Selected Titles in This Series

702 **Ilijas Farah,** Analytic quotients: Theory of liftings for quotients over analytic ideals on the integers, 2000

701 **Paul Selick and Jie Wu,** On natural coalgebra decompositions of tensor algebras and loop suspensions, 2000

700 **Vicente Cortés,** A new construction of homogeneous quaternionic manifolds and related geometric structures, 2000

699 **Alexander Fel'shtyn,** Dynamical zeta functions, Nielsen theory and Reidemeister torsion, 2000

698 **Andrew R. Kustin,** Complexes associated to two vectors and a rectangular matrix, 2000

697 **Deguang Han and David R. Larson,** Frames, bases and group representations, 2000

696 **Donald J. Estep, Mats G. Larson, and Roy D. Williams,** Estimating the error of numerical solutions of systems of reaction-diffusion equations, 2000

695 **Vitaly Bergelson and Randall McCutcheon,** An ergodic IP polynomial Szemerédi theorem, 2000

694 **Alberto Bressan, Graziano Crasta, and Benedetto Piccoli,** Well-posedness of the Cauchy problem for $n \times n$ systems of conservation laws, 2000

693 **Doug Pickrell,** Invariant measures for unitary groups associated to Kac-Moody Lie algebras, 2000

692 **Mara D. Neusel,** Inverse invariant theory and Steenrod operations, 2000

691 **Bruce Hughes and Stratos Prassidis,** Control and relaxation over the circle, 2000

690 **Robert Rumely, Chi Fong Lau, and Robert Varley,** Existence of the sectional capacity, 2000

689 **M. A. Dickmann and F. Miraglia,** Special groups: Boolean-theoretic methods in the theory of quadratic forms, 2000

688 **Piotr Hajłasz and Pekka Koskela,** Sobolev met Poincaré, 2000

687 **Guy David and Stephen Semmes,** Uniform rectifiability and quasiminimizing sets of arbitrary codimension, 2000

686 **L. Gaunce Lewis, Jr.,** Splitting theorems for certain equivariant spectra, 2000

685 **Jean-Luc Joly, Guy Metivier, and Jeffrey Rauch,** Caustics for dissipative semilinear oscillations, 2000

684 **Harvey I. Blau, Bangteng Xu, Z. Arad, E. Fisman, V. Miloslavsky, and M. Muzychuk,** Homogeneous integral table algebras of degree three: A trilogy, 2000

683 **Serge Bouc,** Non-additive exact functors and tensor induction for Mackey functors, 2000

682 **Martin Majewski,** ational homotopical models and uniqueness, 2000

681 **David P. Blecher, Paul S. Muhly, and Vern I. Paulsen,** Categories of operator modules (Morita equivalence and projective modules, 2000

680 **Joachim Zacharias,** Continuous tensor products and Arveson's spectral C^*-algebras, 2000

679 **Y. A. Abramovich and A. K. Kitover,** Inverses of disjointness preserving operators, 2000

678 **Wilhelm Stannat,** The theory of generalized Dirichlet forms and its applications in analysis and stochastics, 1999

677 **Volodymyr V. Lyubashenko,** Squared Hopf algebras, 1999

676 **S. Strelitz,** Asymptotics for solutions of linear differential equations having turning points with applications, 1999

675 **Michael B. Marcus and Jay Rosen,** Renormalized self-intersection local times and Wick power chaos processes, 1999

674 **R. Lawther and D. M. Testerman,** A_1 subgroups of exceptional algebraic groups, 1999

673 **John Lott,** Diffeomorphisms and noncommutative analytic torsion, 1999

(*Continued in the back of this publication*)

Analytic Quotients

Theory of Liftings for Quotients over Analytic Ideals on the Integers

of the
American Mathematical Society

Number 702

Analytic Quotients

Theory of Liftings for Quotients over
Analytic Ideals on the Integers

Ilijas Farah

November 2000 • Volume 148 • Number 702 (second of 5 numbers) • ISSN 0065-9266

American Mathematical Society
Providence, Rhode Island

2000 *Mathematics Subject Classification.*
Primary 03E50, 03E65, 03G05, 03E15, 06E05, 28A05, 54D40, 54C05.

Library of Congress Cataloging-in-Publication Data

Farah, Ilijas, 1966–
  Analytic quotients : theory of liftings for quotients over analytic ideals on the integers / Ilijas Farah.
    p. cm. — (Memoirs of the American Mathematical Society, ISSN 0065-9266 ; no. 702)
  Includes bibliographical references and index.
  ISBN 0-8218-2117-2
  1. Set theory. 2. Algebra, Boolean. 3. Lifting theory. I. Title. II. Series.
QA3 .A57 no. 702
[QA248]
510 s—dc21
[511.3′22]   00-059368

Memoirs of the American Mathematical Society

This journal is devoted entirely to research in pure and applied mathematics.

Subscription information. The 2000 subscription begins with volume 143 and consists of six mailings, each containing one or more numbers. Subscription prices for 2000 are $466 list, $419 institutional member. A late charge of 10% of the subscription price will be imposed on orders received from nonmembers after January 1 of the subscription year. Subscribers outside the United States and India must pay a postage surcharge of $30; subscribers in India must pay a postage surcharge of $43. Expedited delivery to destinations in North America $35; elsewhere $130. Each number may be ordered separately; *please specify number* when ordering an individual number. For prices and titles of recently released numbers, see the New Publications sections of the *Notices of the American Mathematical Society*.

Back number information. For back issues see the *AMS Catalog of Publications*.

Subscriptions and orders should be addressed to the American Mathematical Society, P. O. Box 845904, Boston, MA 02284-5904. *All orders must be accompanied by payment.* Other correspondence should be addressed to Box 6248, Providence, RI 02940-6248.

Copying and reprinting. Individual readers of this publication, and nonprofit libraries acting for them, are permitted to make fair use of the material, such as to copy a chapter for use in teaching or research. Permission is granted to quote brief passages from this publication in reviews, provided the customary acknowledgment of the source is given.

Republication, systematic copying, or multiple reproduction of any material in this publication is permitted only under license from the American Mathematical Society. Requests for such permission should be addressed to the Assistant to the Publisher, American Mathematical Society, P. O. Box 6248, Providence, Rhode Island 02940-6248. Requests can also be made by e-mail to reprint-permission@ams.org.

Memoirs of the American Mathematical Society is published bimonthly (each volume consisting usually of more than one number) by the American Mathematical Society at 201 Charles Street, Providence, RI 02904-2294. Periodicals postage paid at Providence, RI. Postmaster: Send address changes to Memoirs, American Mathematical Society, P. O. Box 6248, Providence, RI 02940-6248.

© 2000 by the American Mathematical Society. All rights reserved.
This publication is indexed in *Science Citation Index*®, *SciSearch*®, *Research Alert*®, *CompuMath Citation Index*®, *Current Contents*®/*Physical, Chemical & Earth Sciences*.
Printed in the United States of America.

∞ The paper used in this book is acid-free and falls within the guidelines established to ensure permanence and durability.
Visit the AMS home page at URL: http://www.ams.org/

10 9 8 7 6 5 4 3 2 1   05 04 03 02 01 00

To the memory of my grandmother, Irena Gajić

Contents

Preface   xiii

Acknowledgments   xv
   A note to the reader   xv

Introduction   1

Chapter 1. Baire-measurable homomorphisms of analytic quotients   5
   1.1. Introduction   5
   1.2. Ideals induced by submeasures   7
   1.3. Preorderings on analytic quotients   11
   1.4. The Radon–Nikodym property   14
   1.5. Asymptotically additive liftings   16
   1.6. The Radon–Nikodym property of Fin and Fin $\times \emptyset$ 19
   1.7. A reformulation of Todorcevic's hypothesis   20
   1.8. Approximate homomorphisms and non-pathological submeasures   25
   1.9. The lifting theorem and the non-lifting theorem   31
   1.10. Permanence properties of quotients   32
   1.11. Simple F_σ P-ideals which are not summable   33
   1.12. The structure of summable ideals   34
   1.13. The structure of density ideals   41
   1.14. Remarks and questions   51

Chapter 2. Open Coloring Axiom and uniformization   55
   2.1. Why Open Coloring Axiom?   56
   2.2. Coherent families of partial functions   59
   2.3. Coherent families of partitions   63
   2.4. Σ_2-reflection for coherent families   65
   2.5. Remarks and questions   66

Chapter 3. Homomorphisms of analytic quotients under OCA   69
   3.1. Introduction   69
   3.2. Homomorphisms without Baire-measurable liftings   71
   3.3. Almost liftings and lifting theorems   72
   3.4. Applications: rigidity of analytic quotients   74
   3.5. Applications: quotients embeddable into analytic quotients   76
   3.6. Automorphism groups   78
   3.7. Homogeneity of analytic quotients   83
   3.8. Almost liftings of embeddings into $\mathcal{P}(\mathbb{N})/\mathrm{Fin}$ 85
   3.9. Almost liftings of embeddings into $\mathcal{P}(\mathbb{N}^2)/\mathrm{Fin} \times \emptyset$ 90
   3.10. Nonmeager hereditary sets   91

3.11.	Approximate homomorphisms; more on stabilizers	92
3.12.	A local version of the OCA lifting theorem	99
3.13.	The proof of the OCA lifting theorem for the analytic P-ideals	106
3.14.	Remarks and questions	114

Chapter 4. Weak Extension Principle — 117

4.1.	Introduction	117
4.2.	Dependence of functions on their variables	118
4.3.	Prime mappings	120
4.4.	Autohomeomorphisms of finite powers of $\beta\omega$ and ω^*	123
4.5.	Čech–Stone remainders of countable ordinals	124
4.6.	Some Parovičenko spaces under wEP	126
4.7.	Remainders of locally compact, countable spaces	130
4.8.	Almost liftings and duality	132
4.9.	OCA and MA imply wEP	135
4.10.	Versions of wEP	139
4.11.	Remarks and questions	140

Chapter 5. Gaps and limits in analytic quotients — 143

5.1.	Introduction	143
5.2.	Gaps in the quotient over Fin	144
5.3.	Gaps in the quotient over $\emptyset \times$ Fin	147
5.4.	Gaps in the quotient over Fin $\times \emptyset$	148
5.5.	The Todorcevic separation property, TSP	151
5.6.	Tukey reductions of nonlinear gaps	153
5.7.	TSP in quotients over analytic P-ideals	155
5.8.	Quotients as reduced products	157
5.9.	Preservation of gaps	160
5.10.	An analytic Hausdorff gap	161
5.11.	Limits in analytic quotients	164
5.12.	A coherent family of functions	164
5.13.	Remarks and questions	165

Bibliography — 167

Index — 172

Index of special symbols — 176

ABSTRACT. We study analytic ideals on the integers and their quotients, mainly concentrating on ideals induced by non-pathological submeasures. This class includes essentially every example of an analytic P-ideal that one can find in the literature. We prove that every Baire-measurable lifting of a homomorphism between quotients over these ideals can be replaced by a completely additive one, and give numerous applications of this result. Concerning arbitrary liftings, we show that the Open Coloring Axiom has exactly the opposite effect than the Continuum Hypothesis, for it implies that every homomorphism has a Baire-measurable *almost lifting*. In particular, two quotients over non-pathological ideals are isomorphic if and only if the corresponding ideals are. Dualization of lifting results leads us to an explicit description of autohomeomorphisms of finite powers of the Čech-Stone remainder $\mathbb{N}^*$ of the integers and to the complete failure of Parovičenko's theorem under OCA. At the end, we show how increasing the complexity of an ideal affects the richness of its gap structure.

Keywords: Analytic ideals, Baire-measurable liftings, completely additive liftings, Boolean algebras, automorphism groups, non-pathological submeasures, Čech–Stone remainders, Hausdorff gaps, Tukey reductions.

Preface

It is with pleasure and gratitude that I offer my acknowledgment to Stevo Todorcevic for an inspiring guidance throughout this enterprise by generously sharing both his deep insight and unique spirit. I would also like to thank: Alan Dow, whose lucid remarks have often led me to improve my results; Fred Galvin, for kindly permitting me to include some of his unpublished results in §22; Alexander Kechris, for many useful remarks upon reading this manuscript; Marion Scheepers, for many hours of discussions; and Boban Velickovic for many hours of discussion during the fall of 1993, and for his later interest in this work. I would also like to thank Bohuslav Balcar, Otmar Spinas, Jianping Zhu and the unknown referee for pointing out some inaccuracies in an earlier version of Chapter V. It was a pleasure to be a part of the Toronto Set Theory Seminar. I would like to thank all of its members for many exciting lectures and conversations afterwards.

I would also like to acknowledge the financial support obtained from the following sources: School of Graduate Studies of the University of Toronto (Connaught Scholarship, 1993/94), Ontario Ministry of Colleges and Universities (Ontario Graduate Scholarship, 1994/95, 1995/96 and 1996/97). My trips to conferences at which some of these results were presented were supported from the NSERC grant of Stevo Todorcevic, with supplements from the Canadian Mathematical Society, the Association for Symbolic Logic, the Serbian Science Foundation and the NSF grants of Tomek Bartoszynski, Claude Laflamme and Marion Scheepers.

Ilijas Farah, Toronto, June 1997

---

[1]This is a preface to author's Ph.D. thesis, [**32**], whose Chapters II and III have eventually evolved into the present monograph.

Acknowledgments

Not only me, but also every reader of this monograph should be grateful to Max Burke. He has carefully read most of Chapters 1 and 3, noticed many errors and simplified some of the proofs. In particular, his remarks have made the long and difficult proof of OCA Lifting Theorem (Theorem 3.3.5) slightly shorter and simpler. Needless to say, I am the only one to blame that this proof is still long and difficult. I would also like to thank Juris Steprans for patiently listening to the proof of this theorem and giving many useful remarks, and to Krzystof Mazur and Vladimir Kanovei for reading parts of the text and giving some useful comments. I would like to thank Stevo Todorcevic again for an additional constructive criticism. Last, but not least, I would like to thank David Fremlin for giving very useful suggestions about the presentation of the material, and for an extensive list of "typos/obscurities/errors."

Between September 1997 and August 1999, when this monograph was revised and some of the results were obtained, I was supported by NSERC. I would like to thank Alan Dow whose NSERC grant provided generous financial support for my trips to conferences during these two years. Some of the results from this monograph were presented at the VIG meeting at UCLA, January 1998, XII Summer Topology Conference, Mexico City, June 1998, Logic Colloquium '98, Prague, August 1998, 5eme Atelier international de theorie des ensembles, Marseille, September 1998, an AMS meeting in Winston–Salem, October 1998 and the Mengenlehre workshop in Oberwolfach, December 1999. I would like to thank the organizers of these meetings for inviting me.

Ilijas Farah, Toronto, June 1999,
and New York City, November 1999.

A note to the reader

This is a revised version of author's Ph.D. thesis [32], obtained under the direction of Professor Stevo Todorcevic at the University of Toronto. Chapters 1 and 3 of this monograph are revised versions to Chapters II and III, respectively, of [32], while Chapters 4 and 5 are based on some related results obtained during the summer and early fall of 1997, shortly after the completion of [32] in June 1997. The result of §5.10 was proved in January 1998, and Chapter 5 was thoroughly revised in July 1998. Chapter 2 was extracted from Chapters 3 and 4 in May 1999 in order to avoid repetitions and present the material in a more friendly way. The version of wEP proved in Theorem 4.9.1 (see also Theorem 4.10.1) is obtained in spring of 1998, and it is stronger than one which appeared in the original version of this monograph. Chapter 4 was thoroughly revised in May 1999.

Some parts of [**32**] not included in this monograph were published in [**33**] (a part of which is included in §1.8), [**29**] (see §5.12), and [**30**]. A survey of results presented here appears in [**35**].

Open problems. Each of the five chapters of this monograph begins with a short introduction and ends with a discussion of the results, possible directions for the further research, and a list of open problems. Additional open problems are scattered throughout the text.

Questions 9.3 and 9.4 from [**32**] have been answered (in negative) by the author in [**36**] (see also §1.11).

Question 9.5 from [**32**] has been answered (in positive) by V. Kanovei and M. Reeken in [**72**] and [**70**] (see also §1.14).

In an earlier version of this monograph I have overlooked the fact that the quotient over $\emptyset \times$ Fin is weakly homogeneous and asked a trivial question; see §3.7 for some less obvious remarks about the homogeneity of analytic quotients.

Theorem 4.11.1 appears in an earlier version of this monograph as a Corollary "with an obvious proof," while Theorem 4.11.2 appears there as a conjecture. The proofs of these two results can be found in [**41**].

The organization of this monograph. The first few sections of Chapter 1 contain the background material. The central result of this chapter is the lifting theorem of §1.9. §§1.10–1.13 contain various applications of this theorem.

Chapter 2 is self-contained, and it contains results needed in some proofs of Chapter 3 and 4. Most of the results of Chapter 2 are well-known, except perhaps for the results of §2.4. Those readers who read only one section from Chapters 2 through 5 may wish to read §2.1.

The main result of Chapter 3 is the OCA lifting theorem, whose proof spreads through §§3.10–3.13. This chapter relies on §§1.2–1.9 and Chapter 2. Basic concepts for this chapter are introduced in §§3.1–3.3. Most readers will want to read §§3.4–§3.7 first; these sections contain some applications of the OCA lifting theorem. §3.8 and §3.9 give proofs of simpler special cases of this theorem.

The main result of Chapter 4 is the isolation of the weak Extension Principle, whose proof is given in §§4.8–4.9. This proof relies on Chapter 2 and §3.9. However, the less technical (and more interesting for a wider audience) part of this chapter, the one concerned with the applications (§§4.1–4.7) is self-contained. In particular, §4.2 and §4.3 are independent from the remaining sections of this text and probably interesting in their own right.

Apart from relying on §1.2 and §1.3, Chapter 5 is self-contained and can be read independently from other chapters. This is the final, and probably the most open-ended of all chapters in this monograph.

Introduction

An ideal $\mathcal{I}$ of sets of integers is analytic if it is given in an explicit and simple way, like for example $\mathcal{I}_{1/n} = \{A : \sum_{n \in A} 1/n < \infty\}$. More generally, if φ is a submeasure supported by $\mathbb{N}$ satisfying certain continuity properties (see §1.2) then
$$\mathcal{I} = \{A : \lim_n \varphi(A \setminus [1, n)) = 0\}$$
is a typical analytic P-ideal (see [**112**], also Theorem 1.2.5). These simple objects, however, give rise to quotients $\mathcal{P}(\mathbb{N})/\mathcal{I}$ which can be quite complex, or *saturated* as they are sometimes called. The basic question that will be considered in this monograph is: How does a change of the ideal $\mathcal{I}$ effect the change of its quotient? Various instances of this question were asked in the literature ([**28**, p. 38–39], [**139**, Problem I.12.11], [**20**, Question 48], [**79**], [**75**]), and this turns out to be a quite subtle question which, first of all, has to be put in a proper setting. For example, the Continuum Hypothesis, via a standard back-and-forth argument, makes many of these quotients isomorphic therefore trivializing the problem (see [**65**] and §3.14). Understanding a connecting map (such as homomorphism) $\Phi \colon \mathcal{P}(\mathbb{N})/\mathcal{I} \to \mathcal{P}(\mathbb{N})/\mathcal{J}$ between quotients requires understanding its *lifting*, namely a map $\Phi_* \colon \mathcal{P}(\mathbb{N}) \to \mathcal{P}(\mathbb{N})$ which induces Φ by making the diagram

$$\begin{array}{ccc} \mathcal{P}(\mathbb{N}) & \xrightarrow{\Phi_*} & \mathcal{P}(\mathbb{N}) \\ \downarrow{\scriptstyle \pi_\mathcal{I}} & & \downarrow{\scriptstyle \pi_\mathcal{J}} \\ \mathcal{P}(\mathbb{N})/\mathcal{I} & \xrightarrow{\Phi} & \mathcal{P}(\mathbb{N})/\mathcal{J} \end{array}$$

commute. (We should remark that sometimes it is customary to require lifting to be additive (see [**57**]), while in our terminology a *lifting* is any map between the underlying structures which induces the given homomorphism of quotients.) The first natural restriction is to consider only those homomorphisms with liftings which are simple either topologically (like continuous or Baire-measurable) or algebraically, like *additive* (ones which themselves are homomorphisms of $\mathcal{P}(\mathbb{N})$) or *completely additive* (ones which moreover respect infinitary operations).

In Chapter 1 we study a rather strong hypothesis of S. Todorcevic about these liftings: Every Baire-measurable lifting of a homomorphism between quotients over analytic P-ideals can be replaced by a completely additive one ([**130**, Problem 1]). This conjecture, whenever true in a given class of ideals, in particular implies that, in this class of ideals, the Baire-measurable embeddability between quotients coincides with the much simpler finite-to-one reduction of the corresponding ideals. One of the main results of this monograph is a positive answer to this conjecture for a large class of so-called non-pathological ideals (§1.9). However, we also prove

---

[1] Received by the editor November 19, 1997.

that the conjecture in general has a negative answer. We use these results to give some answers to the basic question. For example, quotients over non-pathological analytic P-ideals are Baire-isomorphic if and only if the corresponding ideals are. As an interesting application we shall show that, for example, the rather similar ideals $\mathcal{I}_{1/n}$ and $\mathcal{I}_{1/\sqrt{n}} = \{A : \sum_{n \in A} 1/\sqrt{n} < \infty\}$ have nonisomorphic quotients. The diagram which represents the relation of Baire-embeddability of quotients looks as follows (see §1.2 for definitions and §1.14 for a more detailed diagram):

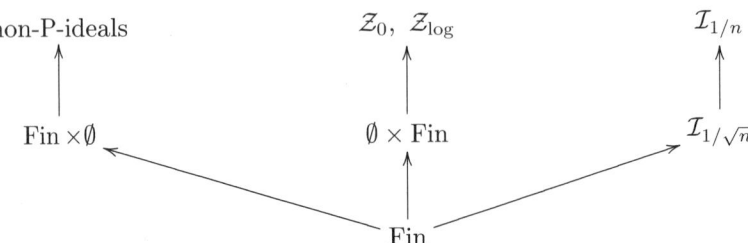

Another line of investigation of the basic problem, which we pursue in Chapter 3, is to consider arbitrary connecting maps but work under an alternative to CH. It turns out that the right alternative is a kind of a two-dimensional Perfect Set Property known under the name of Open Coloring Axiom (OCA), introduced long ago by S. Todorcevic ([**125**]). (We should remark that an axiom called OCA, and several of its variations, was first considered in [**1**], but these axioms do not seem to have the power of what we presently call OCA; see the end of §1.14) This axiom was first shown to be relevant in this context by B. Velickovic ([**142**]) and W. Just ([**64**]). The first result about completely additive liftings between analytic quotients was Shelah's celebrated consistency proof of the statement saying that all automorphisms of $\mathcal{P}(\mathbb{N})/\operatorname{Fin}$ have completely additive liftings ([**106**]). Shelah was solving an old problem of W. Rudin ([**104**]) who showed that under CH every two P-ultrafilters on $\mathbb{N}$ can be mapped one in another by an automorphism of $\mathcal{P}(\mathbb{N})/\operatorname{Fin}$. Clearly, in this situation the number of P-ultrafilters is large while the number of completely additive liftings is small, so there are nontrivial automorphisms of $\mathcal{P}(\mathbb{N})/\operatorname{Fin}$. The situation with the arbitrary homomorphisms rather than automorphisms is by far more complex: If there is a maximal nonprincipal ideal $\mathcal{I}$ on $\mathbb{N}$, then there exists a monomorphism of analytic quotients with no Baire-measurable lifting (see §3.2). Most efforts of Chapter 3 are put in the proof of the result (Theorem 3.3.5) to the effect that under OCA every homomorphism $\Phi \colon \mathcal{P}(\mathbb{N})/\operatorname{Fin} \to \mathcal{P}(\mathbb{N})/\mathcal{I}$ between quotients over analytic P-ideals is an amalgamation of two homomorphisms, Φ_1 and Φ_2:

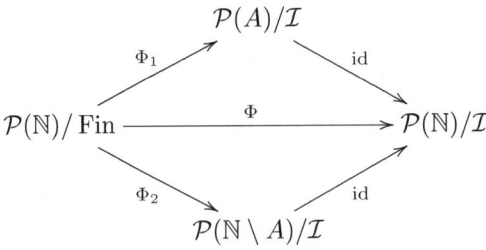

so that Φ_1 has a Baire-measurable lifting, while the kernel of Φ_2 is rather large, in particular not included in any analytic ideal. This implies that under OCA many results from Chapter 1 apply to arbitrary homomorphisms, so the structure of analytic quotients with arbitrary connecting maps resembles the structure of analytic quotients where connecting maps have Baire-measurable liftings. For example, if Φ is an isomorphism of two quotients over analytic P-ideals, then our result implies that Φ has a Baire-measurable lifting. An interesting consequence of this representation theorem is the following metatheorem: If for given two quotients by analytic P-ideals we can prove (without any appeal to additional set-theoretic axioms) that they are isomorphic, then there is such an isomorphism with Baire-measurable lifting. This also shows that the choice of OCA as an alternative to CH is indeed optimal for our purposes. The reader is invited to consult §2.1 for a more thorough discussion of this point.

The key to our approach to Todorcevic's hypothesis is a study of ε-approximate homomorphisms (or *non-exact homomorphisms*) of finite Boolean algebras (see §1.8 and [33]). It is interesting that this can be considered as a part of a general "stability program" initiated long ago by S. Ulam (see [139, §VI.1], also [140, V.4]). The main result of §1.8 is: Non-exact homomorphisms with respect to non-pathological submeasures are, in some sense, stable. Namely, every non-exact homomorphism can be approximated by a real homomorphism up to a constant error. We believe that the results given here are interesting in their own right, in particular because they shed new light on pathological submeasures, which are of significant interest in the study of the famous Control Measure Problem (see [45, 66]).

An initial motivation for the study of automorphisms (and homomorphisms) between analytic quotients came from Topology, in particular from the study of the Čech–Stone remainder $\mathbb{N}^* = \beta\mathbb{N} \setminus \mathbb{N}$ of the integers (see [97]). A dual form of Shelah's consistency result says that all autohomeomorphisms of $\mathbb{N}^*$ can be continuously extended to maps from $\beta\mathbb{N}$ into $\beta\mathbb{N}$, and therefore they are induced by maps of integers into integers, or *trivial* (recall that this fails under CH by [104]). Another classical application of CH in this area is Parovičenko's result ([101]) to the effect that $\mathbb{N}^*$ maps onto a large class of so-called *Parovičenko spaces*, including Čech–Stone remainders of all countable ordinals and their finite powers. That some instances of Parovičenko's conclusion are false under OCA and MA was proved by Just ([64]) and recently by Dow and Hart ([23]). In Chapter 4 we extend and systematize this work by formulating a consequence of OCA and MA that we call *weak Extension Principle* or wEP (see §4.1). For instance, wEP implies that α^* maps onto γ^* if and only if $\alpha \geq \gamma$, whenever α and γ are countable indecomposable ordinals.

We also give a simple description of autohomeomorphisms of finite powers of $\beta\mathbb{N}$, without using any additional set-theoretic axioms. Roughly speaking, there is no continuous map on $(\beta\mathbb{N})^2$ which behaves like the rotation. This required an extension of a result of van Douwen and confirmation of his conjecture about the direction of coordinate axes in finite powers of $\beta\mathbb{N}$ ([17, Conjectures 8.3 and 8.4]), and a combinatorial lemma about the dependence of functions from their variables (Theorem 4.2.1) which may turn out to be interesting in its own right (see [26]). By using wEP, this is extended to describe the autohomeomorphisms of finite powers of $\mathbb{N}^*$.

Although most of the results of Chapter 4 are formulated as consequences of OCA and MA, they also give information about Borel subalgebras of analytic quotients and their homomorphisms with Baire-measurable liftings. Namely, they are statements about turning Baire-measurable liftings of homomorphisms between

Borel subalgebras of $\mathcal{P}(\mathbb{N})/\operatorname{Fin}$ and $\mathcal{P}(\mathbb{N}^2)/\operatorname{Fin}\times\emptyset$ into completely additive liftings. For example, a dual form of above result about mappings of α^* onto γ^* translates as: There is a strictly increasing ω_1 chain of such subalgebras (under the ordering of Baire embeddability) between quotients $\mathcal{P}(\mathbb{N})/\operatorname{Fin}$ and $\mathcal{P}(\mathbb{N}^2)/\operatorname{Fin}\times\emptyset$ (compare with [**76**]).

One of the most striking discoveries in the early Set theory was F. Hausdorff's construction of a *gap* inside the quotient $\mathcal{P}(\mathbb{N})/\operatorname{Fin}$ (see [**54**], [**53**], [**103**]). By another striking result of Kunen ([**81**], see also [**125**, §8]), classical gaps of Hausdorff and Rothberger are the only ones which can be constructed without appealing to additional Set-theoretic axioms. Therefore the structure of gaps in $\mathcal{P}(\mathbb{N})/\operatorname{Fin}$ is well-understood, at least under an assumption such as OCA. Recently Hausdorff gaps were discovered in every analytic quotient ([**95**], [**130**]), and Todorcevic has even proved that Hausdorff gaps are preserved by any embedding with a Baire-measurable lifting of $\mathcal{P}(\mathbb{N})/\operatorname{Fin}$ into some analytic quotient. By a result of Mathias ([**93**]), there is such an embedding of $\mathcal{P}(\mathbb{N})/\operatorname{Fin}$ into any analytic quotient, hence the latter statement is a strengthening of the former. (We should remark that this gap-preservation result was a motivation for posing the conjecture about the completely additive liftings.) In Chapter 5 we extend both Kunen's and Todorcevic's results to the ideals $\operatorname{Fin}\times\emptyset$ and $\emptyset\times\operatorname{Fin}$. We also prove that neither of these results is true for an analytic ideal $\operatorname{Fin}\times\operatorname{Fin}$ or even for some F_σ P-ideals, since for example OCA implies there is an $\langle\omega_2,\omega_2\rangle$ gap in the quotient over $\operatorname{Fin}\times\operatorname{Fin}$ (see Proposition 5.8.5 and §5.10).

Todorcevic ([**126**], see also §5.5) has recently proved a strong variant of classical Hurewicz phenomena ([**56**]) for $\mathcal{P}(\mathbb{N})/\operatorname{Fin}$ which in particular implies that there are no analytic Hausdorff gaps in $\mathcal{P}(\mathbb{N})/\operatorname{Fin}$. We reformulate his result in terms of games and use this to prove its analogue for the quotient over the ideal $\operatorname{Fin}\times\emptyset$. We also prove that Fin is the only analytic P-ideal for which Todorcevic's separation result can be proved, and moreover that there are analytic Hausdorff gaps in some quotients over analytic P-ideals. Gap-phenomena in analytic quotients other than $\mathcal{P}(\mathbb{N})/\operatorname{Fin}$ occur in the domain of *nonlinear gaps*, and we use the language of Tukey reductions (see [**46**]) to analyze such gaps and their preservation.

At the end of our Introduction, let us note what has been omitted here. In Chapters 1 and 3 we have seen that an analytic quotient which can be embedded as a Boolean algebra into $\mathcal{P}(\mathbb{N})/\operatorname{Fin}$ cannot be anything else but $\mathcal{P}(\mathbb{N})/\operatorname{Fin}$. Since this proof made a substantial use of the algebraic structure of analytic quotients, it is natural to ask what happens if we consider quotients as partially ordered sets? This is a subject of the following general program: Investigate the class of all partially ordered sets which can be isomorphically embedded into $\mathcal{P}(\mathbb{N})/\operatorname{Fin}$. The structure of this class of partially ordered sets tends to have impact on a number of mathematical areas. For example, the well-known problem of I. Kaplansky about automatic continuity in Banach Algebras (see [**73**]) has been reduced to a problem of this sort by R. Solovay ([**118**]) and H. Woodin ([**15**]), asking essentially whether an ultrapower of the form $\mathbb{N}^{\mathbb{N}}/\mathcal{U}$ can be embedded into $\mathcal{P}(\mathbb{N})/\operatorname{Fin}$. In Chapter V of [**32**] (which is a revised version of [**30**]) we have applied method of Forcing to describe a general setting for analyzing this and related kind of questions.

CHAPTER 1

Baire-measurable homomorphisms of analytic quotients

1.1. Introduction

In this Chapter we start the study of our basic question which asks how does a change of the ideal $\mathcal{I}$ effect the change of its quotient $\mathcal{P}(\mathbb{N})/\mathcal{I}$ from the point of view of simply definable connecting maps. We decide to interpret "simply definable homomorphism" as "a homomorphism with a Baire-measurable lifting," also called *Baire homomorphism*. This is sufficiently general, because by a famous result of Solovay ([**117**]) every function that can be defined using reals and ordinals as parameters is likely to be Baire-measurable.

Prior to our work, some general results about homomorphisms of analytic quotients and their liftings were obtained by Just ([**62**]). Building on the previous work of Shelah ([**106**]) and Velickovic ([**141**]), Just analyzed homomorphisms between quotients over the ideal $\mathcal{Z}_0$ of sets of density zero and the ideal $\mathcal{Z}_{\log}$ of sets of logarithmic density zero having a Baire-measurable lifting. He proved that every such homomorphism has a lifting which is induced by a sequence of finite functions (see Definition 1.5.1). The major part of [**62**] consists of analysis of such lifting leading to the conclusion that it cannot be an isomorphism, thus answering a long-standing problem of Erdös and Ulam (see also [**65**]). Our proof owes much to this pioneering work.

Completely additive liftings of homomorphisms $\Phi\colon \mathcal{P}(\mathbb{N})/\operatorname{Fin} \to \mathcal{P}(\mathbb{N})/\mathcal{J}$ are always induced by a function $h\colon \mathbb{N} \to \mathbb{N}$ as follows: $\Phi_h(A) = h^{-1}(A)$. Having in mind that the ideal $\mathcal{I}$ is given by a submeasure and that completely additive liftings behave like some sort of "integrals," it is not unreasonable to call the following property of $\mathcal{I}$ the *Radon–Nikodym property of $\mathcal{I}$*: Every homomorphism $\Phi\colon \mathcal{P}(\mathbb{N})/\operatorname{Fin} \to \mathcal{P}(\mathbb{N})/\mathcal{I}$ with a Baire-measurable lifting has a completely additive lifting. In [**130**], Todorcevic posed a hypothesis that the Radon–Nikodym property is shared by all ideals induced by submeasures supported by $\mathbb{N}$, also called *analytic P-ideals* (see §1.2). Note that this hypothesis is obviously very relevant to the basic question because the map h relates the ideals $\mathcal{I}$ and $\ker(\Phi)$. In §1.4 we prove that for the ideals which have the Radon–Nikodym property a version of the basic question where we require the connecting maps to be Baire-measurable, has a sharp answer: quotients over such ideals are Baire-isomorphic if and only if the corresponding ideals are isomorphic (see Proposition 1.10.4).

In §1.5 we extend Just's analysis of liftings to give a weak positive answer to Todorcevic's hypothesis: If $\mathcal{I}$ is an analytic P-ideal, then every Baire homomorphism $\Phi\colon \mathcal{P}(\mathbb{N})/\operatorname{Fin} \to \mathcal{P}(\mathbb{N})/\mathcal{I}$ has a lifting which is induced by a sequence of finite functions, so-called asymptotically additive lifting. Since these finite functions are approximate homomorphisms with respect to the submeasure which induces $\mathcal{I}$, the results of §1.8 (see also [**33**]) can be applied to study these liftings. In §1.6 we show that Radon–Nikodym property of ideals Fin and Fin ×∅ follows from this representation of liftings.

The main result of this Chapter is proved in §1.9 and it shows that a large class of natural ideals shares the Radon–Nikodym property. These are the ideals induced by non-pathological submeasures, and essentially all examples of analytic P-ideals occurring in the literature belong to this class (see e.g., Example 1.2.3). Therefore our result implies that for all previously considered analytic P-ideals the relation of Baire-embeddability of quotients reduces to much simpler relation of finite-to-one reduction, also known under the name of Rudin–Blass ordering. We also prove that some ideals induced by pathological submeasures lack the Radon–Nikodym property. An essential tool needed in §1.9, the analysis of *non-exact homomorphisms*, is provided in §1.8.

In §1.11 we construct some examples of F_σ P-ideals which are not summable (see also [**36**]). In §1.12 we apply the main lifting result of this Chapter to analyze the class of analytic P-ideals which are induced by measures, also called summable ideals. These ideals are of the form

$$\mathcal{I}_f = \left\{ A : \sum_{n \in A} f(n) < \infty \right\}$$

for some $f \colon \mathbb{N} \to \mathbb{R}^+$. We give a complete description of the Baire-embeddability relation between quotients over summable ideals, and show that they form a rather complex structure under this ordering. For example, rather similar ideals $\mathcal{I}_{1/n}$ and $\mathcal{I}_{1/\sqrt{n}}$ have quotients which are not Baire-isomorphic. This should be compared with the fact all quotients over summable ideals are saturated, and therefore isomorphic under CH. In §1.13 we analyze the more complex class of *density* ideals, which contains familiar ideals $\mathcal{Z}_0$ of zero density and $\mathcal{Z}_{\log}$ of logarithmic zero density (see Example 1.2.3). We give a simple criterion for quotients over density ideals to be not Baire-isomorphic. In particular, we reprove Just's result that the quotients over $\mathcal{Z}_0$ and $\mathcal{Z}_{\log}$ are not Baire-isomorphic. In §1.14 we discuss possible directions of the further research on this exciting topic and list some of the most important open problems.

The organization of this chapter. In §1.2 and §1.3 we introduce the basic notions, give examples of analytic P-ideals and results about their structure. The Radon–Nikodym property is introduced in §1.4. A weaker form of Todorcevic's Conjecture is proved in §1.5. It says that a homomorphism into a quotient over an analytic P-ideal which has a Baire-measurable lifting also has an asymptotically additive lifting. The instances of Todorcevic's Conjecture for the ideals Fin and Fin ×∅ are proved in §1.6. This conjecture is proved to be equivalent to a finitary statement about the stability of non-exact homomorphisms in §1.7, and this statement is partially proved, but refuted in general, in §1.8. In §1.9 we isolate the class of non-pathological analytic P-ideals that have the Radon-Nikodym property and construct analytic P-ideals that do not have this property. This is applied in §1.10 to prove permanence properties of certain classes of quotients over non-pathological analytic P-ideals. In §1.11 we give an example of an F_σ-ideal which is not summable. In §1.12 and §1.13 we apply the lifting theorem to study the structure of summable and density ideals, respectively.

1.2. Ideals induced by submeasures

By identifying sets of integers with their characteristic functions, we equip $\mathcal{P}(\mathbb{N})$ with the Cantor-space topology and therefore we can assign the topological complexity to the ideals of sets of integers. In particular, an ideal $\mathcal{I}$ is *analytic* if it is a continuous image of a G_δ subset of the Cantor space. By Fin we denote the ideal of all finite subsets of $\mathbb{N}$, so-called *Fréchet ideal*. Although we will normally consider only the ideals which include Fin, it will be useful to consider the empty set, $\emptyset$, as an ideal, since it will serve as a building block for some important ideals.

DEFINITION 1.2.1. A map $\varphi\colon \mathcal{P}(\mathbb{N}) \to [0, \infty]$ is a *submeasure supported by* $\mathbb{N}$ if
$$\varphi(\emptyset) = 0,$$
$$\varphi(A) \leq \varphi(A \cup B) \leq \varphi(A) + \varphi(B),$$
for all A, B. It is *lower semicontinuous* if for all $A \subseteq \mathbb{N}$ we have
$$\varphi(A) = \lim_{n\to\infty} \varphi(A \cap [1, n]).$$

(This obviously corresponds to φ being lower semicontinuous in the Cantor-set topology on $\mathcal{P}(\mathbb{N})$.)

We can clearly talk about lower semicontinuous submeasures supported by any other countable set, by identifying this set with $\mathbb{N}$. If φ is a lower semicontinuous submeasure supported by $\mathbb{N}$, let $\|\cdot\|_\varphi \colon \mathcal{P}(\mathbb{N}) \to [0, \infty]$ be the submeasure defined by
$$\|A\|_\varphi = \limsup_n \varphi(A \setminus [1, n]) = \lim_n \varphi(A \setminus [1, n])$$
(the second equality follows by the monotonicity of φ). Let
$$\operatorname{Exh}(\varphi) = \{A \subseteq \mathbb{N} : \|A\|_\varphi = 0\},$$
$$\operatorname{Fin}(\varphi) = \{A \subseteq \mathbb{N} : \varphi(A) < \infty\}.$$

It is clear that both $\operatorname{Exh}(\varphi)$ and $\operatorname{Fin}(\varphi)$ are ideals, for an arbitrary submeasure φ.

Sets $A, B \subseteq \mathbb{N}$ are *almost disjoint* ($A \perp B$) if $A \cap B \in \operatorname{Fin}$ and A *almost includes* B ($A \supseteq^* B$) if $B \setminus A \in \operatorname{Fin}$. Sets A and B are *almost equal* ($A =^* B$) if their symmetric difference, $A \triangle B$, is finite. If $\mathcal{I}$ is an ideal, then we define the relations $A \subseteq^{\mathcal{I}} B$, and $A =^{\mathcal{I}} B$ in an analogous way.

An ideal $\mathcal{I}$ is a *P-ideal* if for every sequence $\{A_n\}_{n=0}^\infty$ of sets in $\mathcal{I}$ there is a single set A_∞ in $\mathcal{I}$ such that $A_n \subseteq^* A_\infty$ for all n. Namely, $\mathcal{I}$ is a P-ideal if the partial ordering $\mathcal{I}, \subseteq^*$ is σ-directed. The following result is probably folklore.

LEMMA 1.2.2. *If φ is a lower semicontinuous submeasure supported by $\mathbb{N}$, then $\operatorname{Exh}(\varphi)$ is an $F_{\sigma\delta}$ P-ideal, and $\operatorname{Fin}(\varphi)$ is an F_σ ideal which includes $\operatorname{Exh}(\varphi)$.*

PROOF. The sets $K_n = \{A : \varphi(A) \leq n\}$ are closed, thus $\operatorname{Fin}(\varphi) = \bigcup_n K_n$ is an F_σ ideal. Also, the sets $L_{m,n} = \{A : \varphi(A \setminus m) \leq 1/n\}$ are closed and $\operatorname{Exh}(\varphi) = \bigcap_n \bigcup_m L_{m,n}$ is $F_{\sigma\delta}$.

Since $\operatorname{Fin}(\varphi)$ includes $\operatorname{Exh}(\varphi)$ by the definition, it only remains to see that $\operatorname{Exh}(\varphi)$ is always a P-ideal. Assume A_i ($i \in \mathbb{N}$) is a sequence of sets in $\operatorname{Exh}(\varphi)$. Let n_i ($i \in \mathbb{N}$) be natural numbers such that $\varphi(A_i \setminus n_i) \leq 2^{-i}$ for every i, and let $A_\infty = \bigcup_i (A_i \setminus n_i)$. Fix $n \in \mathbb{N}$ and let m be such that $\varphi(\bigcup_{i \leq n} A_i \setminus m) \leq 2^{-n-1}$. Then $\varphi(A_\infty \setminus m) \leq 2^{-n}$. Since n was arbitrary, A_∞ is in $\operatorname{Exh}(\varphi)$. □

A major part of this book is devoted to study of ideals of the form $\mathrm{Exh}(\varphi)$. For $A \subseteq \mathbb{N}^2$ and $n \in \mathbb{N}$ by A_m we denote the vertical section of A at m:
$$A_m = \{n : \langle m, n \rangle \in A\}.$$
For two ideals $\mathcal{I}$, $\mathcal{J}$ define their *sum*, $\mathcal{I} \oplus \mathcal{J}$, to be the ideal on $\mathbb{N} \times \{0,1\}$ given by:
$$A \in \mathcal{I} \oplus \mathcal{J} \quad \Leftrightarrow \quad \{n : \langle n, 0 \rangle \in A\} \in \mathcal{I} \quad \text{and} \quad \{n : \langle n, 1 \rangle \in A\} \in \mathcal{J}.$$
Similarly, we define the *Fubini product*, $\mathcal{I} \times \mathcal{J}$, of $\mathcal{I}$ and $\mathcal{J}$ to be the ideal on $\mathbb{N} \times \mathbb{N}$ given by:
$$A \in \mathcal{I} \times \mathcal{J} \quad \Leftrightarrow \quad \{i : A_i \notin \mathcal{J}\} \in \mathcal{I}.$$
For $A \subseteq \mathbb{N}$ define $\mathcal{I} \upharpoonright A = \{B \cap A : B \in \mathcal{I}\}$. This notation applies whenever $\mathcal{I}$ is a subset of $\mathcal{P}(\mathbb{N})$, not necessarily an ideal.

By $\forall^\infty$ and $\exists^\infty$ we denote the quantifiers "for all but finitely many" and "there exist infinitely many," respectively. Let us give some examples of analytic ideals.

EXAMPLE 1.2.3. (a) The ideal $\mathrm{Fin} \times \emptyset$ is the Fubini product of the ideals Fin and $\emptyset$, therefore an $A \subseteq \mathbb{N}^2$ is in $\mathrm{Fin} \times \emptyset$ if $\forall^\infty n (A_n = \emptyset)$. It is, in some sense, a minimal non-P-ideal, since it is generated by an increasing countable sequence of sets.

(b) The ideal $\emptyset \times \mathrm{Fin}$ is the Fubini product of $\emptyset$ and Fin, and therefore an $A \subseteq \mathbb{N}^2$ is in $\emptyset \times \mathrm{Fin}$ if $\forall n (A_n \in \mathrm{Fin})$. It is not difficult to see that this is an analytic P-ideal. Perhaps not the easiest way is by noticing that $\emptyset \times \mathrm{Fin} = \mathrm{Exh}(\varphi)$ for the lower semicontinuous submeasure φ defined on $\mathbb{N}^2$ by
$$\varphi(A) = \frac{1}{\min\{n : A_n \neq \emptyset\}}.$$
(Note that we consider that $\mathbb{N} = \{1, 2, 3, \dots\}$.)

(c) *Summable ideals.* If $f \colon \mathbb{N} \to \mathbb{R}^+$ is such that $\sum_{n \in \mathbb{N}} f(n) = \infty$, then let
$$\mathcal{I}_f = \left\{ A : \sum_{n \in A} f(n) < \infty \right\}.$$
This is an F_σ-ideal, since $\mathcal{I}_f = \mathrm{Fin}(\mu_f)$, where
$$\mu_f(A) = \sum_{i \in A} f(i)$$
is a typical lower semicontinuous measure supported by $\mathbb{N}$. But $\mathcal{I}_f$ is also a P-ideal, since $\mathcal{I}_f = \mathrm{Exh}(\mu_f)$, by Cauchy's criterion.

The summable ideals were first introduced by Mathias ([92]) and studied by Mazur ([95]). The summable ideal $\mathcal{I}_{1/n}$ (i.e., the one obtained from $f(n) = 1/n$) was also considered by Erdős and Monk (see [20]). We shall study summable ideals in §1.12.

(d) *Erdős–Ulam-ideals* (or *EU-ideals*). For a function $f \colon \mathbb{N} \to \mathbb{R}^+$ consider the EU-ideal, $\mathcal{EU}_f$, consisting of all sets A of f-*density zero*, i.e., sets such that
$$\lim_{n \to \infty} \frac{\sum_{i \in A \cap [1,n]} f(i)}{\sum_{i=1}^n f(i)} = 0.$$
This is a P-ideal, and it is again not very difficult to check that it is of the form $\mathrm{Exh}(\varphi_f)$, where φ_f is a lower semicontinuous submeasure defined by
$$\varphi_f(A) = \sup_n \frac{\mu_f(A \cap \{1, \dots, n\})}{\mu_f(\{1, \dots, n\})}.$$

(See also Theorem 1.13.3.) Erdős–Ulam ideals were introduced by Just and Krawczyk in [**65**]. Important cases of EU-ideals are the ideal $\mathcal{Z}_0$ of *asymptotic density zero sets* (obtained with $f(n) = 1$):

$$\mathcal{Z}_0 = \left\{ A \subseteq \mathbb{N} : \lim_{n \to \infty} \frac{|A \cap n|}{n} = 0 \right\},$$

and the ideal $\mathcal{Z}_{\log}$ of *logarithmic density zero sets* (obtained when $f(n) = 1/n$):

$$\mathcal{Z}_{\log} = \left\{ A \subseteq \mathbb{N} : \lim_{n \to \infty} \frac{\sum_{i \in A \cap n} 1/i}{\sum_{i < n} 1/i} = 0 \right\}.$$

EU-ideals and their natural generalization, density ideals, are studied in §1.13.

Further examples of natural analytic ideals can be found in ([**92**, p. 206]). We should note that even F_σ ideals, although they form the topologically simplest nontrivial class of ideals, form a very rich and well-studied structure (see [**85**]).

A set $K \subseteq \mathcal{P}(\mathbb{N})$ is *hereditary* if it is closed under taking subsets of its elements. In [**64**], Just has introduced the following important notion:

A set $K \subseteq \mathcal{P}(\mathbb{N})$ is a *closed approximation* to an ideal $\mathcal{I}$ if it is closed, hereditary, and $\mathcal{I} \subseteq \{A : (A \setminus [1,n)) \in K \text{ for some } n\}$.

Every ideal $\mathcal{I} = \text{Fin}(\varphi)$ for a lower semicontinuous submeasure φ is of the form $\{A : (A \setminus [1,n)) \in K \text{ for some } n\}$, for the closed hereditary set $K = \{A : \varphi(A) \leq 1\}$.

LEMMA 1.2.4. *An ideal $\mathcal{I}$ on $\mathbb{N}$ is an F_σ-ideal if and only if it has a closed approximation K such that $\mathcal{I} = \{A : A \setminus [1, n) \in K \text{ for some } n\}$.* □

Lemma 1.2.4 appears in [**84**, Lemma 6.3] and it also follows from [**95**, Lemma 1.2(c)]; see Theorem 1.2.5 below. As pointed out in [**63**], there are many ideals $\mathcal{I}$ which are determined by a countable decreasing sequence of their closed approximations $\{G_n\}$, namely

$$\mathcal{I} = \bigcap_{m=1}^{\infty} \{A : (A \setminus [1, n)) \in G_m \text{ for some } n\}.$$

For example, every ideal of the form $\text{Exh}(\varphi)$ for some lower semicontinuous submeasure φ is of this form (let $G_n = \{A : \varphi(A) \leq 1/n\}$).

All analytic P-ideals given in Example 1.2.3 are of the form $\text{Exh}(\varphi)$ for a suitable φ. The following remarkable theorem shows that all analytic P-ideals are of this form.

THEOREM 1.2.5 (Mazur, Solecki). *Let $\mathcal{I}$ be an ideal on $\mathbb{N}$. Then*
(a) *$\mathcal{I}$ is an F_σ ideal if and only if $\mathcal{I} = \text{Fin}(\varphi)$ for some lower semicontinuous submeasure φ.*
(b) *$\mathcal{I}$ is an analytic P-ideal if and only if $\mathcal{I} = \text{Exh}(\varphi)$ for some lower semicontinuous submeasure φ.*
(c) *$\mathcal{I}$ is an F_σ P-ideal if and only if $\mathcal{I} = \text{Fin}(\varphi) = \text{Exh}(\varphi)$ for some lower semicontinuous submeasure φ.* □

For the proof of (a) see [**95**, Lemma 1.2], and for (b) and (c) see [**114**, Theorem 3.1]. A variant of (b) for "ideals" in $\mathbb{N}^{\mathbb{N}}$ is proved in [**11**]. Every Borel ideal $\mathcal{I}$ is of the form $\text{Fin}(\varphi)$ for some Borel submeasure φ: we can let $\varphi(A) = \infty$ if $A \notin \mathcal{I}$ and $\varphi(A) = 0$ if $A \in \mathcal{I}$. This example seems to suggest that not much can be said about the ideals induced by submeasures that are not lower semicontinuous. In this monograph we will not be dealing with such ideals (with essentially a single exception of the ideal Fin $\times\emptyset$). However, the ideals of the form Fin(φ) and $\mathcal{N}(\varphi) = \{A : \varphi(A) = 0\}$ for a Borel submeasure φ have turned out to be well worth of study (see [**72**, **70**]).

Theorem 1.2.5(b) above was very recently extended by Todorcevic ([**128**]) who has proved that every reasonably definable P-ideal is equal to Exh(φ) for some lower semicontinuous φ. More precisely (for the undefined notions see [**69**]):

THEOREM 1.2.6 (Todorcevic). *Assuming large cardinals, every P-ideal in $L(\mathbb{R})$ is of the form* Exh(φ) *for some lower semicontinuous submeasure φ.* □

Let us also note a curious fact, proved recently by Solecki and Todorcevic ([**116**]): every analytic ideal $\mathcal{I}$ on $\mathbb{N}$ has a *Borel basis*, i.e., there is a Borel (moreover G_δ) subset $\mathcal{B}$ of $\mathcal{P}(\mathbb{N})$ such that $\mathcal{I} = \{A : A \subseteq B \text{ for some } B \in \mathcal{B}\}$. This result has been generalized by Solecki ([**115**]) to the context of analytic ideals of compact sets (see also [**77**]).

Recall that an ideal $\mathcal{I}$ is *nontrivial* if it includes all finite sets, but not all of $\mathcal{P}(\mathbb{N})$. We shall use the notation $f''X$ to denote the image of the set X under the mapping f.

DEFINITION 1.2.7. Two ideals $\mathcal{I}$ and $\mathcal{J}$ are *isomorphic* if and only if there is a $1-1$ partial map $f: \mathbb{N} \to \mathbb{N}$ such that $\mathbb{N} \setminus \text{range}(f) \in \mathcal{J}$, $\mathbb{N} \setminus \text{dom}(f) \in \mathcal{I}$, and

$$A \in \mathcal{I} \quad \Leftrightarrow \quad f''A \in \mathcal{J}$$

for all $A \subseteq \mathbb{N}$.

Note that the only pair of isomorphic ideals for which we cannot choose f to be a permutation of the integers are Fin and Fin $\oplus\mathbb{N}$.

PROPOSITION 1.2.8. *Every countably generated ideal $\mathcal{I}$ is isomorphic either to* Fin *or to* Fin $\times\emptyset$.

PROOF. Let A_n $(n \in \mathbb{N})$ be the generating sequence of $\mathcal{I}$. We can assume that $A_n \subseteq A_{n+1}$ for all n. If the set $A_{n+1} \setminus A_n$ is infinite for infinitely many n, then we can assume (by going to a subsequence of $\{A_n\}$) that it is infinite for all n. Then a bijection $h: \mathbb{N} \to \mathbb{N}^2$ defined so that $h''A_n = [1,n] \times \mathbb{N}$ gives an isomorphism between the ideals $\mathcal{I}$ and Fin $\times\emptyset$.

If the set $A_{n+1} \setminus A_n$ is finite for all but finitely many n, then the ideal is clearly isomorphic to Fin. □

Recall that an *orthogonal* of an ideal $\mathcal{I}$ is the ideal

$$\mathcal{I}^\perp = \{A : A \cap B \text{ is finite for all } B \in \mathcal{I}\}.$$

Note that the ideal $\emptyset \times$ Fin cannot be separated from its orthogonal, Fin $\times\emptyset$, but $\emptyset \times$ Fin and Fin $\times\emptyset$ can be *countably separated*: there is a sequence $\{c_n\}$ of sets of integers such that for every $a \in \emptyset \times$ Fin and every $b \in$ Fin $\times\emptyset$ there is c_n such that $a \subseteq^* c_n$ and $b \cap c_n$ is finite.

Todorcevic ([**126**]; see also Chapter 5) has discovered a quite general fact about these notions of separation not only in the context of $\mathcal{I}$ and $\mathcal{I}^\perp$ but also in general between $\mathcal{I}$ and an arbitrary $\mathcal{B} \subseteq \mathcal{I}^\perp$. He has recently extended this analysis in order to compute possible characters of points in Rosenthal compacta (see [**131**, §7]). We shall need, however (in the proof of Corollary 3.5.3), only the following version of the result which shows that an analytic P-ideal is always countably separated from its orthogonal.

LEMMA 1.2.9 (Todorcevic). *If $\mathcal{I}$ is an analytic P-ideal, then its orthogonal is countably generated.*

PROOF. See [**133**], [**128**, Theorem 2]. This result can also be deduced from the subsequent characterization of analytic P-ideals due to Solecki (Theorem 1.2.5) as follows: Let φ be lower semicontinuous and such that $\mathcal{I} = \text{Exh}(\varphi)$, and let
$$C_n = \{i : \varphi(\{i\}) \geq 1/n\}.$$
We claim that these sets generate $\mathcal{I}^\perp$. Since $\lim_{i \to \infty} \varphi(C_n \setminus [1, i)) \geq 1/n$, each C_n is in $\mathcal{I}^\perp$, so it will suffice to show that every $A \in \mathcal{I}^\perp$ is included in some C_n. Assume that $A \not\subseteq C_n$ for every n, and that A is infinite. Then we can find an infinite $B = \{m_i : i \in \mathbb{N}\}$ included in A such that $\varphi(\{m_i\}) \leq 1/i^2$, and this implies that $B \in \mathcal{I}$, and therefore A is not in $\mathcal{I}^\perp$. This completes the proof. □

Recall that an ideal $\mathcal{I}$ has the *Fréchet property* if $(\mathcal{I}^\perp)^\perp = \mathcal{I}$.

COROLLARY 1.2.10. *If $\mathcal{I}$ is an analytic P-ideal which has the Fréchet property, then $\mathcal{I}$ is isomorphic to* Fin *or to* $\emptyset \times$ Fin.

PROOF. Let C_n be a family which separates $\mathcal{I}$ from $\mathcal{I}^\perp$. We can assume that every C_n is orthogonal to $\mathcal{I}$ (either by the proof of Lemma 1.2.9, or by using the fact that $\mathcal{I}$ is σ-directed under $\subseteq^*$). Therefore $\mathcal{I}^\perp$ is generated by $\{C_n\}$ By Proposition 1.2.8, the ideal $\mathcal{I}^\perp$ is isomorphic to either Fin or Fin $\times \emptyset$, and therefore $\mathcal{I} = (\mathcal{I}^\perp)^\perp$ is isomorphic to either Fin or $\emptyset \times$ Fin. □

Let us state a reformulation of Corollary 1.2.10. Recall that an ideal is *dense* (or *tall*) if every infinite set of integers has an infinite subset which is in the ideal.

COROLLARY 1.2.11. *If $\mathcal{I}$ is an analytic P-ideal not isomorphic to* Fin *or* $\emptyset \times$ Fin, *then $\mathcal{I} \restriction A$ is a dense ideal for some positive set A.* □

1.3. Preorderings on analytic quotients

Let $\mathcal{I}$ be an ideal on $\mathbb{N}$. The elements of the quotient $\mathcal{P}(\mathbb{N})/\mathcal{I}$ are the equivalence classes (here $A \subseteq \mathbb{N}$)
$$[A]_\mathcal{I} = \{C \subseteq \mathbb{N} : A \Delta C \in \mathcal{I}\}.$$
Since this equivalence relation is a congruence, the algebraic operations on the quotient are defined in the natural way. For a homomorphism $\Phi \colon \mathcal{P}(\mathbb{N})/\text{Fin} \to \mathcal{P}(\mathbb{N})/\mathcal{I}$ by Φ_* we denote its *lifting*, i.e., the mapping $\Phi_* \colon \mathcal{P}(\mathbb{N}) \to \mathcal{P}(\mathbb{N})$ such that the diagram

$$\begin{array}{ccc} \mathcal{P}(\mathbb{N}) & \xrightarrow{\Phi_*} & \mathcal{P}(\mathbb{N}) \\ \downarrow{\pi_\mathcal{I}} & & \downarrow{\pi_\mathcal{J}} \\ \mathcal{P}(\mathbb{N})/\mathcal{I} & \xrightarrow{\Phi} & \mathcal{P}(\mathbb{N})/\mathcal{J} \end{array}$$

commutes ($\pi_\mathcal{I}, \pi_\mathcal{J}$ are the natural projections associated to the quotients over $\mathcal{I}$ and $\mathcal{J}$, respectively). Of course, Φ_* is not unique, but we are usually assuming that the homomorphism is given by a lifting.

A set of reals (or more generally, a set in a topological space) is *meager* (or, is of *first category*), if it can be covered by countably many nowhere dense sets. It is *nonmeager* (or, is of *second category*), if it is not meager. A set has the *Property of Baire* (or, is *Baire-measurable*) if it is equal to an open set modulo some meager set. A function is *Baire-measurable* (or simply *Baire*) if the preimage of every open set has the property of Baire.

We will now introduce five preorderings on analytic ideals that will be studied in the remainder of this chapter.

- (O1) (*Rudin–Blass ordering*, see [**86**]) $\mathcal{I} \leq_{\mathrm{RB}} \mathcal{J}$ if there is a finite-to-one $h\colon \mathbb{N} \to \mathbb{N}$ such that $A \in \mathcal{I}$ if and only if $h^{-1}(A) \in \mathcal{J}$. This ordering is also called *finite-to-one reduction*, and denoted by $\leq_f$ in [**112**], [**114**].
- (O2) (*Rudin–Keisler ordering*) $\mathcal{I} \leq_{\mathrm{RK}} \mathcal{J}$ if there is an $h\colon \mathbb{N} \to \mathbb{N}$ such that $A \in \mathcal{I}$ if and only if $h^{-1}(A) \in \mathcal{J}$.
- (O3) (*Baire embeddability*) $\mathcal{I} \leq_{\mathrm{BE}} \mathcal{J}$ if there is a Baire-measurable $F\colon \mathcal{P}(\mathbb{N}) \to \mathcal{P}(\mathbb{N})$ which is a lifting of a monomorphism Φ of the quotient Boolean algebras $\mathcal{P}(\mathbb{N})/\mathcal{I}$ and $\mathcal{P}(\mathbb{N})/\mathcal{J}$:

$$\begin{array}{ccc} \mathcal{P}(\mathbb{N}) & \xrightarrow{F} & \mathcal{P}(\mathbb{N}) \\ \downarrow \pi_{\mathcal{I}} & & \downarrow \pi_{\mathcal{J}} \\ \mathcal{P}(\mathbb{N})/\mathcal{I} & \xrightarrow{\Phi} & \mathcal{P}(\mathbb{N})/\mathcal{J} \end{array}$$

- (O4) $\mathcal{I} \leq_{\mathrm{BE}}^+ \mathcal{J}$ if $\mathcal{I} \leq_{\mathrm{BE}} \mathcal{J} \upharpoonright A$ for some $\mathcal{J}$-positive A.
- (O5) (*Borel cardinality*) $\mathcal{I} \leq_{\mathrm{BC}} \mathcal{J}$ if there is a Borel-measurable map $F\colon \mathcal{P}(\mathbb{N}) \to \mathcal{P}(\mathbb{N})$ which is a lifting of an injection of the set $\mathcal{P}(\mathbb{N})/\mathcal{I}$ into $\mathcal{P}(\mathbb{N})/\mathcal{J}$.

Rudin–Blass and Rudin–Keisler ordering on the maximal ideals and Borel-cardinality of quotients are extensively studied notions (see [**14**], [**75**], [**55**]). We shall be mainly interested in the Baire-embeddability ordering, and the ordering $\leq_{\mathrm{BE}}^+$ will turn out to be important in Chapter 3 (see Corollary 3.4.7).

Probably the first result about preorderings on the analytic ideals was proved by Mathias ([**93**]), and it says that Fin $\leq_{\mathrm{RB}} \mathcal{I}$ for every analytic ideal $\mathcal{I}$. Therefore (by Proposition 1.3.1(a) below) Fin is the minimal analytic ideal in all five preorderings. This was extended later by Jalali–Naini ([**58**]) and Talagrand ([**121**]) who independently gave a characterization of all ideals $\mathcal{I}$ for which Fin $\leq_{\mathrm{RB}} \mathcal{I}$ in terms of Baire category (see also Theorem 3.10.1). Before we continue, let us make some easy observations that will be frequently used below. First we define an *amalgamation* of two homomorphisms. If $\Phi \colon \mathcal{P}(\mathbb{N})/\mathrm{Fin} \to \mathcal{P}(\mathbb{N})/\mathcal{I}$ and $\Psi \colon \mathcal{P}(\mathbb{N})/\mathrm{Fin} \to \mathcal{P}(\mathbb{N})/\mathcal{J}$ are homomorphisms, we define a homomorphism

$$\Phi \oplus \Psi \colon \mathcal{P}(\mathbb{N})/\mathrm{Fin} \to \mathcal{P}(\mathbb{N} \times \{0,1\})/(\mathcal{I} \oplus \mathcal{J})$$

by letting its lifting be $A \mapsto \Phi_*(A) \times \{0\} \cup \Psi_*(A) \times \{1\}$. In other words, $\Phi \oplus \Psi$ is a homomorphism which makes the following diagram commute:

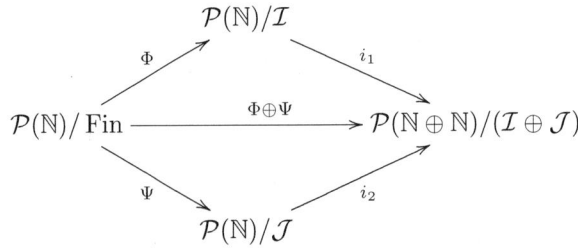

where $i_1 \colon \mathcal{P}(\mathbb{N})/\mathcal{I} \to \mathcal{P}(\mathbb{N} \oplus \mathbb{N})/\mathcal{I} \oplus \mathcal{J}$ is given by its lifting $A \mapsto A \times \{0\}$ and $i_2 \colon \mathcal{P}(\mathbb{N})/\mathcal{J} \to \mathcal{P}(\mathbb{N} \oplus \mathbb{N})/\mathcal{I} \oplus \mathcal{J}$ is given by its lifting $A \mapsto A \times \{1\}$. Note that $\ker(\Phi \oplus \Psi) = \ker(\Phi) \cap \ker(\Psi)$. This operation will be very important in Chapter 3, where it will be used to give a description of arbitrary homomorphisms of quotients over analytic P-ideals under OCA.

We note that (a) below is true for arbitrary ideals on the integers.

PROPOSITION 1.3.1. (a) $\mathcal{I} \leq_{\mathrm{RB}} \mathcal{J}$ implies $\mathcal{I} \leq_{\mathrm{RK}} \mathcal{J}$ implies $\mathcal{I} \leq_{\mathrm{BE}} \mathcal{J}$ implies $\mathcal{I} \leq_{\mathrm{BE}}^+ \mathcal{J}$ implies $\mathcal{I} \leq_{\mathrm{BC}} \mathcal{J}$.
(b) If $\mathcal{J}$ is a P-ideal, then $\mathcal{I} \leq_{\mathrm{RK}} \mathcal{J}$ implies $\mathcal{I} \leq_{\mathrm{RB}} \mathcal{J}$.
(c) The condition $\mathcal{I} \leq_{\mathrm{BE}}^+ \mathcal{J}$ is equivalent to: either $\mathcal{I} \leq_{\mathrm{BE}} \mathcal{J}$ or $\mathcal{I} \oplus \mathrm{Fin} \leq_{\mathrm{BE}} \mathcal{J}$.

PROOF. (a) We will show only that $\mathcal{I} \leq_{\mathrm{RK}} \mathcal{J}$ implies $\mathcal{I} \leq_{\mathrm{BE}} \mathcal{J}$ and that $\mathcal{I} \leq_{\mathrm{BE}}^+ \mathcal{J}$ implies $\mathcal{I} \leq_{\mathrm{BC}} \mathcal{J}$. Assume $h \colon \mathbb{N} \to \mathbb{N}$ is a reduction of $\mathcal{I}$ to $\mathcal{J}$, and define the mapping $\Phi_h \colon \mathcal{P}(\mathbb{N}) \to \mathcal{P}(\mathbb{N})$ by

$$\Phi_h(A) = \bigcup_{n \in A} h^{-1}(\{n\}) = h^{-1}(A).$$

This is a homomorphism of the Boolean algebras $\mathcal{P}(\mathbb{N})$ and $\mathcal{P}(\mathbb{N})$, and therefore, a lifting of the monomorphism of the quotient algebras $\mathcal{P}(\mathbb{N})/\mathcal{I}$ and $\mathcal{P}(\mathbb{N})/\mathcal{J}$ since its kernel is equal to $\mathcal{I}$ by the assumption on h.

Now assume $\mathcal{I} \leq_{\mathrm{BE}}^+ \mathcal{J}$, and let $\Phi_* \colon \mathcal{P}(\mathbb{N}) \to \mathcal{P}(A)$ be a Baire mapping which is a lifting of a homomorphism of $\mathcal{P}(\mathbb{N})/\mathcal{I}$ into $\mathcal{P}(A)/\mathcal{J}$. By Lemma 1.3.2 there is a continuous mapping $\Phi_{**} \colon \mathcal{P}(\mathbb{N}) \to \mathcal{P}(A)$ which is a lifting of the same homomorphism, and this mapping is a lifting of an injection of $\mathcal{P}(\mathbb{N})/\mathcal{I}$ into $\mathcal{P}(A)/\mathcal{J}$.

(b) Assume $h \colon \mathbb{N} \to \mathbb{N}$ is a $\leq_{\mathrm{RK}}$-reduction. Let A be a set in $\mathcal{J}$ which includes modulo finite all $h^{-1}(\{n\})$. Then $h_1 = h \restriction (\mathbb{N} \setminus A)$ is a $\leq_{\mathrm{RB}}$-reduction, because $h_1^{-1}(B) \Delta h^{-1}(B) \subseteq A \in \mathcal{J}$ for all $B \subseteq \mathbb{N}$.

(c) If $\Phi \colon \mathcal{P}(\mathbb{N} \times \{0,1\})/\mathcal{I} \oplus \mathrm{Fin} \to \mathcal{P}(\mathbb{N})/\mathcal{J}$, then $B = \Phi(\mathbb{N} \times \{0\})$ satisfies $\mathcal{I} \leq_{\mathrm{BE}} \mathcal{J} \restriction B$, and therefore $\mathcal{I} \leq_{\mathrm{BE}}^+ \mathcal{J}$. On the other hand, assume $\mathcal{I} \leq_{\mathrm{BE}}^+ \mathcal{J}$ holds, say Φ_1 witnesses $\mathcal{I} \leq_{\mathrm{BE}} \mathcal{J} \restriction B$ for some B. Then if $\mathbb{N} \setminus B$ is in $\mathcal{J}$ there is nothing to prove, and otherwise by Mathias' theorem we have Φ_2 witnessing $\mathrm{Fin} \leq_{\mathrm{BE}} \mathcal{J} \restriction (\mathbb{N} \setminus B)$. Then mapping $\Phi_1 \oplus \Phi_2$ witnesses $\mathcal{I} \oplus \mathrm{Fin} \leq_{\mathrm{BE}} \mathcal{J}$. □

The following well-known fact (see [**141**, p. 132] and [**130**, Theorem 3]) shows that another natural ordering coincides with $\leq_{\mathrm{BE}}$; a hypothesis of Todorcevic (Question 1.14.3) goes one step further.

LEMMA 1.3.2. If a homomorphism $\Phi \colon \mathcal{P}(\mathbb{N})/\mathrm{Fin} \to \mathcal{P}(\mathbb{N})/\mathcal{I}$ has a Baire-measurable lifting then it also has a continuous lifting.

PROOF. Let $\Phi_* \colon \mathcal{P}(\mathbb{N}) \to \mathcal{P}(\mathbb{N})$ be a Baire-measurable lifting of Φ. Then there is a sequence $\{U_i\}$ of dense open subsets of $\mathcal{P}(\mathbb{N})$ such that Φ_* is continuous on $G = \bigcap_{i=1}^{\infty} U_i$. Recursively choose a sequence $1 = n_1 < n_2 < \ldots$ and $t_i \subseteq [n_i, n_{i+1})$ (for $i \in \mathbb{N}$) such that $A \cap [n_i, n_{i+1}) = t_i$ implies $A \in U_j$ for all A and $j = 1, \ldots, i$. Let

$$A_0 = \bigcup_i [n_{2i}, n_{2i+1}), \qquad C_0 = \bigcup_i t_{2i},$$
$$A_1 = \bigcup_i [n_{2i+1}, n_{2i+2}), \qquad C_1 = \bigcup_i t_{2i+1}.$$

Define a mapping Θ by
$$\Theta_0(B) = \Phi_*((B \cap A_0) \cup C_1) \cap \Phi_*(A_0),$$
$$\Theta_1(B) = \Phi_*((B \cap A_1) \cup C_0) \setminus \Phi_*(A_0),$$
$$\Theta(B) = \Theta_0(B) \cup \Theta_1(B).$$

Then Θ is a *stabilization of* Φ_* *by* $\{n_i\}$ *and* $\{t_i\}$. It is continuous because $(B \cap A_\varepsilon) \cup C_{1-\varepsilon}$ is in G for $\varepsilon = 0, 1$. For $B \subseteq A_\varepsilon$ we have $\Phi_*(B) =^{\mathcal{I}} \Phi_*(B \cup C_{1-\varepsilon}) \cap \Phi_*(A_\varepsilon)$. It is therefore easy to see that Ψ is a lifting for Φ. □

The stabilization Θ as defined above and its variations will play a prominent role in proofs of our lifting theorems (see Theorem 1.5.2, §3.11). V. Kanovei and M. Reeken ([**72**]) and D.H. Fremlin ([**47**, Proposition 1C]) have proved an analogue of Lemma 1.3.2 for Lebesgue-measurable liftings: Every homomorphism between two quotients $\mathcal{P}(\mathbb{N})/\mathcal{I}$ that has a Lebesgue-measurable lifting has a continuous lifting.

Let us finish this section by stating several simple facts about the structure of quotients over the ideals of the form $\mathrm{Exh}(\varphi)$. Recall that $\|A\|_\varphi = \liminf_n(A \setminus n) = \lim_n \varphi(A \setminus n)$.

LEMMA 1.3.3. *If φ is a lower semicontinuous submeasure supported by $\mathbb{N}$, then*
(a) $\|\cdot\|_\varphi$ *satisfies the inequality* $\|A \Delta B\|_\varphi \leq \|A\|_\varphi + \|B\|_\varphi$.
(b) $A \Delta B \in \mathrm{Exh}(\varphi)$ *implies* $\|A\|_\varphi = \|B\|_\varphi$.
(c) $\mathcal{P}(\mathbb{N})/\mathrm{Exh}(\varphi)$ *is complete with respect to the metric defined by* $d_\varphi(A, B) = \min(1, \|A \Delta B\|_\varphi)$.

PROOF. Since (a) and (b) are straightforward, we shall prove only (c). Let $\{A_i\}$ be a Cauchy sequence in $\mathcal{P}(\mathbb{N})/\mathcal{I}$. Find $n_1 < n_2 < n_3 < \ldots$ such that
$$\varphi((A_{n_i} \Delta A_{n_j}) \setminus [1, n_i)) < 2^{-i}$$
for all $j < i$. Then $A = \bigcup_i (A_{n_i} \cap [n_i, n_{i+1}))$ is the limit of the sequence $\{A_i\}$. □

The above lemma was proved by Just and Krawczyk (Lemma 3 of [**65**]) in the case when φ is a submeasure corresponding to some EU-ideal. By a result of Solecki (Theorem 3.3 of [**112**]), $\mathrm{Exh}(\varphi)$ is an F_σ ideal if and only if the metric induced by $\|\cdot\|_\varphi$ is discrete. This implies that the quotient over an ideal $\mathcal{I} = \mathrm{Exh}(\varphi)$ can be countably saturated only when $\mathcal{I}$ is an F_σ ideal. The converse is also known to be true—all quotients over F_σ ideals are countably saturated. This was proved in [**65**, §1]; the fact that $\mathcal{P}(\mathbb{N})/\mathrm{Fin}$ is countably saturated is consequence of an earlier result of P. Olin (see [**59**]).

1.4. The Radon–Nikodym property

The notion of Radon–Nikodym property is related to the following conjecture posed by Todorcevic in [**130**, Problem 1] (recall that for $h \colon \mathbb{N} \to \mathbb{N}$ mapping $\Phi_h \colon \mathcal{P}(\mathbb{N}) \to \mathcal{P}(\mathbb{N})$ is defined by $\Phi_h(A) = h^{-1}(A)$).

CONJECTURE 1.4.1 (Todorcevic's conjecture). *Suppose $\mathcal{I}$ is an analytic P-ideal on $\mathbb{N}$ and that Φ is a homomorphism from $\mathcal{P}(\mathbb{N})/\mathrm{Fin}$ into $\mathcal{P}(\mathbb{N})/\mathcal{I}$ with a Baire lifting. Then Φ has a completely additive lifting.*

A positive answer would imply that $\leq_{\mathrm{RB}}$ and $\leq_{\mathrm{BE}}$ coincide on analytic P-ideals, but it would also give much more (see Proposition 1.10.2 below). Note that, since completely additive mappings behave like some sorts of "integrals," a positive answer to Todorcevic's hypothesis would be a form of the Radon–Nikodym property of quotients and their Baire homomorphisms. (Compare with the Radon–Nikodym property of Banach spaces and their vector measures, [**16**, Definition III.1.3]).

DEFINITION 1.4.2. An ideal $\mathcal{I}$ has the *Radon–Nikodym property* if every Baire homomorphism $\Phi\colon \mathcal{P}(\mathbb{N})/\operatorname{Fin} \to \mathcal{P}(\mathbb{N})/\mathcal{I}$ has a completely additive lifting.

Todorcevic's hypothesis can clearly be rephrased as follows.

CONJECTURE 1.4.3 (Todorcevic's Conjecture). *Every analytic P-ideal has the Radon–Nikodym property.*

If $\mathcal{J}$ is not a P-ideal, then there is a homomorphism $\Phi\colon \mathcal{P}(\mathbb{N})/\operatorname{Fin} \to \mathcal{P}(\mathbb{N})/\mathcal{J}$ with completely additive lifting but with no lifting of the form Φ_h for a finite-to-one function h. For example, it is not difficult to see that for $g\colon \mathbb{N} \times \mathbb{N} \to \mathbb{N}$ defined by $g(\langle n,m\rangle) = n$ mapping Φ_g is a lifting of a homomorphism of $\mathcal{P}(\mathbb{N})/\operatorname{Fin}$ into $\mathcal{P}(\mathbb{N}\times\mathbb{N})/\operatorname{Fin}\times\emptyset$ with no lifting of the form Φ_h for a finite-to-one function h.

LEMMA 1.4.4. *The following are equivalent for every ideal $\mathcal{J}$ on the integers:*
(a) *$\mathcal{J}$ has the Radon–Nikodym property.*
(b) *Every Baire homomorphism $\Phi\colon \mathcal{P}(\mathbb{N})/\mathcal{I} \to \mathcal{P}(\mathbb{N})/\mathcal{J}$ has a completely additive lifting for every ideal $\mathcal{I} \supseteq \operatorname{Fin}$.*
(c) *Every Baire monomorphism $\Phi\colon \mathcal{P}(\mathbb{N})/\mathcal{I} \to \mathcal{P}(\mathbb{N})/\mathcal{J}$ has a completely additive lifting for every ideal $\mathcal{I} \supseteq \operatorname{Fin}$.*

PROOF. To see that (a) implies (b), assume (a) and fix $\mathcal{I}$ and Φ as in (b). Since $\mathcal{I} \supseteq \operatorname{Fin}$, the formula
$$\Psi([A]_{\operatorname{Fin}}) = \Phi([A]_\mathcal{I})$$
defines a Baire homomorphism $\Psi\colon \mathcal{P}(\mathbb{N})/\operatorname{Fin} \to \mathcal{P}(\mathbb{N})/\mathcal{J}$. (In the diagram below π stands for the natural projection of $\mathcal{P}(\mathbb{N})/\operatorname{Fin}$ onto $\mathcal{P}(\mathbb{N})/\mathcal{I}$.)

$$\begin{array}{ccc}
\mathcal{P}(\mathbb{N})/\operatorname{Fin} & & \\
{\scriptstyle \pi}\downarrow & \searrow{\scriptstyle \Psi} & \\
\mathcal{P}(\mathbb{N})/\mathcal{I} & \xrightarrow{\Phi} & \mathcal{P}(\mathbb{N})/\mathcal{J}.
\end{array}$$

By the Radon–Nikodym property of $\mathcal{J}$, Ψ has a completely additive lifting. But this mapping is also a completely additive lifting of Φ.

(b) implies (c) is trivial.

To prove that (c) implies (a), assume (a) fails, and let $\Psi\colon \mathcal{P}(\mathbb{N})/\operatorname{Fin} \to \mathcal{P}(\mathbb{N})/\mathcal{J}$ be a Baire homomorphism with no completely additive lifting. Let $\mathcal{I}$ be the kernel of Ψ; then
$$\Phi([A]_\mathcal{I}) = \Psi([A]_{\operatorname{Fin}})$$
defines a Baire monomorphism $\Phi\colon \mathcal{P}(\mathbb{N})/\mathcal{I} \to \mathcal{P}(\mathbb{N})/\mathcal{J}$. Every lifting of Φ is also a lifting of Ψ, and therefore Φ has no completely additive lifting, and (c) fails. □

COROLLARY 1.4.5. *If $\mathcal{J}$ has the Radon–Nikodym property, then $\mathcal{I} \leq_{\operatorname{BE}} \mathcal{J}$ if and only if $\mathcal{I} \leq_{\operatorname{RK}} \mathcal{J}$, for every ideal $\mathcal{I}$ on the integers.* □

It is not clear whether the assumption that $\mathcal{I}$ includes Fin in (b) and (c) is important; some recent results of V. Kanovei and M. Reeken suggest that it is not. Namely, in [**72, 70**] they have proved that in all known cases when $\mathcal{J}$ has the

Radon–Nikodym property, every Baire homomorphism $\Phi\colon \mathcal{P}(\mathbb{N}) \to \mathcal{P}(\mathbb{N})/\mathcal{J}$ has a completely additive lifting.

It would be natural to ask whether after replacing "monomorphism" in (c) of Lemma 1.4.4 with "epimorphism" (or "isomorphism") one obtains an equivalent statement. I would conjecture that it is not so; this would follow from an (expected) positive answer to Question 1.14.4.

Recall that by $\mathcal{I}^*$ we denote the *dual filter* of $\mathcal{I}$, namely the filter of all sets whose complement is in $\mathcal{I}$, and by $\mathcal{I}^+$ we denote the coideal of all *sets*, namely all sets of integers which are not in $\mathcal{I}$. An isomorphism of quotients $\Phi\colon \mathcal{P}(\mathbb{N})/\mathcal{I} \to \mathcal{P}(\mathbb{N})/\mathcal{J}$ is *trivial* if there are sets $A \in \mathcal{I}^*$, $B \in \mathcal{J}^*$ and a bijection $h\colon B \to A$ such that Φ_h is a lifting of Φ. Note that (a) below gives a satisfactory answer to the basic question (see §1.2). The part (b) of Proposition 1.4.6 below for $\mathcal{I} = \mathrm{Fin}$ was first proved by Velickovic ([**141**]). (In Definition 1.2.7 it was defined when two ideals are isomorphic.)

PROPOSITION 1.4.6. *If ideals $\mathcal{I}$ and $\mathcal{J}$ have the Radon–Nikodym property then every Baire isomorphism of quotients $\mathcal{P}(\mathbb{N})/\mathcal{I}$ and $\mathcal{P}(\mathbb{N})/\mathcal{J}$ is trivial. In particular,*

(a) *The quotients over $\mathcal{I}$ and $\mathcal{J}$ are Baire-isomorphic if and only if the corresponding ideals are isomorphic.*
(b) *All Baire automorphisms of $\mathcal{P}(\mathbb{N})/\mathcal{I}$ are trivial.*

PROOF. Let $\Phi\colon \mathcal{P}(\mathbb{N})/\mathcal{I} \to \mathcal{P}(\mathbb{N})/\mathcal{J}$ be a Baire isomorphism and let h_1, h_2 be such that Φ_{h_1} is a lifting of Φ and Φ_{h_2} is a lifting of Φ^{-1}. If $g = h_1 \circ h_2$ is the identity on a set in $\mathcal{I}^*$, then Φ_{h_1} is a lifting of Φ and h_1 witnesses that $\mathcal{I}$ and $\mathcal{J}$ are isomorphic. Let B be the set of all x such that $g(x) \neq x$. Then we can split B into disjoint sets B_1, B_2, B_3 so that $g''B_i$ is disjoint from B_i for $i = 1, 2, 3$ (see e.g., [**14**, Lemma 9.1]). This easily implies each B_i is in $\mathcal{I}$, and therefore the set of fixed points for g is in $\mathcal{I}^*$, as required. Both (a) and (b) follow immediately. □

The main results of this Chapter are that all the natural examples of analytic P-ideals (in particular, all the ideals from Examples 1.2.3) do have the Radon–Nikodym property (Theorem 1.9.1), although in general this is not the case (Theorem 1.9.5).

1.5. Asymptotically additive liftings

In this section we give a partial positive answer to Todorcevic's hypothesis reducing it to a problem of analyzing non-exact homomorphisms, discussed in §1.8 and [**33**]. The following definition essentially appears in [**106**] and [**141**], and Theorem 1.5.2 below was essentially proved by Just ([**62**]) in the case of Erdös–Ulam-ideals (for definition see Examples 1.2.3 or §1.13).

DEFINITION 1.5.1. If $\{n_i\}$ is a strictly increasing sequence of integers, $\{v_i\}$ is a sequence of disjoint finite sets of integers, and H_i is a function mapping subsets of the interval $[n_i, n_{i+1})$ to subsets of v_i, then define $\Psi_H\colon \mathcal{P}(\mathbb{N}) \to \mathcal{P}(\mathbb{N})$ by

$$\Psi_H(X) = \bigcup_{i=1}^{\infty} H_i(X \cap [n_i, n_{i+1})).$$

We say that maps of the form Ψ_H are *asymptotically additive*. This is because we have

$$\Psi_H(s \cup t) = \Psi_H(s) \cup \Psi_H(t)$$

whenever $s < n_i < t$ for some i. Observe that a completely additive lifting, Φ_h, is a special case an asymptotically additive lifting—let $n_i = i$, $H_i(\{i\}) = h^{-1}(i)$ and $H_i(\emptyset) = \emptyset$. Already this form of a lifting is sufficiently simple to be useful in the situations when we do not necessarily have an asymptotically additive lifting. For example, see [130, Theorem 8]. Part (b) of the following theorem will be used in the proof that the ideal Fin $\times \emptyset$ has the Radon–Nikodym property (Theorem 1.6.2).

THEOREM 1.5.2. *If $\mathcal{I}$ is an analytic P-ideal and $\Phi \colon \mathcal{P}(\mathbb{N})/\text{Fin} \to \mathcal{P}(\mathbb{N})/\mathcal{I}$ is a homomorphism which has a Baire lifting, then it has an asymptotically additive lifting. Moreover:*
  (a) *The sequence $\{n_i\}$ can be chosen to be a subsequence of any strictly increasing sequence $\{m_i\}$ of integers given in advance.*
  (b) *The function Ψ_H can be chosen effectively in code for Φ_*.*

PROOF. By Lemma 1.3.2 we can assume Φ has a continuous lifting Φ_*. Note that the proof shows that this Φ_* can be chosen effectively in a code for a given Baire lifting of Φ. Let φ be a lower semicontinuous submeasure such that $\mathcal{I} = \text{Exh}(\varphi)$, as guaranteed by Solecki's theorem (Theorem 1.2.5). We need a notion of a *stabilizer*, first extracted by Just ([63]) (though implicit in an earlier work of Shelah ([106]) and Velickovic ([141])).

DEFINITION 1.5.3. For $n < n'$ let $c \subseteq [n, n')$ be an (n, n')-*stabilizer* of Φ_* if there exists $k \in (n, n')$ such that for all $s, t \subseteq [1, n)$ and all $X, Y \subseteq [n', \infty)$:

(S1) $\varphi((\Phi_*(s \cup c \cup X) \Delta \Phi_*(t \cup c \cup X)) \setminus [1, k)) \leq 2^{-n}$
(S2) $\min(\Phi_*(s \cup c \cup X) \Delta \Phi_*(s \cup c \cup Y)) \geq k$.

CLAIM 1. *For all n and all large enough n' there is an (n, n')-stabilizer.*

PROOF. Fix an n and assume there is no (n, n')-stabilizer for all n'. By the uniform continuity of Φ_*, for all large enough $k > n$ the condition (S2) is satisfied for all $c \subseteq [n, n')$, and therefore (S1) fails for all large enough n'. Recursively pick $c_1 \subseteq c_2 \subseteq \ldots$ and $n = m_0 < m_1 < m_2 < \ldots$ so that $c_j \cap [n, m_i) = c_i$ for $i < j$ and for some $s_i, t_i \subseteq [1, n)$ we have

$$\varphi((\Phi_*(s_i \cup c_i) \Delta \Phi_*(t_i \cup c_i)) \setminus [1, m_{i-1})) > 2^{-n}.$$

Let $\langle \bar{s}, \bar{t} \rangle$ be a pair which appears as $\langle s_i, t_i \rangle$ infinitely often; then $\bar{s} \cup \bigcup_{i=1}^\infty c_i =^* \bar{t} \cup \bigcup_{i=1}^\infty c_i$, but the set

$$F = \Phi_*\left(\bar{s} \cup \bigcup_{i=1}^\infty c_i\right) \Delta \Phi_*\left(\bar{t} \cup \bigcup_{i=1}^\infty c_i\right)$$

is not in $\mathcal{I}$, because $\varphi(F \setminus [1, m)) > 2^{-n}$ for all $m \in \mathbb{N}$. This contradiction finishes the proof. □

By using Claim 1 pick sequences n_i, k_i and $c_i \subseteq [n_i, n_{i+1})$ so that c_i is an (n_i, n_{i+1})-stabilizer of Φ_*, with k_i witnessing this fact. At each step we can pick the pair $\langle c_i, n_{i+1} \rangle$ to be the minimal among all pairs for which c_i is an (n_i, n_{i+1})-stabilizer of Φ_* in some fixed recursive well-ordering of Fin $\times \mathbb{N}$ in order type ω. For $\varepsilon = 0, 1$ let

$$A_\varepsilon = \bigcup_{i=1}^\infty [n_{2i+\varepsilon}, n_{2i+\varepsilon+1}) \qquad C_\varepsilon = \bigcup_{i=1}^\infty c_{2i+\varepsilon+1}.$$

We can assume $\Phi_*(A_\varepsilon) \cap \Phi_*(C_\varepsilon) = \emptyset$ for $\varepsilon = 0, 1$, possibly by replacing Φ_* with the mapping

$$X \mapsto \Phi_*(X) \setminus ((\Phi_*(A_0) \cap \Phi_*(C_0)) \cup (\Phi_*(A_1) \cap \Phi_*(C_1))).$$

This mapping is a lifting of Φ because $A_\varepsilon \cap C_\varepsilon = \emptyset$ for $\varepsilon = 0, 1$. Moreover, it is easily seen from the definitions that c_i is still a (n_i, n_{i+1})-stabilizer for this mapping. Define functions $f_\varepsilon \colon \mathcal{P}(A_\varepsilon) \to \mathcal{P}(\Phi_*(A_\varepsilon))$ for $\varepsilon = 0, 1$ by

$$f_\varepsilon(X) = \Phi_*(X \cup C_\varepsilon) \cap \Phi_*(A_\varepsilon).$$

Again, we can assume $\Phi_*(A_0) \cap \Phi_*(A_1) = \emptyset$. Note that $\Phi_*(A_\varepsilon) \cap \Phi_*(C_\varepsilon) = \emptyset$ implies $f_\varepsilon(\emptyset) = \emptyset$ for $\varepsilon = 0, 1$. Since $C_\varepsilon \cap A_\varepsilon = \emptyset$ for $\varepsilon = 0, 1$ the set $f_\varepsilon(X) \Delta \Phi_*(X)$ is in $\mathcal{I}$ for every $X \subseteq A_\varepsilon$, and therefore the mapping $\bar{\Phi}_*$ defined by

$$\bar{\Phi}_*(X) = f_0(X \cap A_0) \cup f_1(X \cap A_1)$$

is a lifting of Φ.

Define functions $H_i \colon \mathcal{P}([n_i, n_{i+1})) \to \mathcal{P}([k_{i-1}, k_{i+1}))$ by

$$H_i(s) = \bar{\Phi}_*(s) \cap [k_{i-1}, k_{i+1})$$

and let Ψ_H be as in Definition 1.5.1. Note that sets

$$v_i = H_i([n_i, n_{i+1}))$$

are pairwise disjoint, by the assumption that sets $\Phi_*(A_0)$ and $\Phi_*(A_1)$ are disjoint.

CLAIM 2. *The function Ψ_H is a lifting of Φ.*

PROOF. Fix $X \subseteq A_0$; we will prove that

$$\varphi((\bar{\Phi}_*(X) \Delta \Psi_H(X)) \cap [k_{i-1}, k_{i+1})) \leq 2^{-n_i+1}$$

for all even i. Let $x_i = X \cap [n_i, n_{i+1})$; then for some $s < c_{i-1} < x_i < c_{i+1} < Y$ and $s' < c_{i-1} < x_i < c_{i+1} < Y'$ we have

$$\bar{\Phi}_*(X) = f_0(X \cup C_0) = f_0(s \cup c_{i-1} \cup x_i \cup c_{i+1} \cup Y),$$
$$H_i(x_i) = f_0(s' \cup c_{i-1} \cup x_i \cup c_{i+1} \cup Y').$$

Then by (S2) applied to (n_{i+1}, n_{i+2}) we have:

$$\bar{\Phi}_*(X) \cap \{1, \dots, k_{i+1}\} = f_0(s \cup c_{i-1} \cup x_i),$$
$$H_i(x_i) \cap \{1, \dots, k_{i+1}\} = f_0(s' \cup c_{i-1} \cup x_i).$$

Applying (S1) to (n_{i-1}, n_i) we get the desired conclusion. By symmetry the same is true for $X \subseteq A_1$. So for any $X \subseteq \mathbb{N}$ and all i we have $\varphi((\Psi_H(X) \Delta \bar{\Phi}_*(X)) \cap [k_{i-1}, k_{i+1})) \leq 2^{-n_i+1}$, and therefore

$$\varphi((\Psi_H(X) \Delta \Phi_*(X)) \cap [k_{i-1}, \infty)) \leq \sum_{j=i}^{\infty} 2^{-n_j+1} \leq 2^{-n_i+2},$$

and Ψ_H is a lifting of Φ. □

This ends the proof of theorem. For part (b), note that sequences $\{n_i\}$, $\{c_i\}$ were chosen effectively and $\{H_i\}$, $\{v_i\}$ are defined by simple formulas. □

1.6. The Radon–Nikodym property of Fin and Fin $\times \emptyset$

We have already developed enough theory to prove that some ideals have the Radon–Nikodym property. Theorem 1.6.1 below was essentially proved by Velickovic ([**141**]) and Just ([**63**]).

THEOREM 1.6.1. *The ideal* Fin *has the Radon–Nikodym property. Moreover, if Φ is a continuous homomorphism of $\mathcal{P}(\mathbb{N})/$ Fin, then an h for which Φ_h is a lifting of Φ can be chosen effectively in Φ_*.*

THEOREM 1.6.2. *The ideal* Fin $\times \emptyset$ *has the Radon–Nikodym property.*

PROOF OF THEOREM 1.6.1. Let $\Phi \colon \mathcal{P}(\mathbb{N})/\operatorname{Fin} \to \mathcal{P}(\mathbb{N})/\operatorname{Fin}$ be a Baire homomorphism. Then by Theorem 1.5.2, Φ has a lifting of the form Ψ_H with some parameters $\{n_i\}$, $\{v_i\}$ and $\{H_i\}$ chosen effectively in Φ_*. We claim H_i is a homomorphism of the Boolean algebras $\mathcal{P}([n_i, n_{i+1}))$ and $\mathcal{P}(v_i)$ for all but finitely many i.

Note first that $H_i(s) \subseteq v_i$ for all $s \subseteq [n_i, n_{i+1})$. Suppose there are infinitely many i such that for some $s_i, t_i \subseteq [n_i, n_{i+1})$ one of the following happens:

$$H_i(s_i) \cap H_i(t_i) \neq H_i(s_i \cap t_i) \quad \text{or}$$
$$v_i \setminus H_i(s_i) \neq H_i([n_i, n_{i+1}) \setminus s_i).$$

Assume that the first possibility happens infinitely often, say for a pair s_i, t_i for all i in some infinite $A \subseteq \mathbb{N}$. Let $X = \bigcup_{i \in A} s_i$ and $Y = \bigcup_{i \in A} t_i$. Then sets $\Psi_H(X) \cap \Psi_H(Y)$ and $\Psi_H(X \cap Y)$ are not equal modulo finite, because their intersections with v_i are distinct for all $i \in A$. This is a contradiction. The second case is handled in a similar manner.

Therefore we can assume all H_i's are homomorphisms (possibly after deleting finitely many of them). Since $\Psi_H(\mathbb{N}) = \bigcup_{i=1}^\infty v_i$ the set $v_0 = \mathbb{N} \setminus \bigcup_{i=1}^\infty v_i$ is finite. Define a finite-to-one function $h \colon \mathbb{N} \to \mathbb{N}$ as follows:

$$h(m) = \begin{cases} 1, & \text{if } m \in v_0, \\ k, & \text{if } k \in [n_i, n_{i+1}) \text{ and } m \in H_i(\{k\}) \text{ for some } i \geq 1. \end{cases}$$

Then $\Psi_H(X) =^* \Phi_h(X)$ for all X, because:

$$\begin{aligned} k \in \bar{\Phi}_*(X) &\Leftrightarrow & k \in v_i \text{ and } & k \in H_i(X \cap [n_i, n_{i+1})) \\ &\Leftrightarrow & k \in v_i \text{ and } & k \in H_i(\{l\}), \quad \text{for some } l \in X \cap [n_i, n_{i+1}) \\ &\Leftrightarrow & k \in h^{-1}(\{l\}), & \text{for some } l \in X. \end{aligned}$$

This ends the proof, and it is clear that Φ_h is effective in Ψ_H and in Φ_*. $\square$

PROOF OF THEOREM 1.6.2. Let $\Phi \colon \mathcal{P}(\mathbb{N})/\operatorname{Fin} \to \mathcal{P}(\mathbb{N} \times \mathbb{N})/\operatorname{Fin} \times \emptyset$ be a Baire homomorphism. Recall that for a function $f \colon \mathbb{N} \to \mathbb{N}$ we have

$$\Gamma_f = \{\langle m, n \rangle : n \leq f(m)\}.$$

Then, since the ideal Fin $\times \emptyset \upharpoonright \Gamma_f$ is isomorphic to Fin for all f, the mapping

$$\Phi^f(A) = \Phi_*(A) \cap \Gamma_f$$

is a homomorphism of $\mathcal{P}(\mathbb{N})/\operatorname{Fin}$ into $\mathcal{P}(\Gamma_f)/\operatorname{Fin}$ and therefore by Theorem 1.6.1 it has a completely additive lifting, Φ_h, for some $h = h(f) \colon \Gamma_f \to \mathbb{N}$ which is chosen effectively in f, so that the function $f \mapsto h(f)$ is Borel. In particular, the set

$$\mathcal{X} = \{\langle f, h(f) \rangle : f \in \mathbb{N}^\mathbb{N}\}$$

is Borel. Define a partition $[\mathcal{X}]^2 = K_0 \cup K_1$ by letting a pair $\{\langle f, h(f)\rangle, \langle g, h(g)\rangle\}$ in K_0 if
$$h(f)(m) \neq h(g)(m) \quad \text{for some } m \in \Gamma_f \cap \Gamma_g.$$
This partition is open in the natural separable metric topology of $\mathcal{X}$ (see §2.2 for an explanation and terminology), therefore by the Principle of Open Coloring for analytic sets (see Theorem 2.2.12) applied to this partition we have the following two possibilities:

CASE 1. The set $\mathcal{X}$ can be covered by a countable family of K_1-homogeneous subsets. Since the partial order $\langle \mathbb{N}^{\mathbb{N}}, \leq^*\rangle$ is σ-directed there is a K_1-homogeneous set $\mathcal{H}$ such that its projection to $\mathbb{N}^{\mathbb{N}}$ is cofinal in $\langle \mathbb{N}^{\mathbb{N}}, \leq^*\rangle$. Then
$$h = \bigcup_{f \in \mathcal{H}} h(f)$$
is a function because $\mathcal{H}$ is K_1-homogeneous. We claim Φ_h is a lifting of Φ, namely that $\Phi_h(A)\Delta\Phi_*(A) \in \text{Fin} \times \emptyset$ for all A. Assume this is not true, and let A be such that $B = \Phi_h(A)\Delta\Phi_*(A)$ is not in $\text{Fin} \times \emptyset$. Let $f \in \mathcal{H}$ be a function such that $\Gamma_f \cap B$ is not in $\text{Fin} \times \emptyset$; but $h(f) \upharpoonright (\Gamma_f \cap B) = h \upharpoonright (\Gamma_f \cap B)$, and $\Phi_{h(f)}(A)\Delta\Phi_*(A) \in \text{Fin} \times \emptyset$, contradicting the assumption that Φ_h is not a lifting of Φ.

CASE 2. There is a perfect, K_0-homogeneous $\mathcal{H} \subseteq \mathcal{X}$. Then we can find a $g \in \mathbb{N}^{\mathbb{N}}$ such that $\{f \leq g : \langle f, h(f)\rangle \in \mathcal{H}\}$ is perfect. But since $h(f) \upharpoonright \Gamma_f =^* h(g) \upharpoonright \Gamma_f$ for all $f \in \mathcal{H}$, for some $\bar{n}$ the set of $f \in \mathcal{H}$ for which $m > \bar{n}$ implies $h(f)(m, i) = h(g)(m, i)$ for all $\langle m, i\rangle \in \Gamma_f$ is uncountable. Such a set includes an uncountable K_1-homogeneous subset, and therefore Case 2 is not possible, and Case 1 applies. □

1.7. A reformulation of Todorcevic's hypothesis

The results of §1.5 show that a canonical Baire-measurable lifting of a homomorphism between analytic P-ideals is asymptotically additive, i.e., it is obtained by putting together finite maps H_i. In §1.6 we have seen that when in the range we have the quotient $\mathcal{P}(\mathbb{N})/\text{Fin}$, then all but finitely many of H_i are homomorphisms. In general the situation need not be that simple, and these H_i can be only approximate homomorphisms (see Definition 1.7.1 below).

Note that the method of the proof of Theorem 1.6.1 essentially gives that $\emptyset \times \text{Fin}$ has the Radon–Nikodym property. This is because the natural generating submeasure φ for this ideal satisfies the formula $\varphi(A \cup B) = \max\{\varphi(A), \varphi(B)\}$, and this implies that the mapping
$$\Phi_i(s) = \bigcup_{j \in s} H_i(\{j\})$$
is an ε_i-approximation of H_i, if H_i is an ε_i-approximate homomorphism (see Definition 1.7.1 below).

In this section we shall see that Todorcevic's hypothesis turns out to be equivalent to a statement that such non-exact homomorphisms can be approximated by exact homomorphisms modulo a constant error.

Recall that a map $\varphi \colon \mathcal{P}(X) \to [0, \infty]$ is a *submeasure supported by* X if
$\varphi(\emptyset) = 0$,
$\varphi(A) \leq \varphi(A \cup B)$, ($\varphi$ is *monotonic*), and
$\varphi(A \cup B) \leq \varphi(A) + \varphi(B)$ (φ is *subadditive*),

for all $A, B \subseteq X$. We write $\operatorname{supp}(\varphi) = X$. The *norm of* φ is defined by $\|\varphi\| = \varphi(X)$. A submeasure φ is *pathological* if there is $A \subseteq \operatorname{supp}(\varphi)$ such that $\mu(A) < \varphi(A)$ for every measure μ on $X = \operatorname{supp}(\varphi)$ dominated by φ. This corresponds to the notion of an ε-pathological submeasure for $\varepsilon > 0$ (see [**136**]) but it is different from the standard definition of a pathological submeasure as a submeasure which does not dominate a nonzero measure (see [**66**], [**13**], [**122**]). However, we are mainly interested in submeasures with finite supports, and a nonzero submeasure with a finite support always dominates a nonzero measure.

A submeasure is *non-pathological* if it is not pathological, i.e., if it is equal to the pointwise supremum of the measures dominated by itself. For a submeasure φ the function $\hat{\varphi}$ defined by

$$\hat{\varphi}(A) = \sup_{\nu \leq \varphi} \nu(A)$$

(here ν ranges over measures dominated by φ) is a maximal non-pathological submeasure dominated by φ. Let

$$P(\varphi) = \sup_{\hat{\varphi}(A) \neq 0} \frac{\varphi(A)}{\hat{\varphi}(A)}.$$

DEFINITION 1.7.1. If $\mathcal{P}([1, n))$ is equipped with a submeasure φ a mapping $H \colon \mathcal{P}([1, m)) \to \mathcal{P}([1, n))$ is an *ε-approximate homomorphism* (with respect to φ) for some $\varepsilon > 0$ if for all $s, t \subseteq [1, n)$ we have

$$\varphi((H(s) \cup H(t)) \Delta H(s \cup t)) \leq \varepsilon,$$
$$\varphi(H(s^\complement) \Delta H(s)^\complement) \leq \varepsilon.$$

A homomorphism $\Phi \colon \mathcal{P}([1, n)) \to \mathcal{P}([1, m))$ is a *δ-approximation for H*

$$\varphi(\Phi(s) \Delta H(s)) \leq \delta$$

for all $s \subseteq [1, m)$. For a submeasure φ defined on a finite set X let K_φ denote the supremum of all K such that there is an $\varepsilon > 0$ for which some ε-approximate homomorphism $H \colon \mathcal{P}(Y) \to \mathcal{P}(X)$ cannot be $K \cdot \varepsilon$-approximated by a homomorphism.

If $\mathcal{D}$ is a class of metrics, we say that *non-exact homomorphisms with respect to metrics in $\mathcal{D}$ are stable* if there is a universal constant K such that every ε-approximate homomorphism with respect to some metric in $\mathcal{D}$ can be $K\varepsilon$-approximated by an exact homomorphism.

THEOREM 1.7.2. *The following are equivalent*
(1) *The approximate homomorphisms between finite Boolean algebras with respect to arbitrary submeasures are stable,*
(2) *Todorcevic's Conjecture (1.4.1).*

Before proving Theorem 1.7.2, let us prove its special case which shows the idea behind the proof that (1) implies (2) more transparently.

THEOREM 1.7.3. *If the approximate homomorphisms with respect to measures are stable, then all summable ideals have the Radon–Nikodym property.*

PROOF. The following measure supported by $\mathcal{P}(\mathbb{N})$ will be useful

$$\mu_f(A) = \sum_{n \in A} f(n).$$

Note that $\mathcal{I}_f = \mathrm{Exh}(\mu_f)$ (see §1.2). Let $\Phi\colon \mathcal{P}(\mathbb{N})/\mathrm{Fin} \to \mathcal{P}(\mathbb{N})/\mathcal{I}_f$ be a Baire-measurable homomorphism. By Theorem 1.5.2 we can assume that $[n_i, n_{i+1})$, v_i and $H_i\colon \mathcal{P}([n_i, n_{i+1})) \to \mathcal{P}(v_i)$ are such that Ψ_H (see Definition 1.5.1) is a lifting of a homomorphism of $\mathcal{P}(\mathbb{N})/\mathrm{Fin}$ into $\mathcal{P}(\mathbb{N})/\mathcal{I}_g$. Define ε_i to be the minimal ε for which H_i is an ε-approximate homomorphism (for definition see §1.8, and note that ε_i is, being equal to a maximum of a finite set of real numbers, well defined). By $s^\complement$ we shall denote either $[n_i, n_{i+1}) \setminus s$ or $v_i \setminus s$, depending on whether $s \subseteq [n_i, n_{i+1})$ or $s_i \subseteq v_i$ for some i. It will always be clear from the context which case of the definition applies.

CLAIM 1. $\varepsilon = \sum_{i=1}^\infty \varepsilon_i < \infty$.

PROOF. Assume the contrary. For every i there are $s_i, t_i \subseteq [n_i, n_{i+1})$ such that either
$$\mu_f((H_i(s_i) \cup H_i(t_i)) \Delta H_i(s_i \cup t_i)) = \varepsilon_i$$
or
$$\mu_f(H_i(s_i^\complement) \Delta H_i(s_i)^\complement) = \varepsilon_i.$$
Let M_1 be equal to the sum of all ε_i for which the first possibility applies. Then sets $X = \bigcup_i s_i$ and $Y = \bigcup_i t_i$ satisfy
$$\mu_f(\Psi_H(X \cup Y) \Delta (\Psi_H(X) \cup \Psi_H(Y)))) = M_1,$$
therefore M_1 is finite, and the sum of all ε_i for which the second possibility applies is infinite so we can find a set X such that
$$\mu_f(\Psi_H(X)^\complement \Delta \Psi_H(X^\complement))) = \infty.$$
This contradicts to the assumption that Ψ_H is a lifting of a homomorphism and ends the proof. □

By applying the stability assumption to H_i, we obtain a function $h_i\colon v_i \to [n_i, n_{i+1})$ such that Φ_{h_i} is a $K\varepsilon_i$-approximation (with respect to μ_f) of H_i. Let $h = \bigcup_i h_i$. Then for $A \subseteq \mathbb{N}$ we have
$$\mu_f(\Phi_h(A) \Delta \Psi_H(A)) \leq \sum_i K\varepsilon_i < \infty$$
and therefore Φ_h is the required lifting of Φ. □

PROOF OF THEOREM 1.7.2. We shall first prove that (1) implies (2). Let $\Phi\colon \mathcal{P}(\mathbb{N})/\mathrm{Fin} \to \mathcal{P}(\mathbb{N})/\mathcal{I}$ be a homomorphism with a Baire lifting. Then Lemma 1.3.2 implies that Φ has a continuous lifting, so let Φ_* be a fixed continuous lifting of Φ. We shall use the following.

For $A \subseteq \mathbb{N}$ let $\Phi \restriction A$ denote the restriction of Φ to the set $\mathcal{P}(A)$. Let
$$\mathcal{J}_{\mathrm{RN}} = \{A : \Phi \restriction A \text{ has a completely additive lifting }\}.$$
We will prove this is an improper ideal.

CLAIM 2. The ideal $\mathcal{J}_{\mathrm{RN}}$ is a $\mathbf{\Sigma}_2^1$ set of reals.

PROOF. For $A \subseteq \mathbb{N}$ we have $A \in \mathcal{J}_{\mathrm{RN}}$ if and only if
$$\exists h\colon \mathbb{N} \to A \,\forall B \subseteq A \,\forall n \,\exists m \, \varphi((\Phi_*(B) \Delta \Phi_h(B)) \cap [m, \infty)) < 2^{-n},$$
and therefore $\mathcal{J}_{\mathrm{RN}}$ is as required. □

In particular, the assertion of theorem, "the homomorphism Φ has an asymptotically additive lifting," is a $\mathbf{\Sigma}_2^1$-statement, and therefore it will suffice to prove that it follows from Martin's Axiom (see [**82**]), since by Shoenfield's absoluteness theorem ([**110**]) this additional assumption can be removed from the proof.

CLAIM 3. *The ideal $\mathcal{J}_{\mathrm{RN}}$ is nonmeager and therefore it is either improper or it does not have the property of Baire.*

PROOF. By a result of Jalali–Naini ([**58**]) and Talagrand ([**121**]) (Corollary 3.10.2), it will suffice to prove that Fin $\not\leq_{\mathrm{RB}} \mathcal{J}_{\mathrm{RN}}$, namely that for every strictly increasing sequence m_i of integers there is a set in $\mathcal{J}_{\mathrm{RN}}$ which includes infinitely many of the intervals $[m_i, m_{i+1})$. Let m_i be an increasing sequence of integers. By Theorem 1.5.2 this sequence has an infinite subsequence n_i such that the homomorphism Φ has an asymptotically additive lifting, Ψ_H, with parameters $\{n_i\}$, $\{v_i\}$ and $\{H_i\}$ for some sequences $\{v_i\}$ and $\{H_i\}$. For every $i \in \mathbb{N}$ let $m(i)$ be the minimal m such that H_i is not a 2^{-m-1}-approximate homomorphism with respect to φ. We claim that
$$\lim_{i \to \infty} m(i) = \infty.$$
Otherwise we would have an $\bar{m} \in \mathbb{N}$ such that $m(i) = \bar{m}$ infinitely often, and this would give us sets $A, B \subseteq \mathbb{N}$ such that either
$$\varphi((\Psi_H(A \cup B) \Delta (\Psi_H(A) \cup \Psi_H(B))) \cap [k, \infty)) \geq 2^{-\bar{m}}$$
for all k, or
$$\varphi(\Psi_H(A^\complement) \Delta \Psi_H(A)^\complement) \cap [k, \infty)) \geq 2^{-\bar{m}},$$
for all k, contrary to the fact that Ψ_H is a lifting of Φ. (This is proved exactly like in the proof of Theorem 1.6.1, where all but finitely many H_i's were homomorphisms.) Therefore we can find an infinite set C such that the sequence $\{m(i)\}_{i \in C}$ is strictly increasing. Note that, by our assumption, for each $\varphi_i = \varphi \upharpoonright v_i$ we have $P(\varphi_i) \leq M$ and therefore by the stability assumption (1) for each $i \in C$ there is a homomorphism
$$H_i' \colon \mathcal{P}([n_i, n_{i+1})) \to \mathcal{P}(v_i)$$
which is an $M \cdot 2^{-m(i)}$-approximation to H_i. Each H_i' is of the form Φ_{h_i} for some $h_i \colon v_i \to [n_i, n_{i+1})$ and if we let $h = \bigcup_i h_i$ then Φ_h is a witness that $\bigcup_{i \in C}[n_i, n_{i+1})$ is in $\mathcal{J}_{\mathrm{RN}}$. □

Recall that Martin's Axiom implies that all $\mathbf{\Sigma}_2^1$ sets of reals have the property of Baire ([**91**]), and therefore by Claims 2 and 3 the ideal $\mathcal{J}_{\mathrm{RN}}$ is, under MA, improper. In particular, $\mathbb{N} \in \mathcal{J}_{\mathrm{RN}}$ and therefore Φ has a completely additive lifting. As noted before, since this conclusion is absolute, this concludes the proof that (1) implies (2).

Now we prove that (2) implies (1). Let us assume that (1) fails. By using this assumption, we recursively construct

(a) sequences $0 = n_0 < n_1 < n_2 < \ldots$ and $0 = m_0 < m_1 < m_2 < \ldots$,
(b) a submeasure φ_i on $[m_i, m_{i+1})$,
(c) an $1/2^i$-approximate homomorphism $H_i \colon \mathcal{P}([n_i, n_{i+1})) \to \mathcal{P}([m_i, m_{i+1}))$ which cannot be 1-approximated by a homomorphism.

Define a submeasure φ on $\mathbb{N}$ by
$$\varphi(A) = \sum_{i=0}^{\infty} \varphi_i(A \cap [m_i, m_{i+1}))$$

and let $\mathcal{J}$ be $\mathrm{Exh}(\varphi)$. Consider the asymptotically additive mapping Ψ_H determined by the parameters $\{n_i\}$, $\{v_i = [m_i, m_{i+1})\}$, and H_i, as in Definition 1.5.1.

CLAIM 4. *Ψ_H is a lifting of a homomorphism which has no completely additive lifting Φ_h.*

PROOF. It is straightforward to check that Ψ_H is a lifting of a homomorphism from $\mathcal{P}(\mathbb{N})/\mathrm{Fin}$ into $\mathcal{P}(\mathbb{N})/\mathcal{J}$ and it is obviously continuous. Assume h is such that Φ_h is a lifting of the same homomorphism as Ψ_H. Let $n_{k(i)}$ be a subsequence of n_i such that

(1) $h^{-1}([1, n_{k(i)})) \subseteq [1, n_{k(i+1)})$, and
(2) $h''[1, n_{k(i)}) \subseteq [1, n_{k(i+1)})$.

Such a subsequence can be chosen recursively, using the fact that h is finite-to-one. Let $H'_i \colon \mathcal{P}([n_{k(i)}, n_{k(i+1)}))$ be defined by "joining" H_i's, namely by

(3) $H_{k(i)}(s) = \bigcup_{j=k(i)}^{k(i+1)-1} H_j(s)$.

(Clearly, $H_{k(i)}(s) = \Psi_H(s)$.) Let also $v_{k(i)} = \bigcup_{j=k(i)}^{k(i+1)-1} v_j$, so that the range of $H_{k(i)}$ is included in $\mathcal{P}(v_{k(i)})$. Note that the map

$$\Psi_{H_k}(X) = \bigcup_{i=1}^{\infty} H_{k(i)}(X \cap [n_{k(i)}, n_{k(i+1)}))$$

coincides with Ψ_H. Also, it is straightforward to prove that

SUBCLAIM. *The map $H_{k(i)}$ cannot be 1-approximated by a homomorphism.* □

We claim that the set

$$B = \{j : j \in v_{k(i)} \quad \text{and} \quad h(j) \notin [n_{k(i)}, n_{k(i+1)}) \text{ for some } i\}$$

is in the ideal $\mathcal{J}$. Assume otherwise, and let $C = h''B$. Since $B \setminus \Phi_h(C) \in \mathcal{J}$, C is not in $\ker(\Phi_h) = \ker(\Psi_H)$. As usually, let

$$A_0 = \bigcup_{i=1}^{\infty} [n_{k(2i)}, n_{k(2i+1)}) \quad \text{and} \quad A_1 = \bigcup_{i=1}^{\infty} [n_{k(2i+1)}, n_{k(2i+2)}).$$

Assume that $C \cap A_0$ is not in $\ker(\Phi_h)$. Then the set

$$\Phi_h(C \cap A_0) \setminus \bigcap_{i=1}^{\infty} v_{k(2i+1)} = \Phi_h(C \cap A_0) \setminus \Psi_{H_k}(A_0)$$

is $\mathcal{J}$-positive, a contradiction. Therefore $C \cap A_0$ is in $\ker(\Phi_h)$, and an analogous proof shows that $C \cap A_1$ is in $\ker(\Phi_h)$. Therefore, since $B \subseteq \Phi_h(C)$, this proves that $B \in \mathcal{J}$.

Let us now define $H'_{k(i)} \colon \mathcal{P}([n_{k(i)}, n_{k(i+1)})) \to \mathcal{P}(v_{k(i)})$ by

$$H'_{k(i)}(s) = \Phi_h(s) \cap v_{k(i)}.$$

Since the set B defined above is in $\mathcal{J}$, the mapping $H'_{k(i)}$ can be 1/2-approximated by a homomorphism (namely, the restriction of Φ_h) for all but finitely many i. Therefore, by the Subclaim above, $H_{k(i)}$ cannot be 1/2-approximated by $H'_{k(i)}$, and $H'_{k(i)}$ is not a 1/2-approximation to $H_{k(i)}$ for all large enough i.

Let $s_i \subseteq [n_i, n_{i+1})$ be such that $\varphi(H'_i(s_i) \Delta H_i(s_i)) > 1/2$. Since h is finite-to-one, we can find an infinite $D \subseteq \mathbb{N}$ such that $\bigcup_{i \in D} H'_i(s_i) \Delta \Phi_h(\bigcup_{i \in D} s_i)$ is not in $\mathcal{J}$, and therefore $\Phi_h(\bigcup_{i \in D} s_i) \Delta \Psi_H(\bigcup_{i \in D} s_i)$ is also not in $\mathcal{J}$. Therefore Φ_h cannot be a lifting of this homomorphism. □

Therefore the ideal $\mathcal{J}$ does not have the Radon–Nikodym property. □

We shall now give a version Theorem 1.7.2 which will turn out to be more useful. We say that an analytic P-ideal $\mathcal{I}$ is *non-pathological* if it is of the form $\mathrm{Exh}(\varphi)$ for a non-pathological lower semicontinuous submeasure φ.

THEOREM 1.7.4. *The following are equivalent*
(1) *The approximate homomorphisms between finite Boolean algebras with respect to arbitrary non-pathological submeasures are stable,*
(2) *Todorcevic's Conjecture for non-pathological analytic P-ideals.*

PROOF. The proof is a minor modification of the proof of Theorem 1.7.2. In direction (1) implies (2), one only has to notice that if φ is non-pathological, then so is its restriction to a subset of its support, and this follows immediately from the definition of a non-pathological submeasure. In direction (2) implies (1), one only needs to note that if every φ_n is non-pathological, then so is $\sum_n \varphi_n$. □

LEMMA 1.7.5. *If an analytic P-ideal $\mathcal{I}$ is equal to $\mathrm{Exh}(\varphi)$ for a lower semicontinuous submeasure φ such that $K_\varphi < \infty$, then $\mathcal{I}$ is non-pathological.*

PROOF. If $\mathcal{I} = \mathrm{Exh}(\varphi)$ and $P(\varphi) = M < \infty$, then $\hat{\varphi} \leq \varphi \leq M\hat{\varphi}$, therefore $\mathcal{I} = \mathrm{Exh}(\hat{\varphi})$ and $\mathcal{I}$ is non-pathological. □

Therefore we can say that an analytic P-ideal is *pathological* if for every lower semicontinuous submeasure φ such that $\mathcal{I} = \mathrm{Exh}(\varphi)$ we have $P(\varphi) = \infty$.

1.8. Approximate homomorphisms and non-pathological submeasures

Our motivation for studying non-exact homomorphisms is explained in the previous section, but they were studied in their own right, for example in the context of so-called *Hyers–Ulam stability*. See, for example, [**139**, §VI.1], [**140**, V.4], [**44**], also [**67**], [**33**], [**37**], [**72**] for more on non-exact homomorphisms in various contexts.

The upper bound for the constant K_M claimed in [**33**] is 521, which is slightly better than the constant that can be obtained from the proof of Theorem 1.8.1 below. Max Burke and David Fremlin have noticed two mistakes in a proof in [**33**], the correcting of which has caused the numerical bound to increase. Anyway, in [**37**] this bound was improved to 160, and it can be improved further by combining the results of [**72**] and [**37**, §3].

THEOREM 1.8.1. *The non-exact homomorphisms with respect to measures are stable. Namely, there is a universal constant K_M such that $K_\mu \leq K_M$ for every measure μ on some $\mathcal{P}([1, n))$.*

THEOREM 1.8.2. *The non-exact homomorphisms with respect to non-pathological submeasures are stable. Namely, the constant K also satisfies $K_\varphi \leq K_M \cdot P(\varphi)$ for every non-pathological submeasure φ.*

In our proof of Theorem 1.8.1 we shall need a result which says that a family of subsets of some finite set which almost everywhere looks like an ultrafilter must be close to some principal ultrafilter, $\langle k \rangle = \{s \in \mathcal{P}([1, m)) : k \in s\}$. Let ν denote the uniform probabilistic measure on $\mathcal{P}([1,m))$, and let ν^2 denote the product measure on $\mathcal{P}([1,m)) \times \mathcal{P}([1,m))$.

It should be noted that the following theorem remains true even when the condition (A1) is omitted, possibly at the expense of increasing the numerical constant.

THEOREM 1.8.3. *If $\delta > 0$ is small enough (say, $\delta \leq 1/156$) and $A \subseteq \mathcal{P}([1,m))$ satisfies*

(A1) $\nu^2((A \times \mathcal{P}([1,m)) \cup \mathcal{P}([1,m)) \times A) \Delta \{\langle s,t \rangle : s \cup t \in A\}) < \delta$,
(A2) $\nu^2((A \times A) \Delta \{\langle s,t \rangle : s \cap t \in A\}) < \delta$,
(A3) $\nu\{s : s, s^{\complement} \notin A \text{ or } s, s^{\complement} \in A\} < \delta$,

then there is a $k \in \{1,\ldots,m\}$ such that $\nu\{s \in A : k \in s\} > (1-28\delta)/2$. In particular $\nu(A \Delta \langle k \rangle) < 29\delta$.

PROOF. For $I \subseteq \{1,\ldots,m\}$ and s we define
$$s^I = (I \cap s) \cup (I^{\complement} \setminus s),$$
$$A^2(I) = \{s : s, s^I \in A\},$$
$$E_I = \nu(A^2(I)).$$

The number E_I is, in some sense, a measure of how A concentrates on I. For example, if A is an ultrafilter, then E_I is equal to either $1/2$ or 0, depending on whether I is in A or not. Note that (A3) implies

(A) $$E_\emptyset < \delta \quad \text{and} \quad E_{\{1,\ldots,m\}} = \nu(A) > \frac{1}{2} - \delta.$$

The set of all $s \in \mathcal{P}([1,m))$ such that $s, s^{\complement} \notin A$ or $s^I, (s^I)^{\complement} \notin A$ is, by (A3), of measure $< 2\delta$, and for each s outside of this set we either have $s \in A^2(I) \cup A^2(I^{\complement})$ or $s^{\complement} \in A^2(I) \cup A^2(I^{\complement})$. (To see this, consider the possible cases: if $s, s^I \in A$ then $s \in A^2(I)$, if $s, (s^I)^{\complement} \in A$ then $s \in A^2(I^{\complement})$, and so on). Therefore

(B) $$E_I + E_{I^{\complement}} \geq \frac{1-2\delta}{2}.$$

We claim that

(C) $$E_I \cdot E_{I^{\complement}} < 5\delta.$$

To see this, note that the left-hand side is equal to $\nu(A^2(I)) \cdot \nu(A^2(I^{\complement})) = \nu^2(A^2(I) \times A^2(I^{\complement}))$ and that $C = A^2(I) \times A^2(I^{\complement}) = C_0 \cup C_1 \cup C_2$, where

$$C_0 = \{\langle s,t \rangle \in C : s \cap t, s^I \cap t^{I^{\complement}} \in A\},$$
$$C_1 = \{\langle s,t \rangle \in C : s \cap t \notin A\},$$
$$C_2 = \{\langle s,t \rangle \in C : s^I \cap t^{I^{\complement}} \notin A\}.$$

It suffices to show that $\nu^2(C_0) < 3\delta$, $\nu^2(C_1) < \delta$ and $\nu^2(C_2) < \delta$. The latter two inequalities follow immediately from (A2), and C_0 has the same size as the set

$$\{\langle u,v \rangle : \langle (u \cap I) \cup (v \cap I^{\complement}), (v \cap I) \cup (u \cap I^{\complement}) \rangle \in C \quad \text{and} \quad v \cap u, v^{\complement} \cap u \in A\}.$$

To see that the measure of this set is $< 3\delta$, we split it into three pieces, depending on whether $v \notin A$, $v^{\complement} \notin A$, or $v, v^{\complement} \in A$. Each of these pieces has a measure less than δ (by (A2) and (A3)), and therefore $\nu^2(C_2) < 3\delta$. Let

$$\gamma = \sqrt{(1/2 - \delta)^2 - 20\delta}.$$

1.8. APPROXIMATE HOMOMORPHISMS AND NON-PATHOLOGICAL SUBMEASURES

By using (B) and (C), we get $E_I(1/2 - E_I - \delta) < 5\delta$ and therefore

$$(D) \qquad E_I < \frac{1 - 2\delta - 2\gamma}{4} \quad \text{or} \quad \frac{1 - 2\delta + 2\gamma}{4} < E_I.$$

Let $E_k = E_{\{1,\ldots,k\}}$. By (A) there exists a minimal $k \leq m$ such that $E_k > (1 - 2\delta + 2\gamma)/2$, and by (D) for this k we have

$$(E) \qquad E_k - E_{k-1} > \gamma.$$

This means that A concentrates on $\{1,\ldots,k\}$, but not on $\{1,\ldots,k-1\}$. We shall prove that the principal ultrafilter $\langle k \rangle = \{s \in \mathcal{P}([1,m)) : k \in s\}$ is close to A. In order to do so, we have to prove that the set

$$G = \{s \in A^2(\{1,\ldots,k\}) : k \notin s\}$$

is small. Consider the set

$$F = \{s : s \in A \quad \text{and} \quad s \cup \{k\} \notin A\}.$$

Fix $s \in G \setminus A^2(\{1,\ldots,k-1\})$. Then both s and $s^{\{1,\ldots,k\}}$ are in A but $s^{\{1,\ldots,k-1\}}$ is not. Since $k \notin s$ (because $s \in G$), this implies $s^{\{1,\ldots,k-1\}} = s^{\{1,\ldots,k\}} \cup \{k\}$, and therefore $s^{\{1,\ldots,k\}} \in F$. Hence we have

$$(F) \qquad \nu(G) \leq E_{k-1} + \nu(F).$$

We claim that

$$(G) \qquad \nu(F) = \nu\{s : s \in A \quad \text{and} \quad s \cup \{k\} \notin A\} < \frac{8\delta}{1 - 2\delta}.$$

To see this, let $L = \{t : k \notin t \text{ and either } t \text{ or } t \cup \{k\} \text{ is in } A\}$. We have

$$\{t : t, t^C \notin A\} \supseteq \{t : k \notin t \quad \text{and} \quad t, t \cup \{k\}, t^C \setminus \{k\}, t^C \notin A\}$$
$$\supseteq \{t : k \notin t \quad \text{and} \quad t, t^C \setminus \{k\} \notin L\}$$
$$= M, \text{ say,}$$

and therefore by (A3) we have

$$(H) \qquad \nu(L) \geq \frac{\nu\{t : k \notin t\} - \nu(M)}{2} \geq \frac{1 - 2\delta}{4}.$$

Consider sets

$$F_0 = \{\langle s, t \rangle \in F \times L : s \cap t = s \cap (t \cup \{k\}) \notin A\},$$
$$F_1 = \{\langle s, t \rangle \in F \times L : s \cap t = s \cap (t \cup \{k\}) \in A\}.$$

For $t \in L$ define

$$\hat{t} = \begin{cases} t, & \text{if } t \in A, \\ t \cup \{k\}, & \text{otherwise.} \end{cases}$$

Thus (A2) implies $\nu^2\{\langle s, \hat{t} \rangle : \langle s, t \rangle \in F_0\} < \delta$ and $\nu^2\{\langle s \cup \{k\}, t \rangle : \langle s, t \rangle \in F_1\} < \delta$. Since $s \mapsto s \cup \{k\}$ is a $1-1$ mapping on F and $t \mapsto \hat{t}$ is a $1-1$ mapping on L, this implies that $\nu^2(F \times L) = \nu^2(F_0) + \nu^2(F_1) < 2\delta$. Therefore (H) implies that $\nu(F) < 8\delta/(1-2\delta)$, as required. Finally we have

$$\begin{aligned}\nu\{s \in A : k \in s\} &\geq \nu\{s \in A^2(\{1,\dots,k\}) : k \in s\} \\ &= E_k - \nu\{s \in A^2(\{1,\dots,k\}) : k \notin s\} \\ &> E_k - E_{k-1} - \frac{8\delta}{1-2\delta} \quad \text{(by (F) and (G))} \\ &> \gamma - \frac{8\delta}{1-2\delta} \quad \text{(by (E))}.\end{aligned}$$

Therefore

$$\begin{aligned}\nu(A^2(\{k\})) &\geq \nu\{s : k \in s \text{ and } s, s^\complement \cup \{k\} \in A\} \\ &\geq 2\nu\{s \in A : k \in s\} - \frac{1}{2} \\ &> 2\left(\gamma - \frac{8\delta}{1-2\delta}\right) - \frac{1}{2} \\ &> 2(\gamma - 8\delta) - \frac{1}{2}.\end{aligned}$$

We claim that $\nu(A^2(\{k\})) > (1 - 2\gamma - 2\delta)/4$. By the above, this inequality reduces to $10\gamma > 3 + 62\delta$, which is satisfied since $\delta \leq 1/156$ and $\gamma = \sqrt{(1/2 - \delta)^2 - 20\delta}$. By (D), we have $\nu(A^2(\{k\})) = E_{\{k\}} > (1 - 2\delta + 2\gamma)/4$ and

$$\nu\{s \in A : k \in s\} > \nu(A^2(\{k\})) - \nu\{s \in A : s \cup \{k\} \notin A\} > \frac{1 - 28\delta}{2}.$$

Also, $\nu(A \Delta \langle k \rangle) = \nu(A) + \nu(\langle k \rangle) - 2\nu(A \cap \langle k \rangle) \leq 1/2 + \delta/2 + 1/2 - (1 - 28\delta) < 29\delta$. □

REMARK 1.8.4. Note that the proof of Theorem 1.8.3 shows that (under the same assumptions and using the same notation) for all $I \ni k$ we have

$$(I) \qquad \nu\{s : s, s^I \in A\} > \frac{1 - 2\delta + 2\gamma}{4} > \frac{1 - 2\delta + 2\gamma}{4},$$

and $\lim_{\delta \to 0^+} (1 - 2\delta + 2\gamma)/4 = 1$.

PROOF OF THEOREM 1.8.1. Assume $H \colon \mathcal{P}([1,m)) \to \mathcal{P}([1,n))$ is an ε-approximate homomorphism with respect to a measure μ on $\mathcal{P}([1,n))$. For $j \leq n$ let

$$A_j = \{s \in \mathcal{P}([1,m)) : H(s) \ni j\}.$$

Let $\delta = 1/156$, and for $i = 1, 2, 3$ let N_i be the set of all $j \leq n$ for which (Ai) fails for the set A_j. Since for all $s, t \in \mathcal{P}([1,m))$ we have $\mu((H(s) \cup H(t)) \Delta H(s \cup t)) < \varepsilon$, Fubini's theorem applied to the product $(\mathcal{P}([1,m)))^2 \times \mathcal{P}([1,n))$ implies that $\mu(N_1) < \varepsilon/\delta$. To get a bound for $\mu(N_2)$, note that

$$\begin{aligned}\mu((H(s) \cap H(t)) \Delta H(s \cap t)) &= \mu((H(s) \cap H(t))^\complement \Delta H(s \cap t)^\complement) \\ &\leq \mu((H(s)^\complement \cup H(t)^\complement) \Delta (H(s^\complement) \cup H(t^\complement))) \\ &\quad + \mu((H(s^\complement) \cup H(t^\complement)) \Delta H(s^\complement \cup t^\complement)) \\ &\quad + \mu(H((s \cap t)^\complement) \Delta H(s \cap t)^\complement) \\ &\leq 4\varepsilon.\end{aligned}$$

By Fubini's theorem we have $\mu(N_2) < 4\varepsilon/\delta$, and similarly, $\mu(N_3) < \varepsilon/\delta$. Therefore

$$(J) \qquad \mu\left(\bigcup_{i=1}^{3} N_i\right) < \frac{6\varepsilon}{\delta}.$$

For every $j \in \{1, \ldots, n\} \setminus \bigcup_{i=1}^{3} N_i$ the set A_j satisfies assumptions of Theorem 1.8.3, so for such a j let $k(j) \leq m$ be such that $\nu(A_j \cap \langle k(j)\rangle) > 29\delta$. Assume for a moment that $N_i = \emptyset$ for all i, and define a homomorphism $\Phi \colon \mathcal{P}([1,m]) \to \mathcal{P}([1,n])$ by

$$\Phi(s) = \{j : k(j) \in s\}.$$

Let us write $\zeta = (1 - 2\delta + 2\gamma)/4$ and recall that $\zeta \approx 1$ when $\delta \approx 0$.

CLAIM 1. *The homomorphism Φ is a $(10/\zeta + 1)\varepsilon$-approximation to H.*

PROOF. Assume it is not. Thus for some $I \subseteq \{1, \ldots, m\}$ we have

$$\mu(\Phi(I) \Delta H(I)) > \left(\frac{10}{\zeta} + 1\right)\varepsilon.$$

Since $(H(I) \setminus \Phi(I))\Delta(\Phi(I^\complement) \setminus H(I^\complement)) \subseteq H(I) \Delta H(I^\complement)^\complement$ and the measure of the latter set is not bigger than ε, we can assume (possibly by replacing I with $I^\complement$) that

$$\mu(\Phi(I) \setminus H(I)) > \frac{5\varepsilon}{\zeta}.$$

Using the notation defined in the proof of Theorem 1.8.3, by (I) we have

$$\nu\{s : s, s^I \in A_j\} > \zeta$$

for all $j \in \Phi(I) \setminus H(I)$. By Fubini's theorem, there are a $C_1 \subseteq \Phi(I) \setminus H(I)$ and a $\bar{u} \in \mathcal{P}([1,m])$ such that

$$\mu(C_1) \geq \zeta \cdot \frac{5\varepsilon}{\zeta} = 5\varepsilon$$

and $\bar{u}, \bar{u}^I \in \bigcap_{j \in C_1} A_j$. Finally we have

$$0 = \mu(C_1 \cap H(I)) > \mu(C_1 \cap (H(\bar{u}) \cap H(\bar{u}^I))) - 5\varepsilon = \mu(C_1) - 5\varepsilon \geq 0,$$

a contradiction. □

If the set $\bigcup_{i=1}^{3} N_i$ is not empty, then extend the function k obtained using Theorem 1.8.3 by letting $k(j) = 1$ for all j in this set, and define Φ as in Claim above. Then the proof of the Claim shows that

$$\nu\left((\Phi(s)\Delta H(s)) \cap \left(\mathcal{P}([1,n]) \setminus \bigcup_{i=1}^{3} N_i\right)\right) \leq \left(\frac{10}{\zeta} + 1\right)\varepsilon.$$

This shows that Φ is a $(10/\zeta + 1 + 6/\delta)$-approximation of H, for any $\delta \leq 156$. □

The definitions of A_j's and Φ did not depend on either μ or ε, and therefore we have proved the following stronger version of Theorem 1.8.1:

THEOREM 1.8.5. *For every map $H \colon \mathcal{P}([1,m]) \to \mathcal{P}([1,n])$ there is a homomorphism $\Phi \colon \mathcal{P}([1,m]) \to \mathcal{P}([1,n])$ such that for every measure μ on $\{1, \ldots, n\}$ and every ε, if H is an ε-approximate homomorphism with respect to μ, then Φ is an $\varepsilon \cdot K_M$-approximation to H.* □

PROOF OF THEOREM 1.8.1. Let $H\colon \mathcal{P}([1,m)) \to \mathcal{P}([1,n))$ be an ε-approximate (with respect to φ) homomorphism, and let $\Phi\colon \mathcal{P}([1,m)) \to \mathcal{P}([1,n))$ be as guaranteed by Theorem 1.8.5. To prove that Φ is as required, fix $s \in \mathcal{P}([1,m))$, let $u = H(s)\Delta\Phi(s)$ and find a measure $\mu \leq \varphi$ on $\mathcal{P}([1,n))$ such that $P(\varphi) \cdot \mu(u) = \varphi(u)$. Such a μ exists because $\mathcal{P}([1,n))$ is finite, and we have sup = max. Since $\mu \leq \varphi$ and H is an ε-approximate homomorphism with respect to φ, it is an ε-approximate homomorphism with respect to μ as well. Therefore Theorem 1.8.5 implies $\varphi(u) = P(\varphi) \cdot \mu(u) \leq \varepsilon \cdot K_\varphi$. □

Theorem 1.8.1 fails for arbitrary submeasures by the following result.

THEOREM 1.8.6. *The non-exact homomorphisms with respect to arbitrary submeasures are not stable. Namely, for every $M < \infty$ there is a submeasure φ such that $K_\varphi > M$.*

PROOF. For $m > 2^{3M+2}$ let $n = 2^{2^m}$, so we can identify $\{1, \ldots, n\}$ with the set N of all subsets X of $\mathcal{P}([1,m))$. We shall denote the elements of N by X, Y, Z. Define $H\colon \mathcal{P}([1,m)) \to \mathcal{P}([1,n))$ by

$$H(s) = B_s = \{X \in N : s \in X\}.$$

For $k = 1, \ldots, m$ let $\langle k \rangle$ denote the principal ultrafilter $\{s \in \mathcal{P}([1,m)) : k \in s\}$. Define the submeasure $\varphi = \varphi_m$ supported by N by letting

$$\varphi\{\langle 1 \rangle, \langle 2 \rangle, \ldots, \langle m \rangle\} = 0$$

and for $\mathcal{A}$ disjoint from $\{\langle 1 \rangle, \ldots, \langle m \rangle\}$ let

$\varphi(\mathcal{A}) = \min\{|X| : X \in N$ is such that

$Y \cap X \neq \langle k \rangle \cap X$ for all $Y \in \mathcal{A}$ and $k = 1, \ldots, m\}$.

In other words, if for $X \in N$ we define

$$\mathcal{C}_X = \{Y \in N : (Y \Delta \langle k \rangle) \cap X \neq \emptyset \text{ for all } k = 1, \ldots, m\},$$

then $\varphi(\mathcal{A})$ is equal to the smallest size of $X \in N$ such that $\mathcal{A} \subseteq \mathcal{C}_X$.

CLAIM 2. *The function φ is a submeasure.*

PROOF. The monotonicity of φ is trivial, while its subadditivity follows from the formula $\mathcal{C}_X \cup \mathcal{C}_Y \subseteq \mathcal{C}_{X \cup Y}$. □

CLAIM 3. *The function H is a $(3+\varepsilon)$-approximate homomorphism with respect to φ for every $\varepsilon > 0$.*

PROOF. Note that for every $s \in \mathcal{P}([1,m))$ we have $H(s)^\complement \Delta H(s^\complement) \subseteq \mathcal{C}_{\{s,s^\complement\}}$, therefore $\varphi(H(s)^\complement \Delta H(s^\complement)) \leq 2$. Similarly for all s,t we have $(H(s) \cup H(t))\Delta H(s \cup t) \subseteq \mathcal{C}_{\{s,t,s\cup t\}}$, thus $\varphi((H(s) \cup H(t))\Delta H(s \cup t)) \leq 3$. □

If $K_\varphi \leq M$ then H can be $3M$-approximated by a homomorphism Φ. For $s \in \mathcal{P}([1,m))$ let $\mathcal{A}_s = \Phi(s) \Delta H(s)$, let $X_s* \subseteq \mathcal{P}([1,m))$ be of size $3M$ such that $\mathcal{A}_s \subseteq \mathcal{C}_{X_s*}$, and let $X_s = X_s* \cup \{s\}$. Then for $k = 1, \ldots, m$ we have

$$\Phi(\{k\}) \supseteq H(\{k\}) \setminus \mathcal{A}_{\{k\}}.$$

Since any $Y \in H(\{k\})$ satisfying $(Y \Delta \langle k \rangle) \cap X_{\{k\}} = \emptyset$ is not in $\mathcal{A}_{\{k\}}$, we have

$$|\Phi(\{k\})| \geq |\{Y \in H(\{k\}) : (Y \Delta \langle k \rangle) \cap X_{\{x\}} = \emptyset\}|$$
$$= |\{Y \in N : \{k\} \in Y \text{ and } (Y \Delta \langle k \rangle) \cap X_{\{x\}} = \emptyset\}| \geq 2^{2^m - 3M - 2}.$$

The sets $\Phi(\{k\})$ ($k=1,\ldots,m$) are pairwise disjoint, so we have

$$n \geq \left|\bigcup_{k=1}^{m} \Phi(\{k\})\right| \geq m \cdot 2^{2^m - 3M - 2} > 2^{2^m} = n,$$

a contradiction. Therefore $K_\varphi > M$, as promised. □

1.9. The lifting theorem and the non-lifting theorem

In this section we shall use results of §1.5 and §1.8 to prove that for the class of natural analytic P-ideals Todorcevic's hypothesis is true, but that it fails for pathological ideals. An analytic P-ideal $\mathcal{I}$ is pathological if for every submeasure φ such that $\mathcal{I} = \text{Exh}(\varphi)$ we have $P(\varphi) = \infty$. The following lifting theorem for non-pathological analytic P-ideals is the main result of Chapter 1.

THEOREM 1.9.1. *If an analytic P-ideal $\mathcal{I}$ is non-pathological, then it has the Radon–Nikodym property.*

PROOF. This follows immediately from Theorem 1.7.4 and Theorem 1.8.2. □

This result shows that Todorcevic's hypothesis (Conjecture 1.4.1) is "true for all practical purposes"; let us list some of its specific instances.

COROLLARY 1.9.2. *Every summable ideal has the Radon–Nikodym property.*

PROOF. Summable ideals are induced by measures and therefore non-pathological. □

COROLLARY 1.9.3. *Every Erdős–Ulam ideal has the Radon–Nikodym property.*

PROOF. Erdős–Ulam ideals are non-pathological by Theorem 1.13.2 and Theorem 1.13.3. □

COROLLARY 1.9.4. *The ideal $\emptyset \times \text{Fin}$ has the Radon–Nikodym property.*

PROOF. This is because $\emptyset \times \text{Fin} = \text{Exh}(\varphi)$ for a non-pathological submeasure φ defined by $\varphi(A) = \max\{2^{-n} : A \not\subseteq [n, \infty) \times \mathbb{N}\}$. □

The class of non-pathological ideals is rather extensive since it includes essentially all analytic P-ideals occurring in the literature. As we have seen, all P-ideals given in Example 1.2.3 are non-pathological because natural submeasures inducing them satisfy $\varphi = \hat{\varphi}$. A large class of non-pathological ideals was constructed by Louveau and Velickovic ([**87**]), who proved that the structure $\mathcal{P}(\mathbb{N})/\text{Fin}$ under the ordering of almost inclusion embeds into the class of analytic ideals ordered by $\leq_{\text{BC}}$. All the Louveau–Velickovic ideals are non-pathological, because they are of the form $\text{Exh}(\varphi)$ where φ is equal to the supremum of a sequence of orthogonal non-pathological submeasures. In §1.13 we will introduce *density ideals*, a class of non-pathological ideals which includes all Erdős–Ulam ideals. However, not all analytic P-ideals are non-pathological, as the ideal constructed during the course of proving Theorem 1.9.5 shows.

THEOREM 1.9.5 (Non-lifting theorem). *There is an analytic P-ideal without the Radon–Nikodym property. In particular, Todorcevic's hypothesis may be false for ideals generated by some pathological submeasures.*

PROOF. This follows immediately from Theorem 1.7.2 and Theorem 1.8.6. □

The ideal constructed in the proof of Theorem 1.9.5 is F_σ, but there is also an $F_{\sigma\delta}$ ideal which is not F_σ and without the Radon–Nikodym property (recall that these two classes exhaust all analytic P-ideals). If in the proof that $\neg(1)$ implies $\neg(2)$ in Theorem 1.7.2 we define φ to be the supremum of all φ_i's, then the ideal $\mathrm{Exh}(\varphi)$ is not F_σ because $\emptyset \times \mathrm{Fin}$ is Rudin–Blass reducible to it.

Since the homomorphism constructed in the proof of Theorem 1.9.5 is not a monomorphism from $\mathcal{P}(\mathbb{N})/\mathrm{Fin}$ into $\mathcal{P}(\mathbb{N})/\mathcal{J}$, it is natural to ask whether every Baire monomorphism from $\mathcal{P}(\mathbb{N})/\mathrm{Fin}$ into a quotient over an analytic P-ideal has a completely additive lifting. However, the answer is once again in negative. This is because if $\Psi\colon \mathcal{P}(\mathbb{N})/\mathrm{Fin} \to \mathcal{P}(\mathbb{N})/\mathcal{J}$ has no lifting of the form Φ_h and $\Phi\colon \mathcal{P}(\mathbb{N})/\mathrm{Fin} \to \mathcal{P}(\mathbb{N})/\mathrm{Fin}$ is a Baire isomorphism then $\Phi \oplus \Psi$ is a Baire monomorphism of $\mathcal{P}(\mathbb{N})/\mathrm{Fin}$ into $\mathcal{P}(\mathbb{N} \oplus \mathbb{N})/\mathcal{J} \oplus \mathrm{Fin}$ with no completely additive lifting.

1.10. Permanence properties of quotients

In [**130**, Theorem 8], Todorcevic proved that if $\mathcal{J} \leq_{\mathrm{BE}} \mathcal{I}$ and $\mathcal{I}$ is an analytic P-ideal, then so is $\mathcal{J}$; the proof uses Theorem 1.5.2. We shall now see how other results about the structure of liftings imply similar permanence properties of other classes of ideals.

LEMMA 1.10.1. *If $\mathcal{I}$ is an F_σ ideal and the quotient algebra $\mathcal{P}(\mathbb{N})/\mathcal{J}$ is Baire-embeddable into $\mathcal{P}(\mathbb{N})/\mathcal{I}$, then $\mathcal{J}$ is an F_σ-ideal as well.*

PROOF. By Lemma 1.3.2, the embedding has a continuous lifting, and therefore $\mathcal{J}$ is F_σ as a continuous preimage of $\mathcal{I}$. $\square$

The above proof clearly shows that the above lemma is true for $F_{\sigma\delta}$ ideals, and the ideals in any other pointclass closed under continuous preimages.

PROPOSITION 1.10.2. *If $\mathcal{I}_f$ is a summable ideal and $\mathcal{J} \leq_{\mathrm{BE}} \mathcal{I}_f$, then $\mathcal{J}$ is summable as well.*

PROOF. If Φ_h is a lifting of a homomorphism of $\mathcal{P}(\mathbb{N})/\mathcal{J}$ into $\mathcal{P}(\mathbb{N})/\mathcal{I}_f$, then for the function $g(n) = \sum_{h(i)=n} f(i)$ we have $\mathcal{J} = \mathcal{I}_g$. $\square$

PROPOSITION 1.10.3. *If $\mathcal{I}$ is a non-pathological analytic P-ideal and $\mathcal{J} \leq_{\mathrm{BE}} \mathcal{I}$, then $\mathcal{J}$ is a non-pathological P-ideal as well.*

PROOF. Let Φ_h be a lifting of a homomorphism from $\mathcal{P}(\mathbb{N})/\mathcal{J}$ into $\mathcal{P}(\mathbb{N})/\mathcal{I}$. Let $\mathcal{I} = \mathrm{Exh}(\varphi_\mathcal{I})$ for some non-pathological submeasure $\varphi_\mathcal{I}$ (we can assume $\varphi_\mathcal{I}$ is non-pathological by the Fact above). Then the pull-back submeasure $\varphi_\mathcal{J}$ defined by $\varphi_\mathcal{J}(s) = \varphi_\mathcal{I}(\Phi_h(s))$ is non-pathological as well. To see this, pick $s \subseteq \mathbb{N}$ and find a measure $\mu \leq \varphi_\mathcal{I}$ such that $\mu(\Phi_h(s)) = \varphi_\mathcal{I}(\Phi_h(s))$. Then $\nu\colon \mathcal{P}(\mathbb{N}) \to [0,\infty]$ defined by $\nu(A) = \mu(\Phi_h(A))$ is a measure dominated by $\varphi_\mathcal{J}$ and $\varphi_\mathcal{J}(s) = \nu(s)$. Therefore the submeasure $\varphi_\mathcal{J}$ is indeed non-pathological. Also, $\mathrm{Exh}(\varphi_\mathcal{J})$ is equal to the set of all A such that $\Phi_h(A) \in \mathcal{I}$, and therefore to $\mathcal{J}$. $\square$

The following variant of Proposition 1.4.6 gives a strong answer to the basic question for non-pathological analytic P-ideals.

PROPOSITION 1.10.4. *If $\mathcal{I}$ is a non-pathological analytic P-ideal and the quotients $\mathcal{P}(\mathbb{N})/\mathcal{I}$ and $\mathcal{P}(\mathbb{N})/\mathcal{J}$ are Baire-isomorphic, then the ideals $\mathcal{I}$ and $\mathcal{J}$ are isomorphic.*

1.11. Simple F_σ P-ideals which are not summable

PROOF. By Proposition 1.10.3, the ideal $\mathcal{J}$ is non-pathological and therefore both $\mathcal{I}$ and $\mathcal{J}$ have the Radon–Nikodym property, hence the conclusion follows by Proposition 1.4.6. □

1.11. Simple F_σ P-ideals which are not summable

Pathological submeasures have already been used to construct F_σ ideals with peculiar properties. Mazur ([**95**, Theorem 1.9]) constructs an F_σ ideal which is not included in any summable ideal using one of the most frequently rediscovered examples of a pathological submeasure (see [**136**], [**122**], [**66**]). Mazur's ideal is not a P-ideal, but note that our Theorem 1.9.5 gives an example of an F_σ P-ideal which is not summable. This ideal is clearly pathological, and Example 1.11.1 below shows that there can even be a non-pathological F_σ P-ideal which is not summable.

EXAMPLE 1.11.1. A non-pathological F_σ P-ideal $\mathcal{I}$ on $\mathbb{N}$ which is not summable. For $k = 0, 1, 2, \ldots$ let $I_k = [2^k, 2^{k+1})$ and define submeasures $\psi_k \colon \mathcal{P}(I_k) \to \mathbb{R}^+$, $\psi \colon \mathcal{P}(\mathbb{N}) \to \mathbb{R}^+$ and an ideal $\mathcal{I}$ by

$$\psi_k(s) = \frac{\min(k, |s|)}{k^2},$$

$$\psi(A) = \sum_{k=0}^{\infty} \psi_k(A \cap I_k),$$

$$\mathcal{I} = \{A : \psi(A) < \infty\}.$$

We claim that $\mathcal{I}$ is not equal to a summable ideal $\mathcal{I}_f$ for any function f. Assume this is not true, and let $f \colon \mathbb{N} \to \mathbb{R}^+$ be such that $\mathcal{I}_f = \mathcal{I}$. We claim there is an integer M such that

(1) $$\bigcup \{I_k : F(I_k)/\psi(I_k) < M\} \in \mathcal{I}.$$

Assume this fails for all M; then we can pick a sequence of finite disjoint sets w_j such that for all j (recall the notation $\mu_f(A) = \sum_{n \in A} f(n)$):

$$\frac{\mu_f(\bigcup_{i \in w_j} I_i)}{\psi(\bigcup_{i \in w_j} I_i)} \geq j,$$

$$\frac{1}{j^2} \leq \psi\left(\bigcup_{i \in w_j} I_i\right) < \frac{2}{j^2}.$$

This sequence is constructed recursively as follows: If $w_1, \ldots, w_m$ are chosen, let $l \geq m+1$ be such that $2^l \geq \max \bigcup_{j=1}^{m} w_m$, and consider the set A_{m+1} of all $k \geq l$ for which $\mu_f(I_k)/\psi(I_k) \geq m+1$. Then this set is not in the ideal, and, since $\psi(I_k) < 1/(m+1)^2$ for all $k \in A_{m+1}$, we can pick a finite $w_{m+1} \subseteq A_{m+1}$ such that $1/(m+1)^2 \leq \psi(\bigcup_{k \in w_{m+1}} I_k) \leq 2/(m+1)^2$. Let $W = \bigcup_j w_j$. Then the set

$$X = \bigcup_{i \in W} I_i$$

is in $\mathcal{I} \setminus \mathcal{I}_f$. This is because for all j we have

$$\mu_f\left(\bigcup_{i \in w_j} I_j\right) \geq j \cdot \frac{1}{j^2} = \frac{1}{j}, \quad \text{and}$$

$$\psi\left(\bigcup_{i \in w_j} I_j\right) \leq \frac{1}{j^2}.$$

This finishes the proof that there is an integer M satisfying (1). So let M be a fixed integer satisfying (1) and let A be the set of all k for which $\mu_f(I_k)/\psi(I_k) \leq M$; note that $\sum_{k \in A} 1/k = \infty$. Then for $k \in A$ there is an $s_k \subseteq I_k$ of size k such that

$$\mu_f(s_k) \leq \frac{M}{k} \cdot \frac{k}{2^k} = \frac{M}{2^k}$$

and therefore $Y = \bigcup_{k \in A} s_k \in \mathcal{I}_f \setminus \mathcal{I}$ which is again in contradiction with our assumption that $\mathcal{I}$ and $\mathcal{I}_f$ are equal.

The above proof gives a more general fact:

PROPOSITION 1.11.2. *If $\{\varphi_n\}$ is a sequence of submeasures with pairwise disjoint finite supports, then the following are equivalent:*
  (a) *The ideal $\mathrm{Exh}(\sum_n \varphi_n)$ is summable.*
  (b) *There is a sequence of measures $\{\nu_n\}$ such that $\mathrm{supp}(\nu_n) = \mathrm{supp}(\varphi_n)$ and $\sum_n \|\nu_n - \varphi_n\| < \infty$.* □

For a while it was unclear whether there are any F_σ P-ideals more substantially different from summable ideals (see [**75**], [**96**]), in particular which are not of the form $\mathrm{Exh}(\sum_n \varphi_n)$ for some sequence of submeasures with pairwise disjoint finite supports. The discovery of such ideals (see [**113**], [**36**], [**39**]) required importing ideas from the theory of infinite-dimensional Banach spaces, in particular in [**36**] and [**39**] we have used the *Tsirelson space*, an infinite-dimensional Banach space which does not include a copy of c_0 or any ℓ_p for $p \geq 1$ (see [**137**]). Let us note that the ideals of [**113**], [**36**] and [**39**] are non-pathological.

1.12. The structure of summable ideals

Recall that summable ideals are ones of the form

$$\mathcal{I}_f = \{A : \sum_{n \in A} f(n) < \infty\}$$

for some positive function f such that $\sum_n f(n) = \infty$. A summable ideal is *dense* if $\lim_n f(n) = 0$. Since the summable ideals have the Radon–Nikodym property (Theorem 1.9.2), the relation of Baire-embeddability of quotients coincides with the much simpler Rudin–Blass ordering in the realm of summable ideals. This is why the large part of this section is devoted to the study of Rudin–Blass-ordering. In particular, we shall prove the following:

THEOREM 1.12.1. *The structure of all dense summable ideals ordered by $\leq_{\mathrm{BE}}$, or equivalently, by $\leq_{\mathrm{RB}}$,*
  (a) *does not have maximal elements,*
  (b) *does not have minimal elements,*
  (c) *includes an isomorphic copy of $\langle \mathcal{P}(\mathbb{N})/\mathrm{Fin}, \subseteq^* \rangle$*
  (d) *it is a dense ordering, namely for all $a < b$ there is c such that $a < c < b$.*

1.12. THE STRUCTURE OF SUMMABLE IDEALS

The proof of Theorem 1.12.1 will occupy a large part of this section. K. Mazur ([**94**]) has proved (using an idea of Louveau–Velickovic ([**87**])) that the structure of all F_σ quotients ordered by $\leq_{\mathrm{BC}}$ is complicated in the sense that it includes an isomorphic copy of $\langle \mathcal{P}(\mathbb{N})/\mathrm{Fin}, \subseteq^*\rangle$. Theorem 1.12.1 (c) above shows that, with respect to the ordering $\leq_{\mathrm{BE}}$, even the summable ideals have a very rich structure. Note that there are only two different Borel cardinalities of summable quotients (see Claim 1 in proof of Lemma 1.12.4 below), so this result cannot be improved in this direction. We shall now give a characterization of $\leq_{\mathrm{RB}}$ (equivalently, $\leq_{\mathrm{BE}}$) ordering on summable ideals, and for this we introduce one useful piece of notation. For a function $f\colon \mathbb{N} \to \mathbb{R}^+$ let μ_f denote the measure associated to f,

$$\mu_f(s) = \sum_{k \in s} f(k)$$

so that $\mathcal{I}_f = \{A : \mu_f(A) < \infty\} = \{A : \lim_n \mu_f(A \setminus [1,n)) = 0\}$. The following lemma gives a characterization of the $\leq_{\mathrm{RB}}$-ordering on dense summable ideals. Recall that $\mathcal{I}^*$ is the dual filter of an ideal $\mathcal{I}$

LEMMA 1.12.2. *Assume $\mathcal{I}_f$ and $\mathcal{I}_g$ are dense summable ideals. A finite-to-one function $h\colon \mathbb{N} \to \mathbb{N}$ is a reduction of $\mathcal{I}_g$ to $\mathcal{I}_f$ if and only if there are positive real numbers $c \leq C$ such that the set*

$$A[c, C] = \left\{ n : c \leq \frac{\mu_g(h^{-1}(n))}{f(n)} \leq C \right\}$$

is in $\mathcal{I}_f^*$ and $h^{-1}(A)$ is in $\mathcal{I}_g^*$.

PROOF. If h is of this form, then it is easily seen to be a reduction. Assume h is a reduction, yet the above conditions fail. There are three possibilities.

CASE 0. If there are $0 < c \leq C < \infty$ such that $A = A[c, C] \in \mathcal{I}_f^*$ but $h^{-1}(A)$ is not in $\mathcal{I}_g^*$, then $\mathbb{N} \setminus A$ is in $\mathcal{I}_f$ but $h^{-1}(\mathbb{N} \setminus A)$ is not in $\mathcal{I}_g$, which is a contradiction.

CASE 1. For every $c > 0$ the set

$$A[c, \cdot] = \left\{ n : c \leq \frac{\mu_g(h^{-1}(n))}{f(n)} \right\}$$

is not in $\mathcal{I}_f^*$. Therefore we can inductively pick a sequence of disjoint finite sets $p_m \subseteq \mathbb{N} \setminus A[1/m^2, \cdot]$ so that $1 \leq \mu_f(p_m) \leq 2$ for all m. Let $B = \bigcup_{m=1}^\infty p_m$; then B is not in $\mathcal{I}_f$, but $\mu_g(h^{-1}(B)) \leq 2\sum_{m=1}^\infty 1/m^2$ and therefore $h^{-1}(B)$ is in $\mathcal{I}_g$, contradicting the assumption on h.

CASE 2. For every $C < \infty$ the set

$$A[\cdot, C] = \left\{ n : \frac{\mu_g(h^{-1}(n))}{f(n)} \leq C \right\}$$

is not in $\mathcal{I}_f^*$. Therefore we can inductively pick a sequence of disjoint finite sets $p_m \subseteq \mathbb{N} \setminus A[\cdot, m]$ so that $1 \leq \mu_f(p_m) \leq 2$ for all m. Furthermore, we can assume that $\min(p_m)$ is large enough so that f assumes only values less than $1/m^2$ on p_m. Therefore we can find a disjoint partition

$$p_m = \bigcup_{i=1}^{m^2} p_m^i$$

so that $\mu_f(p_m^i) \leq 2/m^2$ for all i. Since $\mu_g(h^{-1}(p_m)) \geq m$, there is $i = i(m) \leq m^2$ such that $\mu_g(h^{-1}(p_m^i)) \geq 1/m$. Let $C = \bigcup_{m=1}^\infty p_m^{i(m)}$. Then $\mu_f(C) \leq \sum_{m=1}^\infty 2/m^2$ but $\mu_g(h^{-1}(C)) \geq \sum_{m=1}^\infty 1/m$, and therefore C is in $\mathcal{I}_f$ but $h^{-1}(C)$ is not in $\mathcal{I}_g$, contradicting the assumption on h. □

Let us now consider other, not necessarily dense, summable ideals. Recall that for an ideal $\mathcal{I}$ its *orthogonal*, $\mathcal{I}^\perp$, is defined by

$$\mathcal{I}^\perp = \{A : A \cap B \text{ is finite for all } B \in \mathcal{I}\}.$$

An ideal $\mathcal{I}$ is *atomic* if is generated by a single set over Fin.

LEMMA 1.12.3. *There are four disjoint classes of summable ideals $\mathcal{I}_f$, and each summable ideal belongs to one of them:*
- (S1) *Atomic ideals: For some $\varepsilon > 0$, the set $A_{f\varepsilon^+} = \{n : f(n) \geq \varepsilon\}$ is infinite and $\mu_f(\mathbb{N} \setminus A_{f\varepsilon^+}) < \infty$.*
- (S2) *Ideals of the form* Fin $\oplus$ Dense: *For some $\varepsilon > 0$ the set $A_{f\varepsilon^+}$ is infinite, $\mu_f(\mathbb{N} \setminus A_{f\varepsilon^+}) = \infty$ and $\lim_{n \notin A_{f\varepsilon^+}} f(n) = 0$.*
- (S3) *Ideals whose orthogonal is isomorphic to* Fin $\times \emptyset$: $A_{f\varepsilon_n^+} \setminus A_{f\varepsilon_{n+1}^+}$ *is infinite for all n for some strictly decreasing sequence $\{\varepsilon_n\}$ converging to 0.*
- (S4) *Dense summable ideals:* $\lim_{n \to \infty} f(n) = 0$

PROOF. Let us first note that all four classes are nonempty. For example, if $f \colon \mathbb{N} \to \mathbb{R}^+$ is given by

$$f(2^n(2m-1)) = \frac{1}{m} \qquad \text{for } m, n \in \mathbb{N},$$

then the ideal $\mathcal{I}_f$ belongs to (S3). Now we prove that these four classes are pairwise disjoint. If $\mathcal{I}$ is atomic, then $\mathcal{I}$ obviously cannot satisfy any of (S2)–(S4). To see that conditions (S2)–(S4) exclude each other, we consider an orthogonal $\mathcal{I}^\perp$ of an ideal $\mathcal{I}$. Let $\mathcal{I}$ be a summable ideal. Then $\mathcal{I}$ is dense if and only if $\mathcal{I}^\perp = $ Fin, $\mathcal{I}$ is in (S3) if and only if $\mathcal{I}^\perp$ is isomorphic to Fin $\times \emptyset$, and $\mathcal{I}$ is of the form Fin $\oplus$ Dense if and only if $\mathcal{I}^\perp$ is an atomic ideal different from Fin but $\mathcal{I}$ itself is not atomic. It remains to prove that each summable ideal $\mathcal{I}_f$ satisfies at least one of (S1)–(S4). If $A_{f\varepsilon^+}$ is finite for all $\varepsilon > 0$ then $\lim f(n) = 0$ and $\mathcal{I}_f$ is dense. Otherwise, for some $\varepsilon > 0$ the set $A_{f\varepsilon^+}$ is infinite. If there is a small enough $\varepsilon > 0$ such that $\lim_{n \notin A_{f\varepsilon^+}} f(n) = 0$ then the ideal belongs either to (S1) or to (S2). Otherwise we can recursively pick a sequence $\{\varepsilon_n\}$ like in (S3). □

LEMMA 1.12.4. *Let $\mathcal{I}_i$ be an ideal in the class (Si), for $i = 1, 2, 3, 4$. Then*
- (a) $\mathcal{I}_1 \leq_{\mathrm{RB}} \mathcal{I}_2 \leq_{\mathrm{RB}} \mathcal{I}_3 \leq_{\mathrm{RB}} \mathcal{I}_2 \leq_{\mathrm{RB}} \mathcal{I}_4$.
- (b) $\mathcal{I}_i \not\leq_{\mathrm{RB}} \mathcal{I}_1$ *for $i = 2, 3, 4$.*
- (c) $\mathcal{I}_4 \not\leq_{\mathrm{RB}} \mathcal{I}_i$ *for $i = 1, 2, 3$.*

PROOF. (a) The first relation is proved in [**93**]. The others will make use of the following.

CLAIM 1. *If $\mathcal{I}_g$ is a dense summable ideal and $\mathcal{I}_f$ is a summable ideal, then there is an $\mathcal{I}_g$-positive set A such that $\mathcal{I}_f \leq_{\mathrm{RB}} \mathcal{I}_g \upharpoonright A$.*

PROOF. Pick an increasing sequence of integers

$$n_1^1 < n_1^2 < \cdots < n_1^{k(1)} < n_2^1 < \cdots < n_2^{k(2)} < n_3^1 < \ldots$$

1.12. THE STRUCTURE OF SUMMABLE IDEALS

such that
$$|\mu_g(\{n_i^1, n_i^2, \ldots, n_i^{k(i)}\}) - f(i)| < \frac{1}{2^i}$$
for all i. This is possible because $\lim_n g(n) = 0$. Then for $A = \{n_i^j : i \in \mathbb{N}, j \leq k(i)\}$ mapping $h \colon A \to \mathbb{N}$ defined by $h(n_i^j) = i$ witnesses that $\mathcal{I}_f \leq_{RB} \mathcal{I}_g \restriction A$. □

Now we prove the remaining relations, that $\mathcal{I}_i \leq_{RB} \mathcal{I}_{i+1}$ for $i = 1, 2, 3$. Fix $i \leq 3$. By the assumptions, we can write $\mathcal{I}_i = \text{Fin} \oplus \mathcal{I}_f$, and by Claim 1 we can find an $\mathcal{I}_{i+1}$-positive set A such that $\mathcal{I}_f \leq_{BE} \mathcal{I}_{i+1} \restriction A$. Combining this reduction with Mathias' theorem $\text{Fin} \leq_{RB} \mathcal{I}_{i+1}$ we get the conclusion.

(b) Assume $h \colon \mathbb{N} \to \mathbb{N}$ is finite-to-one and such that $A \in \mathcal{I}$ if and only if $h^{-1}(A) \in \text{Fin}$. This implies $\mathcal{I} = \text{Fin}$, and therefore $\mathcal{I}$ is not equal to $\mathcal{I}_i$ for $i \geq 2$.

(c) This proof is similar to the proof of (b). □

By Lemma 1.12.4, the only class of summable ideals in which we can expect some structure with respect to $\leq_{RB}$-ordering (or equivalently, $\leq_{BE}$-ordering) are dense summable ideals. The following lemma gives a simple characterization of $\leq_{RB}$-comparability on dense summable ideals, and in particular shows that the complexity of this preordering is $F_{\sigma\delta}$.

LEMMA 1.12.5. *If $\mathcal{I}_f$ and $\mathcal{I}_g$ are dense summable ideals then the following two conditions are equivalent:*

(a) *for every positive integer M there are arbitrarily large positive integers k_0 and $k_1 < k_2$ such that:*
  (A1) $\mu_g([k_1, k_2]) > M \cdot \mu_f([1, k_0])$,
  (A2) $g(k_2) > M \cdot f(k_0)$, *and in particular*
  (A3) $k_0, k_1 > M$.
(b) $\mathcal{I}_f \not\leq_{RB} \mathcal{I}_g$.

PROOF. First of all, note that we can assume function f (and g as well) is monotonic, possibly by composing it with a suitable permutation of $\mathbb{N}$. This operation obviously leads to an ideal isomorphic to $\mathcal{I}_f$.

(a) $\Rightarrow$ (b) If $\mathcal{I}_f <_{RB} \mathcal{I}_g$, then let $h \colon \mathbb{N} \to \mathbb{N}$, $0 < c \leq C < \infty$ and $A[c, C]$ be as guaranteed by Lemma 1.12.2. Let $B = \mathbb{N} \setminus A[c, C]$ and pick $M \geq C + 1$ large enough so that $\mu_g(B \setminus [1, M]) < 1$ and $\mu_f([1, M]) > 1$. Then if k_0, k_1, k_2 satisfy (A1)–(A3) and we let
$$t = [k_1, k_2] \setminus h^{-1}([1, k_0]),$$
we have
$$\mu_g(t) \geq \mu_g([k_1, k_2]) - C\mu_f([1, k_0]) \geq (M - C)\mu_f([1, k_0]) > 1.$$
If $l \in t \cap A[c, C]$ is such that $m = h(l) \geq k_0$ then by (A2) and the definition of $A[c, C]$ we have
$$C \geq \frac{\mu_g(h^{-1}(\{m\}))}{f(m)} \geq \frac{g(l)}{f(m)} \geq \frac{g(k_2)}{f(k_0)} > M,$$
a contradiction. Therefore $h^{-1}(m)$ is disjoint from t for all $m \geq k_0$, and t is included in B. This contradicts to previously proved facts $\mu_g(B \setminus [1, M]) < 1$ and $\mu_g(t) > 1$, together with $t \subseteq [M, \infty)$.

(b) $\Rightarrow$ (a). Assume that (a) fails. Then for some M and all $k_0 > M$ and $k_2 > k_1 > M$ we have

(7) $\qquad \mu_g([k_1, k_2]) > M \cdot \mu_f([1, k_0]) \qquad$ implies $\qquad g(k_2) \leq M \cdot f(k_0).$

Choose some integers $N_0, N_1 > M$ such that
$$\mu_g([M+1, N_0]) > M \cdot \mu_f([1, N_1]).$$
We recursively choose a sequence $N_0 = k_1 \leq k_2 \leq k_3 \leq \ldots$ so that k_{i+1} is the minimal such that
(8) $$\mu_g([k_i, k_{i+1})) > Mf(N_1 + i)$$
for every i. Then we have
$$\mu_g([M+1, k_{i+1})) > M\mu_f([1, N_1]) + M \sum_{j=1}^{i} f(N_1 + j) = M\mu_f([1, N_1 + i])$$
and therefore (7) implies that $g(k_{i+1} - 1) \leq Mf(N_1 + i)$. This implies that
(9) $$\mu_g([k_i, k_{i+1})) = \mu_g([k_i, k_{i+1} - 1)) + g(k_{i+1} - 1) \leq 2Mf(N_1 + i).$$
Let $h \colon [N_0, \infty) \to [N_1 + 1, \infty)$ be a map which collapses the interval $[k_i, k_{i+1})$ to point $N_1 + i$. Since both the domain and the range of h are cofinite and (8) and (9) hold, function h satisfies the conditions of Lemma 1.12.2 with $c = M$ and $C = 2M$. Therefore h is a witness for $\mathcal{I}_g \leq_{\mathrm{RB}} \mathcal{I}_f$. $\square$

We are almost ready to start the proof of Theorem 1.12.1. It will be helpful to visualize summable ideals as follows. If $\mathcal{I}_f$ is a summable ideal, consider the increasing unbounded sequence of real numbers $a_k = a_k^f$ defined by
$$a_k = \sum_{i=1}^{k} f(k).$$
Then $\mathcal{I}_f$ is equal to the family of all A for which the set $\bigcup_{k \in A} [a_{k-1}, a_k)$ has finite Lebesgue measure. On the other hand, every unbounded increasing sequence a_k of positive reals determines an $f = f\{a_k\}$ by $f(k) = a_k - a_{k-1}$.

LEMMA 1.12.6. *If $\{a_k^g\}_{k \geq m}$ is a subsequence of $\{a_k^f\}$ for some positive natural number m, then $\mathcal{I}_g \leq_{\mathrm{RB}} \mathcal{I}_f$.*

PROOF. Since the ideals corresponding to $\{a_k^g\}_{k \geq m}$ and $\{a_k^g\}_{k \geq 1}$ are isomorphic, we can assume $m = 1$. Let $n(k)$ be an increasing sequence such that $a_k^g = a_{n(k)}^f$ for all k. Define $h \colon \mathbb{N} \to \mathbb{N}$ by (let $n(0) = 0$)
$$h^{-1}(k) = [n(k-1), n(k)).$$
Then $\mu_f(h^{-1}(k)) = g(k) = a_k^g - a_{k-1}^g$ (where $a_0^g = 0$), so h is as required. $\square$

LEMMA 1.12.7. *Assume f, g are nonincreasing and such that $\{a_k^g\}$ is a subsequence of $\{a_k^f\}$. Let $\{n(k)\}$ be the increasing sequence determined by $a_k^g = a_{n(k)}^f$. Suppose that for arbitrarily large positive integer N there are $N < k < k'$ such that:*
(B1) $a_{k'}^g = a_{n(k')}^f > N \cdot a_{n(k)}^f$, *and*
(B2) $g(k') > N \cdot f(n(k))$.
Then $\mathcal{I}_g$ is strictly below $\mathcal{I}_f$ in the sense of the preordering $\leq_{\mathrm{RB}}$.

PROOF. By Lemma 1.12.6 we have $\mathcal{I}_g \leq_{\mathrm{RB}} \mathcal{I}_f$. To prove $\mathcal{I}_f \not\leq_{\mathrm{RB}} \mathcal{I}_g$, we have to check that the conditions of Lemma 1.12.5 are satisfied. For $M < \infty$ pick $N > 2M$ such that $a_N^g > 2a_M^g$. If $N < k < k'$ are such that (B1) and (B2) are satisfied, then let $k_0 = n(k)$, $k_1 = M + 1$ and $k_2 = k'$. Let us check (A1)–(A3) are satisfied:

1.12. THE STRUCTURE OF SUMMABLE IDEALS

(A1) $\mu_g([k_1, k_2]) = \mu_g([1, k_2]) - \mu_g([1, M]) > \mu_g([1, k_2])/2$
$= a_{k'}^g/2 > N \cdot a_{n(k)}^f/2 > M \cdot \mu_f([1, k_0])$

(A2) $g(k_2) = g(k') > N \cdot f(n(k)) > M \cdot f(k_0)$.

(A3) This is obvious.

Therefore Lemma 1.12.5 implies the desired conclusion, that $\mathcal{I}_f \not\leq_{\mathrm{RB}} \mathcal{I}_g$. □

We are now ready for the proof of Theorem 1.12.1.

PROOF OF THEOREM 1.12.1 (a). To prove that there are no minimal elements in the structure of dense summable ideals with respect to the preordering $\leq_{\mathrm{RB}}$, fix a dense summable ideal $\mathcal{I}_f$. Since $\lim_n f(n) = 0$, we can assume f is monotonic, possibly by composing it with a suitable permutation of the integers. By Lemma 1.12.7 it suffices to construct a subsequence $\{a_i^g\}$ of $\{a_i^f\}$ which satisfies (B1) and (B2) and such that $a_i^g - a_{i-1}^g$ nonincreasingly converges to zero. We recursively find increasing sequences of positive integers $n(i)$ and $k(i)$ such that the sequence $a_i^g = a_{n(i)}^f$ satisfies (B1) and (B2) for $N = j$ with $k = k(j)$ and $k' = k(j+1)$ for all $j \in \mathbb{N}$. We shall also arrange that the following two conditions be satisfied for all i:

(10) $a_{k(i)+1}^f - a_{k(i)}^f < 1/i^3$, and

(11) $a_{k(i+1)}^f > 2i \cdot a_{k(i)}^f$.

The recursive construction of $\{k(i)\}$ is as follows: If $k(1), k(2), \ldots, k(i)$ are chosen, pick $k(i+1)$ large enough so that (11) holds and

$$f(k(i+1)) < \frac{1}{(i+1)^3}.$$

A sequence $\{k(i)\}$ constructed in this manner satisfies (10) and (11). For every i find integers l_i and $k(i) = n(i,1) < n(i,2) < \cdots < n(i,l_i) = k(i+1)$ such that

(12) $$\frac{1}{i^2} < a_{n(i,j+1)}^f - a_{n(i,j)}^f < \frac{2}{i^2}$$

for all $j = 1, \ldots, l-1$. Note that (10) implies this is possible. Let $n(k)$ be the increasing enumeration of $\{n(i,j) : i \in \mathbb{N}, j \leq l_i\}$ and let g be defined by sequence $a_k^g = a_{n(k)}^f$. Then (12) implies that $a_{k+1}^g - a_k^g$ converges to zero, and by the construction this sequence is monotonic. Given $N > 0$, let $k = k(N)$, $k' = k(N+1)$; then (B1) follows from (11) and (B2) is satisfied because by (12) and (11) we have

$$a_k^g - a_{k-1}^g > \frac{1}{N^2} = N\frac{1}{N^3} > a_{n(k)+1}^f - a_{n(k)}^f,$$

therefore Lemma 1.12.7 implies the desired conclusion. □

PROOF OF THEOREM 1.12.1 (b). To prove that there are no maximal elements in the structure of dense summable ideals with respect to the preordering $\leq_{\mathrm{RB}}$, fix a dense summable ideal $\mathcal{I}_g$. Since $\lim_n g(n) = 0$, we can assume g is monotonic, possibly by composing it with a suitable permutation of the integers. We will find a sequence $\{a_i^f\}$ including $\{a_i^g\}$ so that f is decreasing and Lemma 1.12.7 applies to prove $\mathcal{I}_g$ is strictly below $\mathcal{I}_f$. First pick an increasing sequence of positive integers $\{k(i)\}$ so that

(13) $$a_{k(i+1)}^g > 2i \cdot a_{k(i)}^g$$

for all i. Define a_n^f recursively: assume $a_1^f, \ldots, a_{m(i)}^f$ are defined so that $a_{m(i)}^f = a_{k(i)}^g$. Then for $j \in [k(i), k(i+1))$ partition interval $[a_j^g, a_{j+1}^g)$ into the pieces of equal length less than both

$$\text{(14)} \qquad \frac{1}{i}(a_{k(i+1)}^g - a_{k(i+1)-1}^g) \quad \text{and} \quad \frac{1}{2}(a_{m(i)+1}^f - a_{m(i)-1}^f).$$

Let $a_{m(i)+1}^f, \ldots, a_{m(i+1)}^f$ be an increasing enumeration of endpoints of these intervals; this describes the construction.

Then function $f = f\{a_n^f\}$ is nonincreasing and $\lim_n(a_{n+1}^f - a_n^f) = 0$. To see that the conditions of Lemma 1.12.7 are satisfied, fix $N > 0$ and consider $k = k(N)$, $k' = k(N+1)$. Then (13) implies (B1) and (14) reads as

$$\frac{1}{N}(a_k^g - a_{k-1}^g) > a_{n(k)+1}^f - a_{n(k)}^f$$

which is equivalent to (B2). An application of Lemma 1.12.7 ends the proof. □

PROOF OF THEOREM 1.12.1(c). For $A \subseteq \mathbb{N}$ define $f_A \colon \mathbb{N} \to \mathbb{R}^+$ as follows: Let $0 = n_0^A < n_1^A < n_2^A < \ldots$ be a sequence of integers recursively defined by

$$n_{k+1}^A - n_k^A = \begin{cases} ((2k)!)^2, & k \notin A, \\ 2k((2k)!)^2, & k \in A, \end{cases}$$

and for $i \in \mathbb{N}$ let $k^A(i)$ be the unique k such that $i \in [n_k^A, n_{k+1}^A)$. Let

$$f^A(i) = \begin{cases} 1/(2k)!, & k^A(i) \notin A \\ 1/(2k(2k)!), & k^A(i) \in A. \end{cases}$$

In particular we have $\mu_{f^A}([n_k^A, n_{k+1}^A)) = (2k)!$. If A, B are such that $A \subseteq^* B$, then $\mathcal{I}_{f^A} \leq_{\mathrm{RB}} \mathcal{I}_{f^B}$ because the sequence $\{a_i^{f^A}\}$ is almost included in the sequence $\{a_i^{f^B}\}$, hence Lemma 1.12.7 applies. If $k \in B \setminus A$ is large enough, then $k_1 = n_k^A$, $k_2 = n_{k+1}^A$ and $k_0 = n_k^B$ satisfy (A1)–(A3) of Lemma 1.12.5 for $g = f^A$, $f = f^B$ and $M = k$:

(A1) $\mu_f([n_k^A, n_{k+1}^A)) = (2k)! > k^2(2k-2)! > k\sum_{i=0}^{k-1}(2i)! + 1 = k\sum_{i=1}^{n_k^B} f^B(i) + 1$

(A2) $f^A(n_{k+1}^A) = \dfrac{1}{(2k)!} > k\dfrac{1}{2k(2k)!} = f^B(n_k^B)$.

Therefore if the set $B \setminus A$ is infinite then Lemma 1.12.5 implies $\mathcal{P}(\mathbb{N})/\mathcal{I}_{f^A} \not\leq_{\mathrm{BE}} \mathcal{P}(\mathbb{N})/\mathcal{I}_{f^B}$, therefore the mapping $A \mapsto \mathcal{I}_{f^A}$ is an embedding of $\mathcal{P}(\mathbb{N})/\mathrm{Fin}$ into the class of dense summable ideals ordered by $\leq_{\mathrm{RB}}$, or equivalently, by $\leq_{\mathrm{BE}}$. □

PROOF OF THEOREM 1.12.1(d). We have to prove that if $\mathcal{I}_f \leq_{\mathrm{RB}} \mathcal{I}_g$ and not vice versa, then there is an f' such that $\mathcal{I}_f \leq_{\mathrm{RB}} \mathcal{I}_{f'} \leq_{\mathrm{RB}} \mathcal{I}_g$ and both relations are irreversible. Fix a finite-to-one mapping $h \colon \mathbb{N} \to \mathbb{N}$ witnessing $\mathcal{I}_f \leq_{\mathrm{RB}} \mathcal{I}_g$. Define $f' \colon \mathbb{N} \to \mathbb{R}^+$ by

$$f'(n) = \mu_g(h^{-1}(n)).$$

Then $\mathcal{I}_{f'} = \mathcal{I}_f$, because h is a Rudin–Blass reduction and $\mu_{f'}(A) = \mu_g(h^{-1}(A))$ for all A. Without a loss of generality we may assume that $h^{-1}(k) = [n_k, n_{k+1})$ for some sequence

$$0 = n_0 < n_1 < n_2 < \ldots.$$

For $B \subseteq \mathbb{N}$ let f^B be a mapping defined by

$$\text{dom}(f^B) = (\mathbb{N} \setminus B) \times \{0\} \cup \bigcup_{k \in B} [n_k, n_{k+1}) \times \{1\}$$

$$f^B(i,j) = \begin{cases} f'(i), & i \notin B \text{ and } j = 0, \\ g(i), & i \in \bigcup_{k \in B}[n_k, n_{k+1}) \quad \text{and } j = 0. \end{cases}$$

Let $\mathcal{I}_{f^B}$ be the summable ideal on the index-set $\text{dom}(f^B)$ determined by f^B. Then

$$\mathcal{I}_f \leq_{\text{RB}} \mathcal{I}_{f^A} \leq_{\text{RB}} \mathcal{I}_{f^B} \leq_{\text{RB}} \mathcal{I}_g, \quad \text{for } A \subseteq B \subseteq \mathbb{N}.$$

Let $\mathcal{J} = \{B : \mathcal{I}_{f^B} \leq_{\text{RB}} \mathcal{I}_f\}$ and $\mathcal{F} = \{B : \mathcal{I}_g \leq_{\text{RB}} \mathcal{I}_{f^B}\}$. Then $\mathcal{J}$ and $\mathcal{F}$ are analytic sets (the characterization of orderings $\leq_{\text{RB}}$ and $\leq_{\text{BE}}$ from Lemma 1.12.5 is Borel, and in fact F_σ) which are downwards (respectively. upwards) closed, closed under finite changes, and not equal to $\mathcal{P}(\mathbb{N})$. Therefore these two sets must be meager (by [58] or [121]; see also Theorem 3.10.1 below) hence there is a set $B \in \mathcal{P}(\mathbb{N}) \setminus (\mathcal{F} \cup \mathcal{J})$. Then $\mathcal{I}_{f^B}$ is the required summable ideal. $\square$

Although Theorem 1.12.1 gives many quotients over summable ideals which are pairwise not Baire-isomorphic, let us give an explicit example of a pair of such quotients (see also Corollary 3.4.3). Let $\mathcal{I}_{1/\sqrt{n}} = \{A : \sum_{n \in A} 1/\sqrt{n} < \infty\}$.

PROPOSITION 1.12.8. *There is no isomorphism between quotients over $\mathcal{I}_{1/n}$ and $\mathcal{I}_{1/\sqrt{n}}$ having a Baire lifting.*

PROOF. By the Radon–Nikodym property of summable ideals and Proposition 1.4.6, it suffices to prove there is no bijection $h \colon A \to B$ such that $\mathbb{N} \setminus A \in \mathcal{I}_{1/\sqrt{n}}$, $\mathbb{N} \setminus B \in \mathcal{I}_{1/n}$, and $C \in \mathcal{I}_{1/\sqrt{n}}$ if and only if $h''C \in \mathcal{I}_{1/n}$. By Lemma 1.12.2, we can also assume that for some $0 < p \leq q < \infty$ and all $n \in A$ such function satisfies

$$p \leq \frac{h(n)}{\sqrt{n}} \leq q.$$

This implies that $h(n) \leq q\sqrt{n}$ for all $n \in A$, and since h is $1-1$ this easily implies that $\mathbb{N} \setminus A$ is not in $\mathcal{I}_{1/\sqrt{n}}$, a contradiction. $\square$

Note that Lemma 1.12.3 and Lemma 1.12.5 together show that the ordering $\leq_{\text{BE}}$ on summable quotients does not have the *Schröder–Bernstein property*. Namely, there are nonisomorphic summable quotients $\mathcal{P}(\mathbb{N})/\mathcal{I}_f$ and $\mathcal{P}(\mathbb{N})/\mathcal{I}_g$ such that $\mathcal{P}(\mathbb{N})/\mathcal{I}_f$ is Baire-embeddable into $\mathcal{P}(\mathbb{N})/\mathcal{I}_g$ and vice versa. To see this, first recall that an ideal $\mathcal{I}$ is *dense* if every infinite set of integers has an infinite subset in $\mathcal{I}$. Note that if ideals $\mathcal{I}$ and $\mathcal{J}$ are dense and not isomorphic, then the ideals $\text{Fin} \oplus \mathcal{I}$ and $\text{Fin} \oplus \mathcal{J}$ are not isomorphic as well: Otherwise let $h \colon \mathbb{N} \times \{0,1\} \to \mathbb{N} \times \{0,1\}$ be a $1-1$ mapping such that $A \in \mathcal{I}$ if and only if $h''A \in \mathcal{J}$ and $\mathbb{N} \setminus h''\mathbb{N} \in \mathcal{J}$. It is easy to see that $h''(\mathbb{N} \times \{0\}) \Delta (\mathbb{N} \times \{0\})$ is in $\mathcal{J}$, and this implies that the restriction of h to $\mathbb{N} \times \{1\}$ is an isomorphism between $\mathcal{I}$ and $\mathcal{J}$—a contradiction.

1.13. The structure of density ideals

Sets of integers of *zero density*, i.e., sets A such that

$$\limsup_n \frac{|A \cap [1,n]|}{n} = 0,$$

form a well-known ideal $\mathcal{Z}_0$ which has many interesting properties. For example, the famous result of Szemeredi ([120]) says that every $\mathcal{Z}_0$-positive set contains arbitrarily long arithmetic progressions. Another related classical notion is $\mathcal{Z}_{\log}$, the ideal of sets A of *logarithmic density zero*, i.e., sets such that

$$\limsup_n \frac{\sum_{i \in A \cap [1,n]} 1/i}{\log n} = 0.$$

The question whether quotients over these two ideals are isomorphic was asked by P. Erdös and S. Ulam long ago (see [**139**, Problem I.12.11], [**28**, p. 38–39], [**20**, Question 48]), and solved by Just–Krawczyk ([**65**]) and Just ([**62**], [**64**]), relatively recently. Just and Krawczyk ([**65**]) have generalized the ideals $\mathcal{Z}_0$ and $\mathcal{Z}_{\log}$ in the following way. We say $f \colon \mathbb{N} \to \mathbb{R}^+$ is an *Erdös–Ulam function* if (recall the notation $\mu_f(s) = \sum_{n \in s} f(n)$) we have

$$\mu_f(\mathbb{N}) = \infty \quad \text{and} \quad \lim_n \frac{f(n)}{\mu_f(\{1, \ldots, n\})} = 0.$$

The *Erdös–Ulam submeasure* φ_f and the *Erdös–Ulam-ideal* $\mathcal{EU}_f$ (also called *f-ideal* in [**65**] and [**62**]) on $\mathcal{P}(\mathbb{N})$ associated with an EU-function f are (recall the notation $\|A\|_{\varphi_f} = \lim_n \varphi_f(A \setminus [1, n)))$

$$\varphi_f(A) = \sup_n \frac{\mu_f(A \cap \{1, \ldots, n\})}{\mu_f(\{1, \ldots, n\})},$$
$$\mathcal{EU}_f = \operatorname{Exh}(\varphi_f(A)) = \{A \,:\, \|A\|_{\varphi_f} = 0\}.$$

In particular, $\mathcal{Z}_0 = \mathcal{EU}_1$ and $\mathcal{Z}_{\log} = \mathcal{EU}_{1/n}$.

In this section we shall define a class of *density* ideals which further extends the class of EU-ideals. For a submeasure φ let $\operatorname{supp}(\varphi) = \{n \,:\, \varphi(\{n\}) \neq 0\}$. Submeasures φ and ψ are *orthogonal* if they have disjoint supports. Let

$$\operatorname{at}^+(\varphi) = \sup_{k \in \operatorname{supp}(\varphi)} \varphi(\{k\}),$$
$$\operatorname{at}^-(\varphi) = \inf_{k \in \operatorname{supp}(\varphi)} \varphi(\{k\}).$$

DEFINITION 1.13.1. For a sequence $\mu = \{\mu_i\}$ of orthogonal measures on $\mathbb{N}$ each of which concentrates on some finite set define the submeasure φ_μ by

$$\varphi_\mu = \sup_i \mu_i.$$

Then

$$\mathcal{Z}_\mu = \operatorname{Exh}(\varphi_\mu) = \{A \,:\, \|A\|_{\varphi_\mu} = 0\}$$

is a *density ideal generated by a sequence of measures* (or simply a *density ideal*).

Note that $\emptyset \times \operatorname{Fin}$ is an example of a density ideal which is not an EU-ideal (all EU-ideals are dense). If we drop the requirement that measures concentrate on finite sets then we get an even larger family of ideals including all summable ideals. However, EU-ideals themselves are never F_σ (see Proposition 1.13.14). Since the submeasure φ_μ is defined as the supremum of measures, all density ideals are nonpathological, and therefore by Theorem 1.9.1 their important property immediately follows:

THEOREM 1.13.2. *Every density ideal has the Radon–Nikodym property.* □

1.13. THE STRUCTURE OF DENSITY IDEALS

The following result shows that the class of density ideals in our sense extends the class of EU-ideals in a very natural way. We should note that the seeds of (a) can be found in Oliver's note ([**100**]). The proof of (b) given below is a simplification of my original proof due to Max Burke, and it is included here with his kind permission.

THEOREM 1.13.3. (a) *Every Erdös–Ulam ideal is equal to some density ideal $\mathcal{Z}_\mu$. Moreover, we can assume each μ_i is a probability measure.*
(b) *A density ideal $\mathcal{Z}_\mu$ is an Erdös–Ulam ideal if and only if*
  (D1) $\sup_i \|\mu_i\| < \infty$,
  (D2) $\limsup_i \mathrm{at}^+(\mu_i) = 0$, *and*
  (D3) $\limsup_i \|\mu_i\| > 0$.

PROOF. (a) Consider an ideal $\mathcal{EU}_f$. Since $\lim_n f(n)/\mu_f(\{1,\dots,n\}) = 0$, possibly by changing finitely many values of f, we can assume
$$\sup_{n \geq 2} \frac{f(n)}{\mu_f(\{1,\dots,n\})} < \frac{1}{4}.$$
Pick a sequence $0 = n_0 < n_1 < n_2 < \dots$ so that n_1 is arbitrary and n_{i+1} is the minimal such that $\mu_f(\{n_i+1,\dots,n_{i+1}\}) \geq \mu_f(\{1,\dots,n_i\})$. Then for all i we have
$$(1) \qquad 1 \leq \frac{\mu_f(\{n_i+1,\dots,n_{i+1}\})}{\mu_f(\{1,\dots,n_{i+1}\})} \leq 1 + \frac{f(n_{i+1})}{\mu_f(\{1,\dots,n_{i+1}\})} \leq \frac{5}{4}.$$
Define the measure λ_i with the support $\{n_i+1,\dots,n_{i+1}\}$ by
$$\lambda_i(A) = \frac{\mu_f(A \cap \{n_i+1,\dots,n_{i+1}\})}{\mu_f(\{1,\dots,n_{i+1}\})}.$$
Although λ_i is not a probability measure, it will suffice to prove that $\mathcal{Z}_\lambda = \mathcal{EU}_f$ (because (1) reads as $1 \leq \|\lambda_i\| \leq 5/4$, so taking $\mathcal{Z}_{\lambda'}$ for $\lambda' = \lambda_i/\|\lambda_i\|$ would lead to the identical ideal). Since
$$\lambda_i(A) \leq \frac{\mu_f(A \cap \{1,\dots,n_{i+1}\})}{\mu_f(\{1,\dots,n_{i+1}\})} \leq \varphi_f(A),$$
we have $\varphi_\lambda \leq \varphi_f$.

CLAIM 1. $\|A\|_{\varphi_\lambda} = 0$ *implies* $\|A\|_{\varphi_f} = 0$ *for all* A.

PROOF. Assume $\|A\|_{\varphi_\lambda} = 0$, or in other words, $\lim_p \lambda_p(A) = 0$. Pick $\varepsilon > 0$, and let $i \geq 2$ be such that $\lambda_p(A) < \varepsilon$ for all $p \geq i$. For $m > k$ such that $n_i \leq k$ let j be such that $n_{j-1} < m \leq n_j$, then we have
$$\frac{\mu_f(A \cap [k,m))}{\mu_f([1,m])} \leq \frac{\mu_f(A \cap [n_i, n_j))}{\mu_f([n_{i-1}, n_{j-1}))}$$
$$\leq \max_{i \leq p \leq j-1} \frac{\mu_f(A \cap [n_p, n_{p+1}))}{\mu_f([n_{p-1}, n_p))}$$
$$\leq \frac{5}{4} \max_{i \leq p \leq j-1} \frac{\mu_f(A \cap [n_p, n_{p+1}))}{\mu_f([n_p, n_{p+1}))} \qquad \text{(by (1))}$$
$$< \frac{5}{4}\varepsilon.$$
Therefore
$$\|A\|_{\varphi_f} = \limsup_k \sup_{m > k} \frac{\mu_f(A \cap [k,m))}{\mu_f([1,m])} \leq \frac{5}{4}\varepsilon,$$

and since $\varepsilon > 0$ was arbitrary, we have $\|A\|_{\varphi_f} = 0$. □

We therefore have $\|A\|_{\varphi_f} = 0$ if and only if $\|A\|_{\varphi_\lambda} = 0$, and this implies $\mathcal{EU}_f = \mathrm{Exh}(\varphi_f) = \mathrm{Exh}(\varphi_\lambda) = \mathcal{Z}_\lambda$.

Now we prove Theorem 1.13.3 (b) Observe that (D3) is equivalent to the assertion that $\mathcal{Z}_\mu$ is a proper ideal (i.e., that $\mathbb{N} \notin \mathcal{Z}_\mu$), and every EU-ideal is such. Therefore it will suffice to prove that the conjunction of (D1) and (D2) is equivalent to (a), under the additional assumption that $\mathcal{Z}_\mu$ is a density ideal.

Assume now that (D1) and (D2) hold. We can assume $\|\mu_i\| \leq 1$ for all i (by possibly replacing each μ_i with μ_i/M for a large enough positive real number M). Without a loss of generality we can also assume that there is a sequence $1 = n_0 < n_1 < \ldots$ such that
$$\mathrm{supp}(\mu_i) = D_i = \{n_i + 1, \ldots, n_{i+1}\}.$$

CASE 1. There is an $\varepsilon > 0$ such that the set
$$A_\varepsilon = \{i : \|\mu_i\| \leq \varepsilon\}$$
is either finite or $\lim_{i \in A_\varepsilon} \|\mu_i\| = 0$. Then $\bigcup_{i \in A_\varepsilon} \mathrm{supp}(\mu_i)$ is in $\mathcal{Z}_\mu$, and therefore we can without a loss of generality assume $\|\mu_i\| = 1/2$ for all i. Define $f \colon \mathbb{N} \to \mathbb{R}^+$ by
$$f(k) = 2^i \mu_i(\{k\}), \quad \text{if } n_i < k \leq n_{i+1},$$
$$f(1) = 1.$$

Then $\mu_f(\{n_i + 1, \ldots, n_{i+1}\}) = 2^i$ and $\mu_f(\{1, \ldots, n_i\}) = 2^i$ for all i. Therefore for $n \in D_i$ we have
$$\frac{f(n)}{\mu_f(\{1, \ldots, n\})} \leq \mu_i(\{n\}),$$
and the value of this expression converges to 0 by our assumption (D2). Hence, f is an EU-function. Note that n_{i+1} is the minimal integer $n > n_i$ such that $\mu_f(\{n_i + 1, \ldots, n\}) \geq \mu_f(\{1, \ldots, n_i\})$. Note also that for every set A we have
$$\mu_i(A) = \frac{\mu_f(A \cap \{n_i + 1, \ldots, n_{i+1}\})}{\mu_f(\{1, \ldots, n_i\})},$$
and therefore μ_i is equal to the measure λ_i constructed in the proof of (a) as applied to the ideal $\mathcal{EU}_f$. Hence this proof implies that ideals $\mathcal{Z}_\mu$ and $\mathcal{EU}_f$ coincide.

CASE 2. With A_ε as defined in Case 1, the sequence $\|\mu_i\|$ ($i \in A_\varepsilon$) is infinite and it does not converge to 0 for every $\varepsilon > 0$. Since we can assume $\limsup_i \|\mu_i\| > 0$ (otherwise $\mathcal{Z}_\mu = \mathcal{P}(\mathbb{N})$) there is a sequence $\varepsilon_1 > \varepsilon_2 > \varepsilon_3 > \ldots$ converging to 0 and such that the set
$$A_j = \{i : \varepsilon_j \geq \|\mu_i\| > \varepsilon_{j+1}\}$$
is infinite for all j. By the discussion of Case 1, if
$$B_j = \bigcup_{i \in A_j} D_i$$
then the ideal $\mathcal{I}_j = \mathcal{I} \restriction B_j$ is an EU-ideal. We claim that
$$\mathcal{I} = \{A : A \cap B_j \in \mathcal{I}_j \text{ for all } j\}.$$

The inclusion "⊆" is obvious, so let us prove the reverse inclusion. Assume A is not in $\mathcal{I}$. Then for some j we have that $\limsup_j \mu_j(A) \geq \varepsilon_j$. This implies $A \cap \bigcup_{k=1}^{j} B_k$

is not in $\mathcal{I}$, and therefore $A \cap B_k$ is not in $\mathcal{I}$ (and therefore in $\mathcal{I}_k$) for some $k \leq j$. This shows that replacing μ_i for $i \in A_j$ with

$$\nu_i = 2^{-j} \frac{\mu_i}{\|\mu_i\|}$$

does not change the ideal $\mathcal{I}$. Enumerate $A_j = \{i(j,k) : k = 1, 2, \ldots\}$ and let

$$\nu'_k = \sum_{j=1}^{k} \nu_{i(j,k)}.$$

Then $\mathcal{I} = \mathcal{Z}_{\nu'}$, and Case 1 applies to prove that $\mathcal{I}$ is an EU-ideal.

Back to the proof of (b). To prove the other direction, let us first assume (D2) fails. Let $\varepsilon > 0$ be such that the set A of all i for which there is a $k_i \in \mathrm{supp}(\mu_i)$ such that $\mu_i(\{k_i\}) \geq \varepsilon$ is infinite. Then the set $\{k_i : i \in A\}$ does not have an infinite subset in $\mathcal{I}$. Since EU-ideals are dense, $\mathcal{I}$ is not an EU-ideal.

Let us now assume (D2) is satisfied, yet (D1) fails. By (a), it will suffice to prove there is no sequence ν_j of probability measures for which $\lim_j \mathrm{at}^+(\nu_j) = 0$ such that $\mathrm{Exh}(\sup_i \mu_i) = \mathrm{Exh}(\sup_j \nu_j)$. Assume the contrary, and let

$$\mathrm{supp}(\mu_i) = D_i \quad \text{and} \quad \mathrm{supp}(\nu_i) = E_i.$$

Let

$$S_i = \{j : E_j \cap D_i \neq \emptyset\}.$$

CLAIM 2. *For each m, there are infinitely many i for which there is $t_i \subseteq D_i$ such that $\mu_i(t_i) \geq 1$ but $\sup_j \nu_j(t_i) \leq 1/m$.*

PROOF. We will prove that for infinitely many i there are $a_j \subset E_j \cap D_i$ ($j \in S_i$) such that $\nu_j(a_j) < 1/m$ and $\mu_i(\bigcup_{j \in S_i} a_j) \geq 1$. Take $\delta = 1/(m+1)^2$, so that $(m+1)(1/m - \delta) > 1$. Consider any of the infinitely many i such that $\mu_i(D_i) \geq m+1$ and $\sup_{j \in S_i} \mathrm{at}^+(\nu_j) < \delta$. For $j \in S_i$, partition $E_j \cap D_i$ into (at most) $m+1$ pieces each of measure $< 1/m$. (Take all but (at most) one of the pieces to have measure in the interval $[1/m - \delta, 1/m]$.) Let a_j be the piece of largest μ_i measure. Then

$$\mu_i(a_j) \geq \frac{1}{m+1} \mu_i(E_j \cap D_i),$$

and therefore $\mu_i(t_i) \geq \mu_i(D_i)/(m+1) \geq 1$. □

Using Claim 2, we can choose an increasing sequence of $i(1) < i(2) < \ldots$ so that the sets $S_{i(m)}$ ($m = 1, 2, \ldots$) are pairwise disjoint, and find $t_m \subseteq D_{i(m)}$ so that $\mu_{i(m)}(t_m) \geq 1$ but $\sup_j \nu_j(t_m) \leq 1/m$. Then $\bigcup_i t_i$ is in $\mathcal{Z}_\nu \setminus \mathcal{Z}_\mu$. □

Note that the discussion of Case 2 in the proof of Theorem 1.13.3 (b) gives the following general fact:

LEMMA 1.13.4. *If $\mathcal{I}$ is a density ideal, then there is a sequence $\mu = \{\mu_i\}$ of measures such that $\mathcal{I} = \mathcal{Z}_\mu$ and $\|\mu_i\| \geq 1$ for all i.* □

Theorem 1.13.3 also implies the following.

COROLLARY 1.13.5. *If $\mathcal{I}$ is an Erdös–Ulam ideal and the set A is $\mathcal{I}$-positive, then $\mathcal{I} \restriction A$ is an Erdös–Ulam ideal as well.*

PROOF. By Theorem 1.13.3 (a), ideal $\mathcal{I}$ is equal to some $\mathcal{Z}_\mu$ such that $\|\mu_i\| = 1$ for all i. Then $\mathcal{I} \restriction A$ is equal to $\mathcal{Z}_\nu$, where $\nu = \{\mu_i \restriction A\}$. By Theorem 1.13.3 (b), this is an Erdös–Ulam ideal. □

COROLLARY 1.13.6. *If $\mathbb{N} = \bigcup_n A_n$ is a disjoint partition of $\mathbb{N}$ into infinite sets and $\mathcal{Z}_{\mu_n}$ is an Erdös–Ulam ideal on the set A_n, then*

$$\mathcal{Z} = \{B : B \cap A_n \in \mathcal{Z}_{\mu_n} \text{ for all } n\}$$

is an Erdös–Ulam ideal as well.

PROOF. By Theorem 1.13.3, for every n there is a sequence $\{\mu_{ni}\}_{i \in \mathbb{N}}$ of orthogonal measures supported by A_n such that $\mu_n = \sup_i = \mu_{ni}$, and $\|\mu_{ni}\| = 1/n$ for all i. Then $\mathcal{Z} = \{B : \sup_n \limsup_i \mu_{ni}(B) = 0\} = \{B : \limsup_{n,i \to \infty} \mu_{ni}(B) = 0\}$, and Theorem 1.13.3 implies that $\mathcal{Z}$ is an Erdös–Ulam ideal. □

The following proof of a result stated in [65] was pointed out to me by Max Burke.

COROLLARY 1.13.7. *Under the Continuum Hypothesis all quotients over EU-ideals are homogeneous.*

PROOF. In [65] it was proved that under CH all quotients over EU-ideals are pairwise isomorphic. Corollary 1.13.5 implies that all quotients over EU-ideals are homogeneous under CH. □

Let us state yet another corollary for future reference (see the end of §3.14).

COROLLARY 1.13.8. *Under the Continuum Hypothesis the quotient $\mathcal{P}(\mathbb{N})/\mathcal{Z}_0$ is isomorphic to its infinite power, $(\mathcal{P}(\mathbb{N})/\mathcal{Z}_0)^{\mathbb{N}}$.*

PROOF. By Corollary 1.13.6, a quotient over an Erdös–Ulam ideal is isomorphic to a countable product of quotients over Erdös–Ulam ideals, $\prod_{i=1}^{\infty} \mathcal{P}(\mathbb{N})/\mathcal{Z}_{\mu_i}$. Since by [65] all quotients over Erdös–Ulam ideals are isomorphic under CH, the statement follows. □

LEMMA 1.13.9. *There are four disjoint classes of density ideals $\mathcal{Z}_\mu$ and each density ideal belongs to one of them:*

($\mathcal{Z}$1) *Atomic ideals: $\inf_i \text{at}^-(\mu_i) > 0$.*
($\mathcal{Z}$2) *Nonatomic ideals which are not dense:*
   $\inf_i \text{at}^-(\mu_i) = 0$ *and* $\limsup_i \text{at}^+(\mu_i) > 0$.
($\mathcal{Z}$3) *Erdös–Ulam ideals: $\lim_i \text{at}^+(\mu_i) = 0$ and $\sup_i \|\mu_i\| < \infty$.*
($\mathcal{Z}$4) $\lim_i \text{at}^+(\mu_i) = 0$ *and* $\sup_i \|\mu_i\| = \infty$.

PROOF. It should be clear that these four classes are disjoint, because ideals in ($\mathcal{Z}$3) and ($\mathcal{Z}$4) are always dense, and in Theorem 1.13.3 it was shown that these two classes are disjoint. All four of these classes are nonempty—e.g., the ideal $\emptyset \times \text{Fin}$ is in ($\mathcal{Z}$2). So it will suffice to prove that every ideal $\mathcal{Z}_\mu$ belongs to one of these four classes. If $\mathcal{Z}_\mu$ is not dense, then it belongs to either ($\mathcal{Z}$1) or ($\mathcal{Z}$2), so let us assume it is dense. By Lemma 1.13.4, we can assume $\|\mu_i\| \geq 1$ for all i, and therefore $\mathcal{Z}_\mu$ is in ($\mathcal{Z}$3) or ($\mathcal{Z}$4), depending on whether $\sup_i \|\mu_i\|$ is finite or not. □

The following shows a difference between the EU-ideals and the summable ideals (compare with Theorem 1.12.1), and it moreover shows (together with Theorem 1.13.12 below) that the Schröder–Bernstein property (see remark at the end of §1.12) fails in the realm of quotients over density ideals.

LEMMA 1.13.10. $\mathcal{EU}_f \leq_{\text{RB}} \mathcal{EU}_g$ *for every two Erdös–Ulam ideals $\mathcal{EU}_f$ and $\mathcal{EU}_g$.*

A special case, $\mathcal{Z}_0 \leq_{\text{RB}} \mathcal{Z}_{\log}$, of the above lemma was known to Just (see [**62**, p. 904]). Lemma 1.13.10 fails for density ideals in general. For example, the only ideals $\mathcal{I}$ for which $\mathcal{I} \leq_{\text{RB}} \emptyset \times \text{Fin}$ are Fin and $\emptyset \times \text{Fin}$ itself. This is easy to prove directly, but it follows from a more general result of Kechris ([**75**]).

PROOF OF LEMMA 1.13.10. Let μ_i and ν_i be sequences of measures such that $\mathcal{E}\mathcal{U}_f = \mathcal{Z}_\mu$ and $\mathcal{E}\mathcal{U}_g = \mathcal{Z}_\nu$, as guaranteed by Theorem 1.13.3 (a). Then
$$\lim_n g(n)/\mu_g(\{1,\ldots,n\}) = 0$$
translates as $\lim_i \text{at}^-(\nu_i) = 0$, therefore we can find $0 = m_0 < m_1 < m_2 < \ldots$ such that for every i and all $m_i < j \leq m_{i+1}$ we have
$$(*) \qquad \frac{1}{i}\text{at}^-(\mu_i) \geq \text{at}^+(\nu_j).$$
Let $\psi_i = \sup_{m_i < j \leq m_{i+1}} \nu_j$, $D_i = \text{supp}(\mu_i)$, $E_i = \text{supp}(\nu_i)$ and $F_i = \bigcup_{j=m_i+1}^{m_{i+1}} E_j$.

CLAIM 3. *For every i there is a function $h_i\colon F_i \to D_i$ such that*
$$1 - \frac{2}{i} \leq \frac{\psi_{gi}(h^{-1}(v))}{\lambda_i(v)} \leq 1 + \frac{2}{i}$$
for all $v \subseteq D_i$.

PROOF. For simplicity assume that $D_i = \{1,\ldots,n\}$ and let $\varepsilon = \text{at}^+(\nu_i)$ and $\delta = \text{at}^-(\mu_i)$, so that $(*)$ reads as
$$\frac{\varepsilon}{\delta} \leq \frac{1}{i}.$$
For each j ($m_i < j \leq m_{i+1}$) find pairwise disjoint subsets F_{jl} ($l = 1, \ldots, n-1$) of E_j such that for all $k < n$:
$$|\nu_j(F_{jl} - \mu_i(\{l\}))| \leq 2\varepsilon \quad \text{and} \quad \left|\nu_j\left(\bigcup_{l=1}^k F_{jl}\right) - \mu_i(\{1,\ldots,k\})\right| < 2\varepsilon.$$
These sets are constructed recursively in a straightforward manner. Note that if we set $F_{jn} = E_k \setminus \bigcup_{l=1}^{n-1} F_{jl}$ then the above formulas remain true for $k = n$ because $\nu_j(E_j) = \mu_i(D_i)$. Define $h_i\colon \bigcup_j E_j \to D_i$ by
$$h^{-1}(\{k\}) = \bigcup_{j=m_i+1}^{m_{i+1}} F_{jk}.$$
Then for all $v \subseteq D_i$ and j we have (let $\delta = \text{at}^- \mu_i$):
$$\frac{\nu_j(\bigcup_{k \in v} F_{jk})}{\mu_i(v)} \leq \frac{\mu_i(v) + 2\varepsilon|v|}{\mu_i(v)} \leq 1 + \frac{\varepsilon \cdot |v|}{\delta \cdot |v|} \leq 1 + \frac{2}{i},$$
which implies $\psi_i(h^{-1}(v))/\mu_i(v) \leq 1 + 2/i$. The other inequality is proved in a similar way. □

If h_i ($i \in \mathbb{N}$) are as guaranteed by Claim 3, then the function $h = \bigcup_i h_i$ witnesses that $\mathcal{E}\mathcal{U}_f \leq_{\text{RB}} \mathcal{E}\mathcal{U}_g$. □

As mentioned before, a question of Erdős and Ulam whether the quotients over $\mathcal{Z}_0$ and $\mathcal{Z}_{\log}$ are isomorphic was the motivation for the introduction of Erdős–Ulam ideals. In [**65**], Just and Krawczyk used CH to prove that all quotients over Erdős–Ulam ideals are isomorphic. Just ([**62**], [**64**]) proved that under a different set-theoretic axiom the quotients over $\mathcal{Z}_0$ and $\mathcal{Z}_{\log}$ are not isomorphic. (Let us remark here that by Lemma 1.13.10 this is best possible, since $\mathcal{Z}_0 \leq_{\mathrm{RB}} \mathcal{Z}_{\log}$ and $\mathcal{Z}_{\log} \leq_{\mathrm{RB}} \mathcal{Z}_0$.) We shall now show how easily our methods give many pairwise Baire-nonisomorphic quotients over density ideals, in particular $\mathcal{Z}_0$ and $\mathcal{Z}_{\log}$. In connection with methods of Chapter 3, this gives another proof of Just's result (see Corollary 3.4.4).

DEFINITION 1.13.11. Fix a density ideal $\mathcal{Z}_\mu$ such that $\lim_i \mathrm{at}^+(\mu_i) = 0$. For a $\delta > 0$ define the functions $F_\mu, G_{\mu\delta} \colon \mathbb{N} \to \mathbb{N}$ by

$$F_\mu(n) = |\mathrm{supp}(\mu_n)|,$$
$$G_{\mu\delta}(n) = \max\{j : \mu_j(s) \geq \delta \text{ for some } s \text{ of size } \leq n\}.$$

The function $G_{\mu\delta}$ is well-defined because $\lim_i \mathrm{at}^+(\mu_i) = 0$.

For example, if $\mathcal{Z}_0 = \mathcal{Z}_\mu$ and $\mathcal{Z}_{\log} = \mathcal{Z}_\nu$ for natural sequences of measures $\{\mu_i\}$ and $\{\nu_i\}$ induced by EU-functions like in Theorem 1.13.3 (a), then for $\delta > 0$ we have $F_\mu(n) = 2^n$, $G_{\mu\delta}(n) = O(\log(n))$, $F_\nu(n) = O(a^{2^i})$ (for some $a > 1$) and $G_{\nu\delta}(n) \leq K \log(n)$, for a fixed $K > 0$.

THEOREM 1.13.12. *If the ideals $\mathcal{Z}_\mu$ and $\mathcal{Z}_\nu$ satisfy*

$$\lim_i \mathrm{at}^+(\mu_i) = \lim_i \mathrm{at}^+(\nu_i) = 0 \quad \text{and} \quad G_{\nu\delta} \circ F_\mu = o(n)$$

for all $\delta > 0$, then their quotients are not Baire isomorphic.

PROPOSITION 1.13.13. *Quotients over $\mathcal{Z}_0$ and $\mathcal{Z}_{\log}$ are not Baire-isomorphic.*

PROOF. By the above, for $\mathcal{Z}_0 = \mathcal{Z}_\mu$ and $\mathcal{Z}_{\log} = \mathcal{Z}_\nu$ we have

$$G_{\nu\delta} \circ F_\mu(n) \leq K\log(2^n) = o(n)$$

for all $\delta > 0$, and therefore Theorem 1.13.12 implies the desired conclusion. $\square$

PROOF OF THEOREM 1.13.12. Assume the contrary, that the quotients are Baire-isomorphic. Since the density ideals have the Radon–Nikodym property, by Proposition 1.10.4 ideals $\mathcal{Z}_\mu$ and $\mathcal{Z}_\nu$ are isomorphic as well. Let h be a $1-1$ partial function witnessing the isomorphism, such that $A \in \mathcal{Z}_\mu$ if and only if $h''A \in \mathcal{Z}_\nu$. Let $D_i = \mathrm{supp}(\mu_i)$.

CLAIM 4. *There is a $\delta > 0$ such that for all but finitely many i there is j for which $\nu_j(h''D_i) > \delta$.*

PROOF. Assume otherwise, that for every m there are infinitely many i such that for all j we have $\nu_j(h''D_i) < 1/m$. Since $\mathrm{supp}(\nu_j)$ is finite for every j, we can find a sparse enough subsequence of $\{D_i\}$ whose union A is such that $\limsup_j \nu_j(h''A) = 0$, which contradicts the choice of h. $\square$

For $i \in \mathbb{N}$ let $J(i)$ denote the minimal j guaranteed by Claim 4. Then

CLAIM 5. *There is an m such that the function J is at most m-to-1 (i.e., the size of $J^{-1}(k)$ never exceeds m). In particular, $J(n) \neq o(n)$.*

PROOF. Assume otherwise, that for every m there is a k for which $|J^{-1}(k)| \geq m$. Since $\lim_k \mathrm{at}^+(\nu_k) = 0$, we can find such k and $s_k \subseteq \mathrm{supp}(\nu_k)$ so that $\nu_k(s_k) \geq \delta$ and $\varphi_\mu(h^{-1}(s_k)) \leq 2/m$. Now it is easy to find an infinite set A such that the set $\bigcup_{k \in A} s_k$ is not in $\mathcal{Z}_\nu$ but its h-preimage is in $\mathcal{Z}_\mu$. $\square$

Now observe that $J(n) \leq G_{\nu\delta} \circ F_\mu(n)$, and therefore by Claim 5 the function $G_{\nu\delta} \circ F_\mu$ cannot be $o(n)$. $\square$

Let us now relate summable and density ideals. Proposition 1.13.14 below shows that these two classes are quite distant in the sense of the $\leq_{\mathrm{RB}}$-ordering. A much stronger result, to the effect that these two classes are distant in the sense of the coarser Borel-cardinality ordering $\leq_{\mathrm{BC}}$, is proved in [**40**].

PROPOSITION 1.13.14. *If $\mathcal{I}_f$ and $\mathcal{Z}_\mu$ are both nonatomic, then $\mathcal{I}_f$ and $\mathcal{Z}_\mu$ are $\leq_{\mathrm{RB}}$-incomparable. Therefore their quotients are $\leq_{\mathrm{BE}}$-incomparable.*

Before starting a proof of Proposition 1.13.14, let us prove a lemma.

LEMMA 1.13.15. $\emptyset \times \mathrm{Fin} \leq_{\mathrm{RB}} \mathcal{Z}_\mu$ *for every density ideal $\mathcal{Z}_\mu$.*

Lemma 1.13.15 can be proved by using a result of Solecki ([**112**, Theorem 3.3]): An analytic P-ideal $\mathcal{I}$ is not F_σ if and only if $\emptyset \times \mathrm{Fin} \leq_{\mathrm{RB}} \mathcal{I}$. Note also that another equivalent statement is: There is a partition of $\mathbb{N}$ into disjoint $\mathcal{I}$-positive sets F_n ($n \in \mathbb{N}$) such that $\mathcal{I}$ is equal to $\{B : B \cap F_n \in \mathcal{I} \text{ for all } n\}$. We shall prove a more general version of Lemma 1.13.15.

PROPOSITION 1.13.16. *Assume $\varphi = \sup_i \varphi_i$, where $\{\varphi_i\}$ is a sequence of submeasures with disjoint supports such that $\limsup_i \|\varphi_i\| > 0$ and $\lim_i \mathrm{at}^+(\varphi_i) = 0$. Then $\emptyset \times \mathrm{Fin} \leq_{\mathrm{RB}} \mathrm{Exh}(\varphi)$.*

PROOF. It will suffice to prove that $\emptyset \times \mathrm{Fin} \leq_{\mathrm{RB}} \mathrm{Exh}(\varphi) \upharpoonright A$ for some $\mathcal{I}$-positive set A. By restricting $\mathcal{I} = \mathrm{Exh}(\varphi)$ to a positive set, we can assume $s_i = \mathrm{supp}(\varphi_i)$ is finite and $\|\varphi_i\| \leq 1$ for all i. By Lemma 1.13.4, we can moreover assume $\|\varphi_i\| = 1$. Condition $\lim_i \mathrm{at}^+(\varphi_i) = 0$ is equivalent to the assertion that $\mathcal{I}$ is a dense ideal, therefore it is not destroyed by applying Lemma 1.13.4. By going to a subsequence of φ_i, we can moreover assume that $\mathrm{at}^+(\varphi_i) < 2^{-2i}$, and use this to find a disjoint partition $s_i = t_i^1 \dot\cup t_i^2 \dot\cup \ldots \dot\cup t_i^i$ such that

$$\left|\varphi_i(t_i^j) - \frac{1}{2^j}\right| \leq \frac{1}{2^i}$$

for all $j \leq i$. Let $A_j = \bigcup_{i=j}^\infty t_i^j$. Then each A_j is $\mathcal{I}$ positive and we have

$$B \in \mathcal{I} \quad \Leftrightarrow \quad B \cap A_j \in \mathcal{I} \text{ for all } j,$$

so if $h_j \colon A_j \to \{j\} \times \mathbb{N}$ is an RB-reduction of $\mathcal{I} \upharpoonright A_j$ to Fin, then $h = \bigcup_j h_j$ is an RB-reduction of $\mathcal{I}$ to $\emptyset \times \mathrm{Fin}$. $\square$

PROOF OF PROPOSITION 1.13.14. One direction follows from Lemma 1.13.15, because an F_σ ideal cannot be RB-reducible to $\emptyset \times \mathrm{Fin}$, since the latter ideal is not F_σ. Let us now prove another direction. Assume the contrary, and let h be such that $A \in \mathcal{I}_f$ if and only if $h^{-1}(A) \in \mathcal{Z}_\mu$. Find a sequence $1 = n_1 < n_2 < n_3 < \ldots$ and $t_i \subseteq \{n_i, \ldots, n_{i+1} - 1\}$ such that for all i, j (let $D_k = \mathrm{supp}(\mu_k)$):

(1) $1/i \leq \mu_f(t_i) < 2/i$
(2) $h^{-1}(t_i) \cap D_k \neq \emptyset$ and $h^{-1}(t_j) \cap D_k \neq \emptyset$ for some k implies $i = j$.

The recursive construction of these sequences proceeds as follows:

Assume $n_1, \ldots, n_{k-1}$ and $t_1, \ldots, t_{k-1}$ are already chosen. Since the set

$$E = \bigcup_{l=1}^{k-1} t_l \cup h^{-1}(t_l)$$

is finite, there are only finitely many j such that $D_j \cap E \neq \emptyset$, therefore we can pick an n_k so that $\{1, \ldots, n_k - 1\}$ includes E and all such D_j's. Now choose t_k satisfying (1) (this is possible because $\mathcal{Z}_\mu$ is a proper density ideal) and such that $\min(t_k) \geq n_k$. This describes the recursive construction, and it is easy to see that (1) and (2) will be satisfied.

Assume that t_i ($i \in \mathbb{N}$) are chosen to satisfy (1) and (2). Then $\bigcup_i t_i$ is, by (1), not in $\mathcal{I}_f$, and therefore $\lim_j \sup_i \mu_j(h^{-1}(t_j)) = \varepsilon > 0$. By (2)

$$\limsup_j \mu_j \left(\bigcup_i t_i \right) = \limsup_j (\mu_j(t_i)),$$

and there are subsequences t'_i of t_i and μ'_i of μ_i such that $\mu'_i(t'_i) > \varepsilon/2$ or all i, in particular $\bigcup_{i \in A} t'_i$ is $\mathcal{I}_f$-positive whenever A is infinite. But by (1) there is an infinite set A such that $\bigcup_{i \in A} t'_i$ is in $\mathcal{I}_f$—a contradiction. □

Classes of density and summable ideals do not exhaust all non-pathological analytic P-ideals—for example, consider ideals of the form $\mathcal{I}_f \oplus \mathcal{Z}_\mu$. To end this Chapter, let us give an example of a non-pathological analytic P-ideal which is substantially different from both density and summable ideals.

EXAMPLE 1.13.17. Let $\mathcal{I}$ be the ideal on the set $\{0,1\}^{<\mathbb{N}}$ of all finite sequences of $\{0,1\}$ defined as follows. A submeasure φ with $\mathrm{supp}(\varphi) = \{0,1\}^{<\mathbb{N}}$ is defined by:

$$\varphi(A) = \sup_{x \in \{0,1\}^{\mathbb{N}}} \sum_{x \restriction n \in A} \frac{1}{n}.$$

This is obviously a non-pathological submeasure. Let $\mathcal{J}_{\mathrm{br}}$ be the ideal on $\{0,1\}^{<\mathbb{N}}$ generated by its branches, and let $\langle \mathcal{J}_{\mathrm{br}}, \mathcal{I} \rangle$ denote the ideal generated by $\mathcal{I}$ and $\mathcal{J}_{\mathrm{br}}$. For $t \in \{0,1\}^{<\mathbb{N}}$ let $[t]$ be the set of all $s \in \{0,1\}^{<\mathbb{N}}$ end-extending t.

CLAIM 6. *The following are equivalent for $A \subseteq \{0,1\}^{<\mathbb{N}}$:*

(1) $A \notin \langle \mathcal{J}_{\mathrm{br}}, \mathcal{I} \rangle$.
(2) $\mathcal{I} \restriction A$ *is not summable.*
(3) *There is a $B \in \mathcal{I}^+ \restriction A$ such that $\mathcal{I} \restriction B$ is a proper density ideal.*

PROOF. For a set in $A \in \mathcal{J}_{\mathrm{br}}$ a function $f \colon A \to \mathbb{R}^+$ by $f(A) = \sum_{x \restriction n \in A} 1/n$ satisfies $\mathcal{I} \restriction A = \mathcal{I}_f \restriction A$. Therefore if (1) fails, (2) fails as well. Since (3) obviously implies (2), it remains only to prove (1) implies (3). So let A be such that for every $B \in \mathcal{J}_{\mathrm{br}}$ set $A \setminus B$ is not in $\mathcal{I}$, and let

$$T(A) = \{ t \in \{0,1\}^{<\mathbb{N}} : A \cap [t] \notin \langle \mathcal{J}_{\mathrm{br}}, \mathcal{I} \rangle \}$$

The tree $T(A)$ is infinite, so by König's lemma it has an infinite branch C. Then for some $\varepsilon > 0$ we have (let $\{0,1\}^{\leq n}$ be the set of all $t \in \{0,1\}^{<\mathbb{N}}$ of length $\leq n$):

$$\lim_{n \to \infty} \varphi((A \setminus C) \setminus \{0,1\}^{\leq n}) \geq \varepsilon.$$

Now recursively pick a sequence s_k of finite chains of $\{0,1\}^{<\mathbb{N}}$ included in $A \setminus C$ such that $\varepsilon/2 \leq \varphi(s_k) \leq \varepsilon$ for all k and every $t \in s_k$ is incomparable with every $u \in s_l$ for $l \neq k$. This is done as follows: If $s_1, \ldots, s_k$ are already chosen and satisfy the conditions, pick an n such that $\bigcup_{i=1}^k s_i \subseteq \{0,1\}^{\leq n}$. Let $t \in C$ be of length $> n$ and such that $[t] \cap (A \setminus C) \notin \langle \mathcal{J}_{\mathrm{br}}, \mathcal{I} \rangle$. Find $s \subseteq [t] \cap (A \setminus C)$ such that $\varphi(s) > \varepsilon/2$, and let $s_k \subseteq s$ be a finite chain such that $\varphi(s_k) \geq \varepsilon/2$. This describes the construction.

If s_k are as above, then $\bigcup_k s_k$ is not in $\mathcal{I}$, $\mu_k = \varphi \restriction s_k$ is a measure for every k, $\lim_k \mathrm{at}^-(\mu_k) = 0$, and $\varphi(D \cap \bigcup_k s_k) = \sup_k \mu_k(D)$, therefore $\mathcal{I} \restriction \bigcup_k s_k$ is a proper density ideal. □

By Claim 6, every A such that $A \cap B$ is $\mathcal{I}$-positive for infinitely many branches B of $\{0,1\}^{<\mathbb{N}}$ has subsets A_0, A_1 such that $\mathcal{I} \restriction A_0$ is a dense summable ideal and $\mathcal{I} \restriction A_1$ is a proper density ideal. □

To conclude this section we state an appropriate open problem. An ideal $\mathcal{J}$ is c_0-like if $\mathcal{J} = \mathrm{Exh}(\sup_i \varphi_i)$ for some sequence of pairwise orthogonal submeasures φ_i. Thus every density ideal is c_0-like, but c_0-like ideals form a larger class, including, for example, many pathological ideals.

QUESTION 1.13.18. *Is there an analytic P-ideal $\mathcal{I}$ such that*
 (i) *$\mathcal{I}$ is not isomorphic to* Fin,
 (ii) *If $\mathcal{J} \leq_{\mathrm{RB}} \mathcal{I}$ is F_σ, then $\mathcal{J}$ is isomorphic to* Fin, *and*
 (iii) *If $\mathcal{J} \leq_{\mathrm{RB}} \mathcal{I}$ is c_0-like, then $\mathcal{J} \leq_{\mathrm{RB}} \emptyset \times$ Fin.*

I am inclined towards conjecturing a negative answer to the above, but this may be simply because of the lack of interesting examples of analytic P-ideals (compare with Kechris–Mazur conjecture, see [**36**]). Motivation for asking Question 1.13.18 comes from the "basis problem for turbulent actions," and the reader may consult [**40**] for the background and a closely related Question 12.10.

1.14. Remarks and questions

To end this section let us point out to, in our opinion, most important open questions related to this work.

QUESTION 1.14.1. *Is there a simple characterization of analytic P-ideals with the Radon–Nikodym property?*

We do not know whether all pathological analytic P-ideals lack the Radon–Nikodym property. Let us recall Question 11 from [**33**]. This is a question of a finitary nature closely related to Question 1.14.1. For a submeasure φ (recall that $\hat{\varphi}$ is the maximal non-pathological submeasure dominated by φ; see §1.8 or §1.9) let

$$C(\varphi) = \frac{\|\varphi\| - \|\hat{\varphi}\|}{\|\varphi\|}.$$

Then $0 \leq C(\varphi) \leq 1$.

QUESTION 1.14.2. *Is $\lim_{t \to 1^-} \inf\{K_\varphi : C(\varphi) = t\} = \infty$?*

A positive answer to this question can be interpreted as "for every sufficiently pathological submeasure there is a non-exact homomorphism which cannot be approximated by a homomorphism." Therefore, such a positive answer would in turn imply that the class of pathological ideals coincides with the class of analytic P-ideals which lack the Radon–Nikodym property (by a construction of Theorem 1.9.5). The following weaker form of Todorcevic's conjecture would, if true, still be very useful.

QUESTION 1.14.3. *Is it true that every isomorphism of quotients over analytic P-ideals which has a Baire-measurable lifting has a completely additive lifting?*

Note that a positive answer would imply that basic question has a satisfactory answer in the realm of analytic P-ideals: Two quotients over analytic P-ideals would be isomorphic if and only if the corresponding ideals are isomorphic. It is interesting that this question has a finitary version. Let us say that an ε-approximate homomorphism $H\colon \mathcal{P}([1,m)) \to \mathcal{P}([1,n))$ (see §1.8) is an *ε-approximate epimorphism* if for every $s \in \mathcal{P}([1,n))$ there is a $t \in \mathcal{P}([1,m))$ such that $\varphi(H(t)\Delta s) \leq \varepsilon$.

QUESTION 1.14.4. *Is there a universal constant K such that every ε-approximate epimorphism can be $K \cdot \varepsilon$-approximated by a homomorphism?*

The proofs of Theorems 1.9.1 and 1.9.5 easily give that Question 1.14.4 has a positive answer if and only if Question 1.14.3 does.

Some of results of Chapter 1 can be summarized in the following diagram which represents the ordering of Baire-embeddability between analytic quotients:

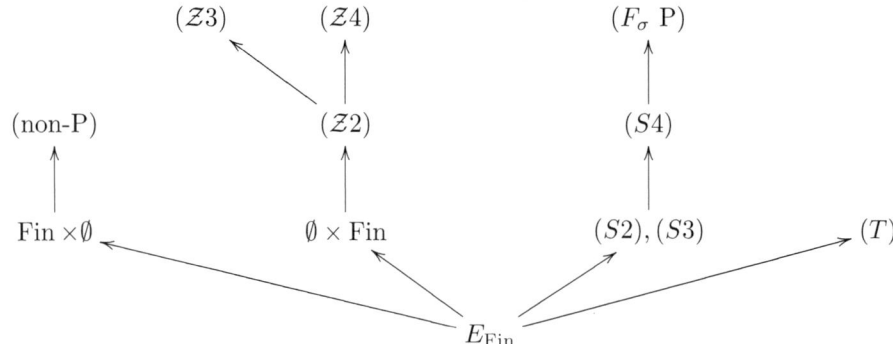

The ordering $\leq_{\mathrm{BE}}$.

(non-P) Analytic ideals which are not P-ideals.
  ($\mathcal{Z}$2) Nonatomic density ideals which are not dense (Lemma 1.13.4).
  ($\mathcal{Z}$3) Erdös–Ulam ideals (§1.13).
  ($\mathcal{Z}$4) Density ideals $\mathcal{Z}_\mu$ for which $\sup_i \|\mu_i\| = \infty$ (Lemma 1.13.4).
 (F_σ P) F_σ P-ideals which are not summable occurring in Theorem 1.9.5 and Example 1.11.1.
  (S2) Summable ideals of the form Fin $\oplus$ Proper (Lemma 1.12.3).
  (S3) Summable ideals whose orthogonal is countably generated (Lemma 1.12.3).
  (S4) Proper summable ideals (Lemma 1.12.3).
  (T) F_σ P-ideals which are not $\leq_{\mathrm{BE}}$-comparable to either $\emptyset \times$ Fin, Fin $\times \emptyset$ or any summable ideal other than Fin ([**36**], [**39**]).

Let us now explain the above diagram. It is not difficult to prove that the ideals Fin $\times \emptyset$ and $\emptyset \times$ Fin are immediate successors of Fin in $\leq_{\mathrm{BE}}$-ordering, because by the Radon–Nikodym property it reduces to the proof that they are immediate successors of Fin in the much simpler $\leq_{\mathrm{RB}}$-ordering. This fact also follows from a more general result of Kechris ([**75**]). By a result of Solecki ([**112**, Theorem 2.1]), Fin $\times \emptyset \leq_{\mathrm{RB}} \mathcal{I}$ for every analytic ideal $\mathcal{I}$ which is not a P-ideal. By Lemma 1.13.15, $\emptyset \times$ Fin is Rudin–Blass reducible to every density ideal. By a result of Solecki ([**112**, Theorem 3.3]), it is moreover Rudin–Blass reducible to any analytic P-ideal which is not F_σ. Since the ideal $\emptyset \times$ Fin belongs to the class ($\mathcal{Z}$2), it is the minimal element of this class in the Rudin–Blass ordering. By Lemma 1.13.10 for every pair $\mathcal{Z}_\mu$ and $\mathcal{Z}_\nu$ of

Erdős–Ulam ideals we have $\mathcal{Z}_\mu \leq_{\mathrm{RB}} \mathcal{Z}_\nu$. On the other hand, by Theorem 1.13.12, there are many nonisomorphic quotients in ($\mathcal{Z}$3). Every ideal in ($\mathcal{Z}i$) ($i = 2, 3, 4$) is, by Proposition 1.13.14, $\leq_{\mathrm{RB}}$-incomparable to every ideal in (Sj) ($j = 2, 3, 4$). The fact that every ideal in ($\mathcal{Z}$2) is Rudin–Blass incomparable with every ideal in ($\mathcal{Z}$3) is proved exactly like Lemma 1.12.2 (the discussion of cases 1 and 2) or Theorem 1.13.3 (b) (Claim 2). By Lemma 1.12.4, for all ideals $\mathcal{I}, \mathcal{J}$ in classes (S2) and (S3) we have $\mathcal{I} \leq_{\mathrm{RB}} \mathcal{J}$ (compare this with Remark at the end of §1.12, by which many of quotients over these ideals are not Baire-isomorphic). Proposition 1.10.2 easily implies that no ideal is strictly between Fin and (S2)∪(S3) in the sense of $\leq_{\mathrm{RB}}$, or equivalently $\leq_{\mathrm{BE}}$, ordering. By Theorem 1.12.1, the partial ordering $\langle (S4), \leq_{\mathrm{BE}} \rangle$, isomorphic to $\langle (S4), \leq_{\mathrm{RB}} \rangle$, is complicated; for example, it includes an isomorphic copy of $\mathcal{P}(\mathbb{N})/\mathrm{Fin}$. By a result of Solecki, all ideals belonging to class (T) are F_σ P-ideals. It was conjectured by Kechris and Mazur that such ideals do not exist (see [**75**], [**96**]), and an example of such an ideal was recently found by the author (see [**36**], [**39**]).

It is interesting that we get a similar picture under any of the five other pre-orderings defined in §1.2 and also under the ordering $\leq_{\mathrm{EM}}$ (see §3.4) if we assume OCA. For example, one gets a very similar picture even in the rather distant ordering of Borel-cardinality of quotients; see [**75**].

Let us turn to the problem of determining which analytic ideals which are not necessarily P-ideals have the Radon–Nikodym property? Recall that by a result of Mazur ([**95**]; see Theorem 1.2.5) every F_σ ideal is of the form

$$\mathrm{Fin}(\varphi) = \{A : \varphi(A \setminus n) < \infty\}$$

for some lower semicontinuous submeasure $\varphi \colon \mathcal{P}(\mathbb{N}) \to [0, \infty]$. Therefore it is natural to say that an F_σ ideal is *non-pathological* if it is of the form $\mathrm{Fin}(\varphi)$ for a non-pathological lower semicontinuous submeasure φ. The following result from ([**70**], [**72**]) nicely supplements Theorem 1.9.1.

THEOREM 1.14.5 (Kanovei–Reeken). *All non-pathological F_σ ideals have the Radon–Nikodym property.* □

In particular, this result confirms a conjecture from [**32**] that the F_σ ideals constructed by Mazur in ([**94**]) have the Radon–Nikodym property. It also gives an alternative proof of Theorem 1.6.2, the Radon–Nikodym property of the ideal $\mathrm{Fin} \times \emptyset$.

Let us describe another important class of analytic ideals. Recall that an ordinal α is *additively indecomposable* (or simply *indecomposable*) if α cannot be represented as the sum of two strictly smaller ordinals. For an indecomposable countable ordinal α let an *ordinal ideal* $\mathcal{I}_\alpha$ be the ideal of all subsets of α of strictly smaller order type. Note that the Fubini product $\mathrm{Fin} \times \mathrm{Fin}$ is isomorphic to $\mathcal{I}_{\omega^2}$. All these ideals have the property that $\mathcal{I}_\alpha \upharpoonright A$ is isomorphic to $\mathcal{I}_\alpha$ for every positive set A; we say that such ideals are *homogeneous*. The following answers a question from [**32**] and it was proved in [**72**], where the ideals $\mathcal{I}_\alpha$ are called *generalized Fréchet ideals*.

THEOREM 1.14.6 (Kanovei–Reeken). *The ideals $\mathcal{I}_\alpha$ have the Radon–Nikodym property. In particular, $\mathcal{I}_{\omega^2} \approx \mathrm{Fin} \times \mathrm{Fin}$ has the Radon–Nikodym property.* □

An important role in proofs of these two Kanovei–Reeken theorems plays the notion of a *Fubini* submeasure. These are exactly the submeasures to which the proof of Theorem 1.8.2 applies. Let us say a word about the relationship between

Fubini and non-pathological submeasures. From results of Christensen ([**13**]) it follows that every Fubini submeasure is non-pathological. On the other hand, it was proved in [**72**] that every lower semicontinuous non-pathological submeasure is Fubini.

Theorem 1.14.6 may be related to a question of Galvin ([**50**]) concerning partially ordered sets $P(\alpha) = \langle \mathcal{P}(\alpha)/\mathcal{I}_\alpha, \subseteq^{\mathcal{I}_\alpha}\rangle$. Theorem 1.14.6 implies, for example, that there is no definable Boolean algebra monomorphism of $P(\omega^3)$ into $P(\omega^\omega)$. Galvin ([**50**]) asked whether there is (provably without using any additional set-theoretic axioms) a *strictly increasing* mapping from $P(\omega^3)$ into $P(\omega^\omega)$ (see [**32**, §3], [**30**] for more details).

A related class of topological ordinal ideals was suggested by W. Weiss ([**144**]). Let $\mathcal{W}_\alpha$ be the family of all subsets of α which do not include a subset which is homeomorphic to α in the ordinal topology. Then an application of Ramsey's theorem shows that $\mathcal{W}_\alpha$ is an ideal if and only if α is a countable ordinal of the form ω^{ω^β} (i.e., a *multiplicatively indecomposable* ordinal). In [**70**] the authors prove that all of these ideals have the Radon–Nikodym property, answering a question of author.

A similar theory of liftings can be developed in contexts other than the quotients of Boolean algebras; see [**38**] for a rather general setting. For example, analogous results to lifting theorems have been proved for Baire homomorphisms between analytic quotients considered as groups (see [**38**], [**72**], [**70**], also [**71**]). Theorems 1.14.5 and 1.14.6 were first proved in this context. Again, in the heart of all these results lies the Ulam stability of non-exact (group) homomorphisms (see [**37**]).

Groups have "less" structure than Boolean algebras, and this fact has three curious consequences:

(a) The results about liftings of group homomorphisms tend to be stronger than the results about liftings of Boolean-algebraic homomorphisms (see [**37**, §4], [**70**]).

(b) The proofs of the stability of group homomorphisms turn out to be simpler than the proofs of stability of Boolean-algebraic homomorphisms (compare §1.8 and [**37**]).

(c) It is impossible to develop a theory of group homomorphisms with arbitrary liftings analogous to Chapter 3 (see [**38**, §7]).

Even in the context of the ordering of the Borel-cardinality of quotients, where we do not have any algebraic structure at all, it is possible to say a lot about the structure of the liftings—see [**40**].

CHAPTER 2

Open Coloring Axiom and uniformization

In this short chapter we gather some results about the uniformization of coherent families needed in proofs of lifting theorems of Chapter 3 and Chapter 4. Although the needed prerequisites on the Open Coloring Axiom are given in this chapter, it is not intended to be an introduction into OCA; see [**125**, §8] or [**135**, §10] instead.

Let us recall the statement of the Open Coloring Axiom. Let X be a separable metric space, and by $[X]^2$ denote the family of all unordered pairs of its elements,

$$[X]^2 = \{\{x,y\} : x \neq y \text{ and } x, y \in X\}.$$

Subsets of $[X]^2$ can be naturally identified with the symmetric subsets of $X \times X$ minus the diagonal. A partition (or *coloring*) $[X]^2 = K_0 \cup K_1$ is *open* if K_0, when identified with a symmetric subset of $X \times X$, is open in the product topology. We say that a subset Y of X is K_i-*homogeneous* if $[Y]^2$ is included in K_i ($i = 0, 1$).

(**OCA**) If X is a separable metric space and $[X]^2 = K_0 \cup K_1$ is an open partition, then X either has an uncountable K_0-homogeneous subset or it can be covered by a countable family of K_1-homogeneous sets.

We shall usually say that a set covered by countably many K_1-homogeneous sets is σ-K_1-*homogeneous*. We should first say a word to clarify our use of the phrase "open coloring." The spaces like $\mathcal{P}(\mathbb{N})$, $\mathbb{N}^{\mathbb{N}}$, $\text{Fin}^{\mathbb{N}}$, and the finite products of such spaces, will always be considered with their natural separable metric product topology. In order to be able to apply OCA to some partition $[X]^2 = K_0 \cup K_1$, it suffices to know that there is a separable metric topology τ on X which makes K_0 open. For example, for $X \subseteq \mathcal{P}(\mathbb{N})$ and for each $x \in X$ we fix an $f_x \in \mathbb{N}^{\mathbb{N}}$ consider the partition $[X]^2 = K_0 \cup K_1$ defined by

$$\{x,y\} \in K_0 \text{ if and only if } f_x(n) \neq f_y(n) \text{ for some } n \in x \cap y.$$

This K_0 is not necessarily open in the topology inherited from $\mathcal{P}(\mathbb{N})$, but if we identify X with a subspace of $\mathcal{P}(\mathbb{N}) \times \mathbb{N}^{\mathbb{N}}$ via the embedding $x \mapsto \langle x, f_x \rangle$, then this partition becomes open. We shall use this fact tacitly quite often and say only that the partition $K_0 \cup K_1$ is open, meaning that it is open in some separable metric topology.

Indeed, all of our open partitions will be variations of the above, and it can be proved (see (c) of Proposition 2.2.11) that every open partition is isomorphic to one of this form. (Two partitions $K_0 \cup K_1$ and $L_0 \cup L_1$ are *isomorphic* if there is a bijection between their domains which sends pairs in K_0 into pairs in L_0.)

We shall frequently say that a coloring $[X]^2 = K_0 \cup K_1$ is *open* if K_0 is equal to a union of countably many rectangles, $K_0 = \bigcup_i U_i \times V_i$. This is equivalent to K_0 being open in some separable metric (not necessarily Hausdorff) topology on X (see Proposition 2.2.11).

OCA has a strong influence on essentially any structure closely related to the real line (see e.g., [**125**, §8], [**135**, §10]). In the case when X is an analytic set, OCA is true even if we replace the word "uncountable" with "perfect" (see Theorem 2.2.12). This fact was already used in §1.5 to prove the Radon–Nikodym property of Fin $\times \emptyset$.

A word on the evolution of OCA. The roots of OCA can be traced back to the work of Baumgartner [**3, 4**] who has discovered the phenomenon that assuming the Continuum Hypothesis one can force, in a ccc fashion, uncountable homogeneous sets for certain open partitions on sets of reals. For example, one such partition considered by Baumgartner is the incomparability relation of $\langle \mathcal{P}(\mathbb{N}), \subset \rangle$ (see the proof of Theorem 2.2.1). This idea was pushed further by Abraham, Rubin and Shelah who have isolated several axioms stated in terms of the existence of large homogeneous sets for open partitions ([**1**]). One of these axioms was called OCA and another, SOCA, had the asymmetric form of what we presently call OCA. However, SOCA, as well as all other axioms of [**1**], applies only to sets of reals of size $\aleph_1$ lacking thus the following crucial component of Todorcevic's OCA:

If X is not σ-K_1-homogeneous, then it has a subset of size $\aleph_1$ which is not σ-K_1-homogeneous.

This new reflection component of OCA does not appear (nor is hinted at) in the work of Baumgartner ([**3, 4**]) or Abraham, Rubin and Shelah ([**1**]). It is in fact this component of OCA that is responsible for many of its fundamental consequences needed for our purpose here, e.g., its effect on the gap spectrum of $\mathcal{P}(\mathbb{N})/\text{Fin}$ (Corollary 5.2.5). Furthermore, the axioms of [**1**], even when taken in conjunction with MA, are consistent with the existence of various 'nonreflecting' objects incompatible with Todorcevic's OCA, such as for example $\langle \omega_2, \omega_2 \rangle$-gaps in $\mathcal{P}(\mathbb{N})/\text{Fin}$. This can be easily shown e.g., by using the methods of [**81**]. A discussion on the evolution of OCA can also be found in [**125**, 8.17], the place where this axiom has been first introduced.

The organization of this chapter. In §2.1 we give a brief survey of the results of this and the following chapter and put these results in a broader context. In §2.2 we prove standard results about OCA and uniformization of coherent families of functions. These results are extended to uniformization of coherent families of partitions in §2.3, and in §2.4 we note how the properties of the uniformizing function (or a partition) reflect to the members of the coherent family. These results will be very useful in the proof of the OCA lifting theorem (Theorem 3.3.5). In the concluding §2.5 we give some remarks and an open question.

2.1. Why Open Coloring Axiom?

In Chapter 3 we study arbitrary homomorphisms between analytic quotients. We have already remarked that the Continuum Hypothesis oversimplifies the problem by making many analytic quotients isomorphic regardless of the structure of the corresponding ideals (i.e., they are isomorphic simply because they are *saturated*; see [**12**]). Therefore, in order to hope for any nontrivial result about the rigidity of analytic quotients in the presence of the Axiom of Choice, we have to work under an alternative to CH. It turns out that the right alternative is a version of two-dimensional Perfect-Set Property known under the name of Open Coloring Axiom (OCA). This axiom, since its first exposition by Todorcevic in [**125**], has found many applications, especially in problems related to the structure of real numbers.

The original purpose for introducing OCA was a study of Hausdorff-completeness of the structure $\mathcal{P}(\mathbb{N})/\operatorname{Fin}$ (more on this in Chapter 5). However, it turned out that OCA implies that $\mathcal{P}(\mathbb{N})/\operatorname{Fin}$ is complete in many other ways (see e.g., [34]). We should remark that an axiom called OCA (and several of its variations) have been previously considered in [1], but these axioms do not seem to be as useful to us as the axiom we presently call OCA.

We will prove that the theory of Baire-measurable liftings, as developed in Chapter 3, can be extended to arbitrary liftings assuming OCA (modulo some obvious limitations—see §3.2). The reader may be inclined to think of these results as relative consistency results which are merely saying what happens in yet another model of ZFC. We assert that this is not quite the case. Consider, for example, the following result:

THEOREM 2.1.1. *For every two given analytic P-ideals $\mathcal{I}$ and $\mathcal{J}$ the following are equivalent:*

(1) *There is a transitive model M of a large enough fragment of ZFC containing $\mathcal{I}$, $\mathcal{J}$, and all countable ordinals such that the quotients over $\mathcal{I}$ and $\mathcal{J}$ are nonisomorphic in M.*
(2) *The quotients over $\mathcal{I}$ and $\mathcal{J}$ are nonisomorphic in every transitive model of a large enough fragment of ZFC containing $\mathcal{I}$, $\mathcal{J}$ and all countable ordinals and satisfying OCA and MA.*

PROOF. It suffices to prove that (1) implies (2). Let $\mathcal{I}, \mathcal{J}$ be two analytic (and therefore $F_{\sigma\delta}$) P-ideals. Then statement "There exists a continuous map $F\colon \mathcal{P}(\mathbb{N}) \to \mathcal{P}(\mathbb{N})$ which is a lifting of an isomorphism between $\mathcal{P}(\mathbb{N})/\mathcal{I}$ and $\mathcal{P}(\mathbb{N})/\mathcal{J}$" is, being $\mathbf{\Sigma}^1_2$, subject to Shoenfield's Absoluteness Theorem ([110]) and therefore the nontrivial direction follows from Theorem 3.4.1. □

Roughly, Theorem 2.1.1 is saying that if the usual axioms of Set theory are not strong enough to show that the quotients over given two analytic P-ideals are isomorphic, then OCA and MA imply that they are not.

In fact, a more precise version of this result (see Theorem 3.4.1) states that two quotients over analytic P-ideals are isomorphic under OCA and MA only when there is a simple lifting witnessing this. This is an instance of a frequently encountered phenomenon in mathematics, that connecting maps between definable structures are usually definable themselves (see [46]). A result analogous to Theorem 2.1.1 in the case of embeddings rather than isomorphisms is the following:

THEOREM 2.1.2. *For every two given analytic P-ideals $\mathcal{I}$ and $\mathcal{J}$ the following are equivalent:*

(1) *There is a transitive model M of a large enough fragment of ZFC containing $\mathcal{I}$, $\mathcal{J}$, and all countable ordinals such that the quotient over $\mathcal{I}$ does not embed into the quotient over $\mathcal{J}$ in M.*
(2) *The quotient over $\mathcal{I}$ does not embed into the quotient over $\mathcal{J}$ in every transitive model of a large enough fragment of ZFC containing $\mathcal{I}$, $\mathcal{J}$ and all countable ordinals and satisfying OCA and MA.*

Theorem 2.1.2 follows from Proposition 3.2.2 and Corollary 3.4.7 proved in this Chapter, followed by an application of Shoenfield's absoluteness theorem ([110]). Consider now a topological interpretation of the similar phenomenon ($X^* = \beta X \setminus X$ is the Čech–Stone remainder of X):

THEOREM 2.1.3. *Assume X, Y are countable, locally compact, topological spaces[1]. Then the following are equivalent:*

(1) *There is a transitive model M of a large enough fragment of ZFC containing X, Y, and all countable ordinals such that X^* and Y^* are not isomorphic in M.*
(2) *X^* and Y^* are not isomorphic in every transitive model of a large enough fragment of ZFC containing X, Y and all countable ordinals and satisfying OCA and MA.* □

(This can be deduced from Corollary 4.7.2 using Shoenfield's absoluteness theorem, in the same way that Theorem 2.1.1 is deduced from Theorem 3.4.1.) A result analogous to Theorem 2.1.1 is also true in this context: If the usual axioms of Set theory are not strong enough to prove that X^* maps onto Y^*, then OCA and MA imply that it does not (follows from Corollary 4.7.4). A deeper principle lying behind these two results, called the *weak Extension Principle*, or *wEP*, (see §4.1), roughly says that maps between Čech–Stone remainders can often be extended to maps between the corresponding compactifications. Clearly, the dual statement merely says that the existence of a given homomorphism between clopen algebras is witnessed by a simple lifting. The following metatheorem summarizes a part of the above discussion (where $\mathcal{I}$ and $\mathcal{J}$ are analytic P-ideals, and X and Y are countable locally compact spaces).

THEOREM 2.1.4. *Assume φ is a statement in the language of Set theory saying that*

(1) *The quotients over $\mathcal{I}$ and $\mathcal{J}$ are not isomorphic, or*
(2) *The quotient over $\mathcal{I}$ does not embed into the quotient over $\mathcal{J}$, or*
(3) *The remainders X^* and Y^* are not isomorphic, or*
(4) *The remainder X^* does not map onto the remainder Y^*.*

Then φ is either false in every transitive model of a large enough fragment of ZFC which contains all countable ordinals or φ is true in every transitive model of a large enough fragment of ZFC which contains all countable ordinals and satisfies OCA and MA. □

A remarkable consequence of this result is that all sentences that can be put into one of the forms (1)–(4) above and which are not outright false hold *simultaneously* in a single model! It is well-known that the effect of the Continuum Hypothesis is quite the opposite from the described effect of OCA and MA. Under CH all quotients over, say, F_σ ideals are saturated and therefore isomorphic (see [65], also §1.2 and §3.14), but the liftings which make them isomorphic are very far from being simply definable. Similarly, under CH many remainders of rather different topological spaces are homeomorphic to $\mathbb{N}^*$, and $\mathbb{N}^*$ maps onto any compact space of weight $\leq 2^{\aleph_0}$ (see [101], also §4.1). Both statements fail badly under OCA and MA. A metamathematical explanation of this effect of CH is provided by a result of Woodin ([145]; see also [146] for a complementary ¬CH-result).

THEOREM 2.1.5 (Woodin). *Assuming large cardinals, if a Σ_1^2-statement is true in one forcing extension of the universe then it is also true in any other forcing extension which satisfies CH.* □

---
[1] I.e., countable limit ordinals

Note that the negations of (1), (2) and the negations of the duals of (3), (4) of Theorem 2.1.4 are all of the form Σ_1^2: each one of these statements asserts the existence of a set of reals (a lifting) which satisfies certain projective statement. Theorem 2.1.4, together with Theorem 2.1.5, show that if we want to know the relationship between two analytic quotients (whether they are provably/consistently isomorphic, or whether one provably/consistently embeds into another) we need to use only two different tests, the CH-test and the (OCA and MA)-test. Note that, while the CH-test requires substantial large cardinal assumptions, the (OCA and MA)-test does not require any.

Let us emphasize that Theorem 2.1.4 summarizes only a part of what is true in this context. Another major point is that under OCA and MA the relation between two structures (analytic quotients, or Čech–Stone remainders) has to be witnessed by a simple lifting, and the assertion that such a lifting exists is of much lower complexity—Σ_2^1, instead of Σ_1^2, and therefore absolute by Shoenfield's theorem (see [**110**])! (Sometimes the complexity of this statement is as low as the complexity of the involved ideals, see Lemma 1.12.5.) This brings the theory of analytic quotients as close as possible to their natural setting in the "Borel world," as discussed in Chapter 1. All this seems to suggest that OCA and MA is, in some sense, the optimal test for studying not only analytic quotients, but also many related structures. The class of statements to which Theorem 2.1.4 applies can be extended to treat other ideals (Theorem 3.3.6), finite and infinite powers of Čech–Stone remainders (§4.1 and [**41**], respectively) a larger class of topological spaces (§4.9, §4.10), preservation of Hausdorff gaps by embeddings (§5.9), ... We believe that this is only the beginning of a very promising line of research.

2.2. Coherent families of partial functions

Let us recall some results of Todorcevic about OCA and uniformization of coherent families. There results will be slightly generalized in the following section.

A family $\mathcal{F}$ of partial functions on some countable set S is *trivial* if there is a function $h\colon S \to S$ such that

$$f(n) = h(n) \text{ for all but finitely many } n \in \mathrm{dom}(f), \text{ for every } f \in \mathcal{F}.$$

We also say that h *trivializes* $\mathcal{F}$, or that $\mathcal{F}$ is *trivialized* by h. The family $\mathcal{F}$ is σ-*trivial* if it can be covered by countably many trivial subfamilies.

Two partial functions f, g on some countable set are *incompatible* if

$$f(i) \neq g(i) \text{ for some } i \in \mathrm{dom}(f) \cap \mathrm{dom}(g).$$

Equivalently, f and g are incompatible if $f \cup g$ is not a function.

THEOREM 2.2.1 (Todorcevic). *Assume OCA. Then every family $\mathcal{F}$ of partial functions on some countable set is either σ-trivial or it includes an uncountable subset of pairwise incompatible functions.*

PROOF. We can assume that the domain of each function in $\mathcal{F}$ is a subset of the natural numbers. Define a partition $[\mathcal{F}]^2 = K_0 \cup K_1$ by letting $\{f_A, f_B\}$ in K_0 if f_A and f_B are incompatible, i.e., if

$$f_A(n) \neq f_B(n) \text{ for some } n \in A \cap B.$$

If we identify functions with sets of ordered pairs and consider $\mathcal{F}$ as a subset of $\mathcal{P}(\mathbb{N}) \times \mathcal{P}(\mathbb{N})$, then this partition is open in the Cantor-set topology, and we may apply OCA. (Alternatively, note that $K_0 = \bigcup_i U_{i0} \times U_{i1}$, for $U_{ij} = \{f_A =: i \in$

$\mathrm{dom}(f)$, $f_A(i) = j\}$, and use Proposition 2.2.11.) Therefore OCA implies that either there is an uncountable set of pairwise incompatible functions, or $\mathcal{F} = \bigcup_i \mathcal{X}_i$, where each $\mathcal{X}_i$ is K_1-homogeneous. But then the union of all functions in $\mathcal{X}_i$ is a function which trivializes $\mathcal{X}_i$. Therefore $\mathcal{F}$ is σ-trivial. □

The open partition used in the above proof has played an important role in most of the later applications of OCA found in the literature; for example, see [**142**] and Chapter 3. This fact is not surprising, since it is essentially the only open partition (see the proof of (c) implies (b), Proposition 2.2.11). We will state several consequences of Theorem 2.2.1, after recalling some terminology.

A *coherent family of partial functions indexed by* $\mathcal{A}$, for some $\mathcal{A} \subseteq \mathcal{P}(\mathbb{N})$, is a family h_A $(A \in \mathcal{A})$ such that

(2) $h_A \colon A \to \mathbb{N}$, and
(3) $h_A(n) = h_B(n)$ for all but finitely many $n \in A \cap B$, for all A, B in $\mathcal{A}$.

Recall that a partial ordering $\mathcal{P}$ is *directed* if for every pair of elements $p, q \in \mathcal{P}$ there is $r \geq p, q$. It is σ-*directed* if for every sequence p_n $(n \in \mathbb{N})$ of elements of $\mathcal{P}$ there is $r \in \mathcal{P}$ such that $r \geq p_n$ for all n. More generally, a partially ordered set is κ-*directed* if all of its subsets of size at most κ are bounded.

LEMMA 2.2.2. (a) *If a directed partial ordering $\mathcal{Q}$ is partitioned into finitely many pieces, then one of these pieces is cofinal.*
(b) *If a σ-directed partial ordering $\mathcal{Q}$ is partitioned into countably many pieces, then one of these pieces is cofinal.*

PROOF. (a) Assume $\mathcal{Q}_i \subseteq \mathcal{Q}$ for $1 \leq i \leq n$. If $\mathcal{Q}_i$ is not cofinal in $\mathcal{Q}$, pick a $q_i \in \mathcal{Q}$ such that $q_i \not\leq q$ for every $q \in \mathcal{Q}_i$. By the directedness, there is a $p \in \mathcal{Q}$ such that $q_i \leq p$ for all i; but then $p \notin \mathcal{Q}_1 \cup \cdots \cup \mathcal{Q}_n$.
The proof of (b) is analogous. □

In this section, we consider subsets of $\mathcal{P}(\mathbb{N})$ as ordered by $\subseteq^*$, the inclusion modulo finite. Therefore we can talk about σ-directed, $\aleph_1$-directed subsets of $\mathcal{P}(\mathbb{N})$. Variants of the following lemma are given in Lemma 3.8.5 and Proposition 3.13.3 (K_1 is defined as in the proof of Theorem 2.2.1, i.e., by $\{f, g\} \in K_0$ if $f(n) = g(n)$ for all $n \in \mathrm{dom}(f) \cap \mathrm{dom}(g)$).

LEMMA 2.2.3. *Let $\mathcal{F}$ be a coherent family of partial functions indexed by a σ-directed family $\mathcal{A}$. Then the following are equivalent*

(1) $\mathcal{F}$ *is trivial,*
(2) $\mathcal{F}$ *is σ-K_1-homogeneous.*

PROOF. (2) $\Leftarrow$ (1). If h trivializes $\mathcal{F} = \{h_A : A \in \mathcal{A}\}$, then for $s \in \mathrm{Fin}$ and $p \colon s \to \mathbb{N}$ the set

$$\mathcal{F}_{s,p} = \{h_A : p \subseteq h_A \text{ and } (h_A \upharpoonright (\mathbb{N} \setminus s)) \subseteq h\}$$

is K_1-homogeneous.
(1) $\Leftarrow$ (2). If $\mathcal{F}$ is σ-K_1-homogeneous, then by Lemma 2.2.2 for at least one of the K_1-homogeneous pieces, say $\mathcal{H}$, is the set

$$\{A \in \mathcal{A} : h_A \in \mathcal{H}\}$$

is cofinal in $\langle \mathcal{A}, \subseteq^* \rangle$. Then $h = \bigcup \mathcal{H}$ is a function which trivializes $\mathcal{F}$. □

COROLLARY 2.2.4. *Assume OCA. Then every coherent family of partial functions indexed by a σ-directed set is either trivial or it includes an uncountable subset of pairwise incompatible functions.*

PROOF. By Theorem 2.2.1, the family $\mathcal{F}$ either includes an uncountable subset of pairwise compatible functions or it is σ-trivial. But by Lemma 2.2.2, if $\mathcal{F}$ is σ-trivial then it is trivial, and this concludes the proof. □

COROLLARY 2.2.5. *Assume OCA. Then every coherent family $\mathcal{F}$ of partial functions indexed by an $\aleph_1$-directed set $\mathcal{A}$ is trivial.*

PROOF. By Corollary 2.2.4, it suffices to note that every bounded subset of a coherent family of partial functions is trivial. □

We shall consider the set $\mathbb{N}^\mathbb{N} = \{f : f \colon \mathbb{N} \to \mathbb{N}\}$ with two natural orderings. The first one is $\leq$, the pointwise dominance: $f \leq g$ if $f(m) \leq g(m)$ for all m. The second one is the ordering of *eventual dominance*, $\leq^*$, namely the ordering $\leq$ taken modulo Fin: we let $f \leq^* g$ if $f(m) \leq g(m)$ for all but finitely many m. Note that $\langle \mathbb{N}^\mathbb{N}, \leq^* \rangle$ is σ-directed. By $\mathfrak{b}$ we denote the *bounding number*: the minimal cardinal κ such that there is a subset of $\mathbb{N}^\mathbb{N}$ of size κ which is unbounded in the ordering $\leq^*$. Equivalently, $\mathfrak{b}$ is the minimal cardinal such that $\emptyset \times$ Fin is not $\mathfrak{b}^+$-directed.

If a family of partial functions $\mathcal{F}$ is indexed by $\mathcal{A} = \{\Gamma_f : f \in \mathbb{N}^\mathbb{N}\}$, then we say that $\mathcal{F}$ is *indexed by* $\mathbb{N}^\mathbb{N}$ and denote a typical element of $\mathcal{F}$ by h_f instead of h_{Γ_f}.

THEOREM 2.2.6 (Todorcevic). *OCA implies that $\mathfrak{b} = \omega_2$.* □

PROOF. This follows by [**125**, Theorem 3.4] and [**125**, Theorem 8.5]. See also Corollary on page 87 of [**135**]. □

The following was proved in [**125**, Theorem 8.7].

COROLLARY 2.2.7. *OCA implies that every coherent family $\mathcal{F}$ of partial functions indexed by $\mathbb{N}^\mathbb{N}$ is trivial.*

PROOF. This follows immediately by Theorem 2.2.6 and Corollary 2.2.5. □

COROLLARY 2.2.8. *Assume OCA and MA. Then every coherent family of functions indexed by an analytic P-ideal is trivial.*

PROOF. A result of Todorcevic (see Lemma 5.6.3) implies that every analytic P-ideal is κ-directed for every cardinal κ less than the additivity of the Lebesgue measure. Since Martin's Axiom implies that the latter is at least $\aleph_2$ (by [**91**], see also [**82**], or [**135**, Proposition 7.3]), the result follows from Corollary 2.2.5. □

So far we have considered only the coherent families of functions with range equal to $\{0,1\}$, and we shall now see that this was not a loss of generality. The statement (1) below has been considered before, in particular in connection with Milnor's additivity axiom for strong homology (see [**90**], [**25**], [**125**, 8.7], [**129**]). It is also known that (1) does not follow from MA ([**129**]) and that it is inconsistent with CH ([**90**]).

The following is a consequence of a result of Todorcevic reproduced in [**25**, Theorem 4.1] (or rather, its proof).

THEOREM 2.2.9. *The following are equivalent:*

(1) *Every coherent family of functions with range $\mathbb{N}$ indexed by $\mathbb{N}^\mathbb{N}$ is is trivial.*

(2) *Every coherent family of functions with range $\{0,1\}$ indexed by $\mathbb{N}^{\mathbb{N}}$ is trivial.*

PROOF. It will clearly suffice to prove that (2) implies (1). Assume (1) fails, and let g_f ($f \in \mathbb{N}^{\mathbb{N}}$) be a nontrivial coherent family of functions whose ranges are included in $\mathbb{N}$. For $f \in \mathbb{N}^{\mathbb{N}}$ define $\hat{f} \in \mathbb{N}^{\mathbb{N}}$ by

$$\hat{f}(m) = 2^{f(m)+1} - 1.$$

Family $\hat{f}$ ($f \in \mathbb{N}^{\mathbb{N}}$) is clearly cofinal in $\mathbb{N}^{\mathbb{N}}$, therefore to refute (2) it will suffice to construct a nontrivial family of coherent functions $h_{\hat{f}} \colon \Gamma_{\hat{f}} \to \{0,1\}$. Let

$$A_{\hat{f}} = \{\langle m, 2^n(2g_f(m,n)+1)\rangle : m \in \mathbb{N}\} \cap \Gamma_{\hat{f}},$$

and let $h_f \colon \Gamma_{\hat{f}} \to \{0,1\}$ be defined by $h_f^{-1}(0) = A_{\hat{f}}$ and $h_f^{-1}(1) = \Gamma_{\hat{f}} \setminus A_{\hat{f}}$. We claim that this is a nontrivial coherent family.

Let us introduce some notation. For $g_f \colon \Gamma_f \to \mathbb{N}$ and $g_{f'} \colon \Gamma_{f'} \to \mathbb{N}$ let

$$\Delta'(g_f, g_{f'}) = \min\{\bar{n} : g_f(n,i) = g_{f'}(n,i)$$
$$\text{whenever } \bar{n} \leq n \text{ and } i \leq \min(f(n), f'(n))\},$$

This is clearly well-defined if functions g_f are coherent. Family $h_{\hat{f}}$ ($f \in \mathbb{N}^{\mathbb{N}}$) is coherent because $\Delta'(h_{\hat{f}}, h_{\hat{f'}}) \leq \Delta'(g_f, g_{f'})$. To see that this family is nontrivial, assume for the sake of getting a contradiction that it is trivialized by some $\hat{F} \colon \mathbb{N}^2 \to \{0,1\}$. Define $F \colon \mathbb{N}^2 \to \mathbb{N}$ by

$$F(m,n) = \min\{i : F(m, 2^n(2i+1)) = 0\},$$

if such i exists and $F(m,n) = 0$ otherwise. Then $\Delta'(g_f, F \upharpoonright \Gamma_f) \leq \Delta'(h_{\hat{f}}, \hat{F} \upharpoonright \Gamma_{\hat{f}})$, and therefore F trivializes family $\{g_f\}$. □

COROLLARY 2.2.10. *OCA implies that every coherent family of functions with range $\mathbb{N}$ indexed by $\mathbb{N}^{\mathbb{N}}$ is trivial.*

PROOF. This follows by Corollary 2.2.7 Theorem 2.2.9. □

A nontrivial coherent family indexed by a P-ideal can be obtained from a Hausdorff gap (see [**5**, pages 96–98] and Chapter 5, in particular §5.12). A uniformization theorem for partial functions into the reals can be found in [**135**, Theorem 10.6]; see also [**135**, Theorems 10.3* and 10.3**] for consequences of Corollary 2.2.5. Related results about coherent families of partial functions from countable subsets of an arbitrary cardinal κ can be found in [**134**] and [**21**].

The remaining part of this section, in which we give two reformulations of OCA, will not be used later in this text. A subset $A \subseteq [X]^2$ is a *rectangle* if it is of the form $A = U \otimes V = \{\{x,y\} : x \in U, y \in V \text{ and } x \neq y\}$. If $\mathcal{A} \subseteq \mathcal{P}(\mathbb{N})$ we say that a family $\mathcal{F}$ of functions $f_A \colon A \to \mathbb{N}$ ($A \in \mathcal{A}$) is *indexed by* $\mathcal{A}$. Such family is *trivial* if there is $g \colon \mathbb{N} \to \mathbb{N}$ such that $f_A =^* g \upharpoonright A$ for all $A \in \mathcal{A}$. We say that g *trivializes* $\mathcal{F}$. We say that $\mathcal{F}$ is *σ-trivial* if it can be partitioned into countably many trivial subfamilies. In the following result, only the implication from (c) to (a) and (d) is due to the author; all the other implications are Todorcevic's.

PROPOSITION 2.2.11. *The following are equivalent:*
(a) *OCA.*
(b) *For every partition $[X]^2 = K_0 \cup K_1$ such that K_0 is equal to the union of countably many rectangles, X either has an uncountable K_0-homogeneous subset or it can be covered by countably many K_1-homogeneous sets.*

(c) *Every family of partial functions $\{f_A \colon A \to \{0,1\}\}$ indexed by some $\mathcal{A} \subseteq \mathcal{P}(\mathbb{N})$ is either σ-trivial or it includes an uncountable subfamily $\mathcal{F}'$ such that for all $a \neq b$ in $\mathcal{A}$ we have $f_A(n) \neq f_B(n)$ for some $n \in A \cap B$.*
(d) *The same as (c), but with $f_A \colon A \to \mathbb{N}$.*

PROOF. Note that if K_0 is open and X is separable metric, then K_0 is equal to some union of countably many rectangles; therefore (a) implies (b). To show the reverse implication, assume that $K_0 \subseteq [X]^2$ is such that $K_0 = \bigcup_i U_{0i} \oplus U_{1i}$. Define $f \colon X \to \mathcal{P}(\{0,1\} \times \mathbb{N})$ be defined by
$$f(x) = \{\langle i, j \rangle : x \in U_{ij}\}.$$
Let $Y = f''X$ and define a partition $[Y]^2 = L_0 \cup L_1$ by letting $\{a,b\} \in L_0$ if $\langle 0, i \rangle \in a$ and $\langle 1, i \rangle \in b$, or $\langle 1, i \rangle \in a$ and $\langle 0, i \rangle \in b$ for some i. This partition is open (when $\mathcal{P}(\{0,1\} \times \mathbb{N})$ is given its natural Cantor-set topology). Note that the f-preimages of L_1-homogeneous sets are K_1-homogeneous, and therefore if Y is σ-L_1-homogeneous then X is σ-K_1-homogeneous. So by OCA we can assume that there is an uncountable L_0-homogeneous subset of Y; let y_ξ ($\xi < \omega_1$) be its $1-1$ enumeration. Pick x_ξ such that $f(x_\xi) = y_\xi$; then the set $\{x_\xi : \xi < \omega_1\}$ is K_0-homogeneous, and therefore (b) is true.

(a) implies (d) will be proved in Theorem 2.2.1.

(d) implies (c) is trivial.

To see that (c) implies (b), we fix a partition $[X]^2 = K_0 \cup K_1$ such that $K_0 = \bigcup_i U_{0i} \oplus U_{1i}$ and prove that one of the alternatives given by OCA is true. Note that each set $U_{0i} \cap U_{1i}$ is K_0-homogeneous, and therefore we can assume that all these sets are countable. Since the set $\bigcup_i U_{0i} \cap U_{1i}$ is countable, it is σ-K_1 homogeneous and we can assume that it is empty without a loss of generality. For $x \in X$ define $a(x) \subseteq \mathbb{N}$ and $f_{a(x)} \colon a(x) \to \{0,1\}$ by
$$a(x) = \{i : x \in U_{0i} \cup U_{1i}\}$$
$$f_{a(x)}(i) = j, \text{ if } x \in U_{ji}.$$
The two alternatives given by (c) clearly translate into two OCA-alternatives. □

We shall frequently say that a coloring is *open* whenever it is equal to a union of countably many rectangles.

Variations on (c) of Proposition 2.2.11 are the main theme of this and the following section, and its slight weakening is equivalent to a statement about gaps in $\mathcal{P}(\mathbb{N})/\mathrm{Fin}$ (see the remark after the proof of Theorem 5.2.4).

One of the reasons why OCA can be considered a natural axiom is the fact that it has a definable version, the following Principle of Open Coloring for analytic sets.

THEOREM 2.2.12. *Assume that X is an analytic set of real numbers and that $[X]^2 = K_0 \cup K_1$ is its open partition. Then exactly one of the following applies:*

(1) *X has a nonempty perfect K_0-homogeneous subset, or*
(2) *X is σ-K_1-homogeneous.*

PROOF. See [**135**, Theorem 10.1], also [**43**]. □

2.3. Coherent families of partitions

By $[A]^{<\mathbb{N}}$ we denote the family of all finite subsets of a set A. A *coherent family of partitions* indexed by $\mathbb{N}^\mathbb{N}$ is a family P_f ($f \in \mathbb{N}^\mathbb{N}$) such that

(1) $P_f \colon [\Gamma_f]^{<\mathbb{N}} \to \mathbb{N}$, and
(2) $P_f(s) = P_g(s)$ for all but finitely many $s \subseteq \Gamma_f \cap \Gamma_g$, for all $f, g \in \mathbb{N}^{\mathbb{N}}$.

Let us write
$$E_n = [n, \infty) \times \mathbb{N}.$$
A coherent family $\mathcal{F}$ of partitions is *trivial* if there is an n and a partition
$$P \colon [E_n]^{<\mathbb{N}} \to \mathbb{N}$$
such that

(3) $P_f(s) = P(s)$ for all but finitely many $s \subseteq [\Gamma_f \cap E_n]^{<\mathbb{N}}$, for every $f \in \mathbb{N}^{\mathbb{N}}$.

We also say that P *trivializes* $\mathcal{F}$ on E_n.

THEOREM 2.3.1 (The OCA uniformization theorem). *Assume OCA. Then every coherent family of partitions indexed by $\mathbb{N}^{\mathbb{N}}$ is trivial.*

PROOF. Let $\{P_f \colon [\Gamma_f]^{<\mathbb{N}} \to \mathbb{N}\}$ be a coherent family of partitions. Define a partition $[\mathbb{N}^{\mathbb{N}}]^2 = K_0 \cup K_1$ by letting $\{f, g\}$ into K_0 if
$$P_f(s) \neq P_g(s) \text{ for some } s \subseteq \Gamma_f \cap \Gamma_g.$$
This partition is open, if we identify each f with the pair $\langle f, P_f \rangle$. (Alternatively, note that $K_0 = \bigcup_{s, m \neq n} U_{s,m} \times U_{s,n}$, for $U_{s,m} = \{f : s \in \mathrm{dom}(f), f(s) = m\}$, and apply Proposition 2.2.11.) By Lemma 2.2.3 (or rather, its version for coherent families of partitions) and Theorem 2.2.6, every subset of $\mathbb{N}^{\mathbb{N}}$ of size $\aleph_1$ is σ-K_1-homogeneous. Therefore there are no uncountable K_0-homogeneous sets. By OCA, $\mathbb{N}^{\mathbb{N}}$ is σ-K_1-homogeneous. By Lemma 2.2.2, there is a K_1-homogeneous subset $\mathcal{H}$ of $\mathbb{N}^{\mathbb{N}}$ which is cofinal in the $\leq^*$-ordering. By Kunen's lemma $\mathcal{H}$ is cofinal in $\mathbb{N}^{[n,\infty)}$ for some n. Then
$$P = \bigcup_{f \in \mathcal{H}} P_f$$
is a function whose domain includes $[[n, \infty) \times \mathbb{N}]^{<\mathbb{N}}$. Note that for every $f \in \mathbb{N}^{\mathbb{N}}$ there is $g \in \mathcal{H}$ such that $\Gamma_f \cap ([n, \infty) \times \mathbb{N}) \subseteq \Gamma_g$, and that $P \upharpoonright [\Gamma_g]^{<\mathbb{N}} = P_g$. This proves that P trivializes $\{P_f\}$ on $[[n, \infty) \times \mathbb{N}]^{<\mathbb{N}}$, and completes the proof. □

We can define coherent families of partitions indexed by various directed sets.

COROLLARY 2.3.2. *Assume OCA. If*
$$h_{gf} \colon [\Gamma_f]^{<\mathbb{N}} \times [\Gamma_g]^{<\mathbb{N}} \to \mathbb{N}$$
is a coherent family of partitions indexed by $(\mathbb{N}^{\mathbb{N}})^2$, then it is trivial.

PROOF. This follows from Theorem 2.3.1 by an obvious coding procedure. □

Let us note two simple lemmas for future reference; we omit the obvious proofs.

LEMMA 2.3.3. *A family of partitions indexed by an ideal (or, a set directed under $\subseteq$) $\mathcal{I}$ is coherent if and only if the coherence condition holds for all $A \subseteq A'$ in $\mathcal{I}$.* □

A family $\mathcal{A}$ *generates* the ideal $\mathcal{I}$ if for every $B \in \mathcal{I}$ there is an $A \in \mathcal{A}$ such that $B \subseteq A$.

LEMMA 2.3.4. *If a coherent family $\mathcal{F}$ of partitions is indexed by a set $\mathcal{A}$ which generates an ideal $\mathcal{I}$, then $\mathcal{F}$ can be extended to a coherent family of partitions indexed by $\mathcal{I}$ which is trivial if and only if $\mathcal{F}$ is trivial.* □

2.4. Σ_2-reflection for coherent families

The following two lemmas are extracted from proofs of lifting theorems.

LEMMA 2.4.1. *Assume $\mathcal{F} = \{h_f \colon \Gamma_f \to \mathbb{N}\}$ is a trivial coherent family indexed by $\mathbb{N}^{\mathbb{N}}$ and $[\mathbb{N}^2]^{<\mathbb{N}} = K_0 \cup K_1$ is a partition such that*

(1) $h_f^{-1}(\{i\})$ *is K_0-homogeneous for all f and i.*

Then we can choose the trivializing function h for $\mathcal{F}$ so that $h^{-1}(\{i\})$ is K_0-homogeneous for all i.

LEMMA 2.4.2. *Assume $\mathcal{F} = \{h_f \colon \Gamma_f \to \mathbb{N}\}$ is a trivial coherent family indexed by $\mathbb{N}^{\mathbb{N}}$ and $[\mathbb{N}^2]^{<\mathbb{N}} = K_0 \cup K_1$ is a partition such that*

(1) $h_f^{-1}(\{i\})$ *is either empty or a maximal K_0-homogeneous subset of Γ_f, for every f and every i.*

Then we can choose the trivializing function h for $\mathcal{F}$ so that $h^{-1}(\{i\})$ is a maximal K_0-homogeneous subset of $\mathrm{dom}(h)$ for all i.

We shall prove these two lemmas at the end of this section. The "real" result behind them seems to be Theorem 2.4.4 below. Recall that a formula φ is Σ_2 if it is of the form

$$\exists x \forall y \psi(x, y)$$

for some quantifier-free formula ψ.

A proof of the following lemma, as well as its strengthening, can be found in [**80**].

LEMMA 2.4.3. *If a subset $\mathcal{H}$ of $\mathbb{N}^{\mathbb{N}}$ is cofinal in the $\leq^*$-ordering, then there is $n \in \mathbb{N}$ such that*

$$\{f \restriction [n, \infty) : f \in \mathcal{H}\}$$

is cofinal in $\mathbb{N}^{[n,\infty)}$, ordered by the strict dominance, $\leq$. □

THEOREM 2.4.4 (Σ_2-reflection for coherent families). *Assume $R \colon [\mathbb{N}^2]^{<\mathbb{N}} \to \mathbb{N}$ is a partition, and*

$$P_f \colon [\Gamma_f]^{<\mathbb{N}} \to \mathbb{N}$$

is a coherent family of partitions indexed by $\mathbb{N}^{\mathbb{N}}$. If this family is trivial, then we can choose the trivializing partition $P \colon [[n, \infty) \times \mathbb{N}]^{<\mathbb{N}} \to \mathbb{N}$ so that every Σ_2-sentence—possibly with parameters—true in the structure (let $A = [n, \infty) \times \mathbb{N}$)

$$\langle A, \mathbb{N}, R \restriction [A]^{<\mathbb{N}}, P \restriction [A]^{<\mathbb{N}} \rangle$$

is also true in (let $A_f = A \cap \Gamma_f$)

$$\langle A_f, \mathbb{N}, R \restriction [A_f]^{<\mathbb{N}}, P_f \restriction [A_f]^{<\mathbb{N}} \rangle$$

for $\leq^*$-cofinally many $f \in \mathbb{N}^{\mathbb{N}}$.

PROOF. Let P' be any trivializing partition. By Lemma 2.2.2 and Lemma 2.4.3, there is an n such that the set

$$\mathcal{H} = \{f : P_f(s) = P'(s) \text{ for all } s \subseteq \Gamma_f \cap ([n, \infty \times \mathbb{N})\}$$

is cofinal in $\mathbb{N}^{[n,\infty)}$. Let $A = [n, \infty) \times \mathbb{N}$. Then if

$$(\exists i_1, \ldots, i_k)(\forall j_1, \ldots, j_l) \varphi(i_1, \ldots, i_k, j_1, \ldots, j_l, \vec{\mathbf{m}})$$

is a formula with parameters $\vec{\mathbf{m}}$ in $A \cup \mathbb{N}$, let $\mathbf{i_1}, \ldots, \mathbf{i_k} \in A \cup \mathbb{N}$ be such that the formula

$$(\forall j_1, \ldots, j_l) \varphi(\mathbf{i_1}, \ldots, \mathbf{i_k}, j_1, \ldots, j_l, \vec{\mathbf{m}})$$

is satisfied in

$$\langle A, \mathbb{N}, R \upharpoonright [A]^{<\mathbb{N}}, P_0 \upharpoonright [A]^{<\mathbb{N}} \rangle.$$

Then every $f \in \mathcal{H}$ such that $\mathbf{i_1}, \ldots, \mathbf{i_k}, \vec{\mathbf{m}} \in \Gamma_f$ satisfies the same formula. □

To prove Lemma 2.4.1 and Lemma 2.4.2 we shall need only a version of Theorem 2.4.4 for coherent families of functions.

PROOF OF LEMMA 2.4.1. For every $k \in \mathbb{N}$ the statement "there is a k-element set which is not K_0-homogeneous but the uniformizing function h is constant on this set" is Σ_1. If h is as guaranteed by Theorem 2.4.4 and $h^{-1}(\{m\})$ is not a K_0-homogeneous set, there is a subset of this set of size k which is not in K_0. Then $h_f^{-1}(\{m\})$ is not K_0-homogeneous for cofinally many f. □

PROOF OF LEMMA 2.4.2. For a fixed pair $\langle i, j \rangle \neq \langle i', j' \rangle$ and every $k \in \mathbb{N}$ the statement "$m = h(i,j) \neq h(i',j')$, and for every k-element subset s of $h^{-1}(\{m\})$ the set $s \cup \{\langle i,j \rangle, \langle i',j' \rangle\}$ is in K_0" is Π_1. If h is as guaranteed by Theorem 2.4.4 and $h^{-1}(\{m\})$ is nonempty and not a maximal K_0-homogeneous set for some k, we can find $\langle i', j' \rangle \notin h^{-1}(\{m\})$ such that $\{\langle i', j' \rangle\} \cup h^{-1}(\{m\})$ is K_0-homogeneous. Fix $\langle i, j \rangle \in h^{-1}(\{m\})$ and note that by Theorem 2.4.4 the set $h_f^{-1}(\{m\})$ will not be maximal K_0-homogeneous for cofinally many f. □

2.5. Remarks and questions

It is natural to ask whether the results of this Chapter can be generalized to families coherent modulo an arbitrary analytic P-ideal $\mathcal{I}$. We shall show that this is not the case. Let us say that a family $\mathcal{F}$ of partial functions is *coherent modulo* $\mathcal{I}$ if for all $f, g \in \mathcal{F}$ the set $\{n \in \text{dom}(f) \cap \text{dom}(g) : f(n) \neq g(n)\}$ is in $\mathcal{I}$. This family is *trivial* if there is a single function $h \colon \mathbb{N} \to \mathbb{N}$ such that $\{n \in \text{dom}(f) : f(n) \neq h(n)\}$ is in $\mathcal{I}$ for every $f \in \mathcal{F}$. It is σ-*directed* if the family $\{\text{dom}(f) : f \in \mathcal{F}\}$ is σ-directed modulo Fin.

THEOREM 2.5.1. *There are F_σ P-ideals $\mathcal{I}$ and $\mathcal{C}$ and a family $\mathcal{F}$ of functions coherent modulo $\mathcal{I}$ indexed by $\mathcal{C}$ such that*

(1) *$\mathcal{F}$ is σ-directed*
(2) *$\mathcal{F}$ is not trivial,*
(3) *every subfamily of $\mathcal{F}$ of size less than the additivity of Lebesgue measure is trivial, moreover*
(4) *$\mathcal{F}$ is an analytic subset of $\mathcal{P}(\mathbb{N}) \times \{0, 1\}$.*

PROOF. This is a reformulation of Theorem 5.10.2, where it will be proved that there are F_σ P-ideals $\mathcal{I}, \mathcal{A}$, and $\mathcal{B}$ such that

(a) $A \cap B \in \mathcal{I}$, for all $A \in \mathcal{A}$ and all $B \in \mathcal{B}$ ($\mathcal{A}$ and $\mathcal{B}$ are $\mathcal{I}$-*orthogonal*),
(b) there is no $C \subseteq \mathbb{N}$ such that $A \setminus C \in \mathcal{I}$ for all $A \in \mathcal{A}$ and $B \cap C \in \mathcal{I}$ for all $B \in \mathcal{B}$ ($\mathcal{A}$ and $\mathcal{B}$ are not *separated* in $\mathcal{P}(\mathbb{N})/\mathcal{I}$).

By Mazur's theorem (Theorem 1.2.5(a)), let φ_0 and φ_1 be lower semicontinuous submeasures such that $\mathcal{A} = \mathrm{Fin}(\varphi_0)$ and $\mathcal{B} = \mathrm{Fin}(\varphi_1)$. Let

$$\mathcal{K}_0 = \{A \subseteq \mathbb{N} : \varphi_0(A) \leq 1\},$$
$$\mathcal{K}_1 = \{B \subseteq \mathbb{N} : \varphi(B) \leq 1\},$$

and let $\mathcal{D} = \{\langle A, B\rangle : A \in \mathcal{K}_0, B \in \mathcal{K}_1, A \cap B = \emptyset\}$. Finally, let $\mathcal{F}$ be the family of all $f_{\langle A,B\rangle}$ for $\langle A, B\rangle \in \mathcal{D}$ defined by

$$f_{\langle A,B\rangle}^{-1}(\{0\}) = A, \qquad f_{\langle A,B\rangle}^{-1}(\{1\}) = B.$$

Then $\mathcal{F}$ is clearly closed. It is also coherent, by (a). Since $\mathrm{dom}(f)$ ($f \in \mathcal{F}$) generates the F_σ P-ideal generated by $\mathcal{A}$ and $\mathcal{B}$, $\mathcal{F}$ is σ-directed. By (b), $\mathcal{F}$ is not trivial. □

Let us now show that Theorem 2.4.4 cannot be extended to cover the coherent families indexed by other analytic P-ideals.

PROPOSITION 2.5.2. *If $\mathcal{I}$ is an analytic P-ideal different from* Fin *and* $\emptyset \times$ Fin, *then there is a trivial coherent family of functions* $\{h_A \colon A \to \mathbb{N} \cup \{*\}\}$ *indexed by $\mathcal{I}$ and a partition $[\mathbb{N}]^{<\mathbb{N}} = K_0$ such that $h_A^{-1}(\{j\})$ is K_0-homogeneous for all $j \in \mathbb{N} \cup \{*\}$, but for no trivializing function h the set $h^{-1}(\{*\})$ is K_0-homogeneous.*

PROOF. By Corollary 1.2.11, we can assume that the ideal $\mathcal{I}$ is dense. By Solecki's theorem (Theorem 1.2.5), let φ be a lower semicontinuous submeasure such that $\mathcal{I} = \mathrm{Exh}(\varphi)$; note that $\lim_i \varphi(\{i\}) = 0$, since $\mathcal{I}$ is dense. We can assume that $\lim_i \varphi(\mathbb{N} \setminus [1,k)) \geq 2$. Let $K_0 = \{s : \varphi(s) \leq 1\}$. For $A \in \mathcal{I}$ let k_A be the minimal such that $\varphi(A \setminus [1,k_A)) \leq 1$, and let h_A be equal to the identity on $A \cap [1, k_A)$ and to the constant $*$ on $A \setminus [1, k_A)$. Then since $h_A(n)$ is equal to $*$ for all but finitely many $n \in A$, the constructed family is coherent. Also, $h_A^{-1}(\{i\})$ is K_1-homogeneous for every $i \in \mathbb{N} \cup \{*\}$. But if some h trivializes $\{h_A\}$ then the complement of $h^{-1}(\{*\})$ has to be in $\mathcal{I}$. Therefore $\varphi(h^{-1}(\{*\})) > 1$, and $h^{-1}(\{*\})$ cannot be K_0-homogeneous. □

The uniformization and reflection results of this chapter will play an important role in the proof of the OCA lifting theorem (Theorem 3.3.5). The main result of §3.8 also relies on uniformization of a particular coherent family of functions indexed by a P-ideal. However, the proof that there are no K_0-homogeneous sets is much more complex in this case, in particular it requires invoking Martin's Axiom.

PROBLEM 2.5.3. *Generalize uniformization results of this Chapter so that they have Lemma 3.13.4 as a consequence.*

Let us note that Σ_2-reflection is the most that one can hope for in Theorem 2.4.4, since already the Π_2-reflection fails in this context. Namely, there is a partition R of $\mathbb{N}^2$ and a Π_2-sentence φ which is true in $\langle [n,\infty) \times \mathbb{N}, R\rangle$ but fails in all $\langle \Gamma_f \cap ([n,\infty) \times \mathbb{N}, R\rangle$ for all f and n. These R and φ are constructed using the fact that in $[n,\infty) \times \mathbb{N}$ every vertical section is infinite, while in Γ_f no vertical section is infinite.

OCA is a consequence of the Proper Forcing Axiom, PFA, and the structure of the continuum in all known models of set theory in which OCA is true resembles the structure of the continuum under PFA (modulo the obvious differences implied by the fact that OCA does not have any large cardinal strength); see [31]. An important and difficult open question about OCA along these lines is the following (see [127]):

QUESTION 2.5.4. *Does OCA imply that the continuum has size $\aleph_2$, or equivalently, that every subset of $\mathbb{N}^{\mathbb{N}}$ of size less than continuum is bounded in the ordering of eventual dominance?*

One may say that this subject has started with Shelah's ground-breaking consistency proof ([**106**]), of the statement "All automorphisms of $\mathcal{P}(\mathbb{N})/\mathrm{Fin}$ are trivial." In this proof Shelah introduced *oracle-chain condition* forcing construction, and his ideas were later systematized by others (see [**63**], [**10**]). Later on, the axiomatic approach was suggested by Shelah and Steprans ([**107**]) who adapted the original proof into the PFA-context. After Todorcevic has isolated the Open Coloring Axiom ([**125**]), Velickovic ([**142**]) used OCA and MA to prove that all automorphisms of $\mathcal{P}(\mathbb{N})/\mathrm{Fin}$ are trivial. We believe that Chapter 3 and Chapter 4 of this monograph show the advantages of the axiomatic approach. This gives an additional interest to the following question (see e.g., [**45**] or [**10**] for terminology, and note that the word "lifting" has different meaning than elsewhere in this monograph):

QUESTION 2.5.5. *Does OCA (or PFA) imply that the Lebesgue measure algebra has no Borel lifting?*

CHAPTER 3

Homomorphisms of analytic quotients under OCA

3.1. Introduction

The results of this Chapter can be put in proper mathematical context only in the light of §2.1, and the reader may wish to consult this section before proceeding further. We prove, working in the context of OCA (see §2.2) and familiar Martin's Axiom (MA), that arbitrary homomorphisms of quotients over analytic P-ideals are close to ones with Baire-measurable liftings. Most efforts of this Chapter are put in the proof of the result (Theorem 3.3.5) to the effect that under OCA every homomorphism $\Phi\colon \mathcal{P}(\mathbb{N})/\operatorname{Fin} \to \mathcal{P}(\mathbb{N})/\mathcal{I}$ between quotients over analytic P-ideals is obtained as an amalgamation of two homomorphisms, Φ_1 and Φ_2:

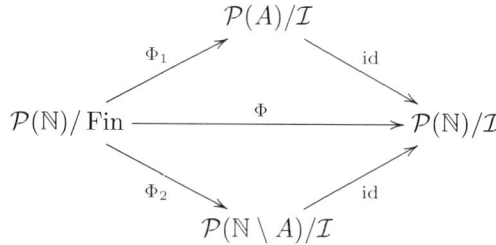

so that Φ_1 has a Baire-measurable lifting, while the kernel of Φ_2 is nonmeager (even ccc over Fin; see §3.3), and therefore not included in any analytic ideal. We shall refer to this result as the *OCA lifting theorem*. An analogous result is proved in the case when $\mathcal{I}$ is a countably generated ideal (Theorem 3.3.6).

A homomorphism Φ is an amalgamation of two homomorphisms Φ_1 and Φ_2 if and only if it has a Baire-measurable almost lifting, i.e., a map F such that $F(A)\Delta\Phi_*(A)$ is in $\mathcal{I}$ for all A contained in some nonmeager ideal (Lemma 3.3.4). The OCA lifting theorem immediately implies that (under OCA and MA) every isomorphism between two analytic P-ideals has a Baire-measurable lifting. By the rigidity results of Chapter 1, this fact has a number of consequences. For example, under OCA and MA the ideals Fin, Fin $\times \emptyset$, $\emptyset \times$ Fin, $\mathcal{I}_{1/n}$, $\mathcal{I}_{1/\sqrt{n}}$, $\mathcal{Z}_0$ and $\mathcal{Z}_{\log}$ have pairwise nonisomorphic quotients (Corollary 3.4.5). More generally, in the class of non-pathological analytic P-ideals our basic question has a satisfactory answer: Two quotients are isomorphic if and only if the corresponding ideals are isomorphic themselves. Moreover, a version of the basic question stated in terms of the embeddability (instead of isomorphism) relation between quotients has a satisfactory answer as well: $\mathcal{P}(\mathbb{N})/\mathcal{I}$ embeds into $\mathcal{P}(\mathbb{N})/\mathcal{J}$ if and only if there is a reduction of $\mathcal{I}$ to $\mathcal{J}$ (Corollary 3.4.7).

Some of the rigidity phenomena extend to the arbitrary ideals whose quotients embed into some quotient over an analytic P-ideal. If $\mathcal{P}(\mathbb{N})/\mathcal{I}$ embeds into $\mathcal{P}(\mathbb{N})/\mathcal{J}$, and $\mathcal{J}$ is an analytic P-ideal, then $\mathcal{I}$ is an amalgamation of an analytic P-ideal (reducible to $\mathcal{J}$) and a nonmeager ideal (Corollary 3.5.1). This implies that there are many quotients $\mathcal{P}(\mathbb{N})/\mathcal{I}$ that are not embeddable into any quotient over an analytic P-ideal (Corollary 3.5.4). We also prove that if $\mathcal{P}(\mathbb{N})/\mathcal{I}$ embeds into $\mathcal{P}(\mathbb{N})/\operatorname{Fin}$, then $\mathcal{I}$ and its orthogonal, $\mathcal{I}^\perp$, do not form a gap (Corollary 3.5.2). Needles to say, all of these results require OCA and MA.

An another consequence of the OCA lifting theorem is that all automorphisms of a quotient over a non-pathological analytic P-ideal are induced by $1-1$ mappings of the integers (this was proved by Shelah for $\mathcal{P}(\mathbb{N})/\operatorname{Fin}$ in [**106**]). Automorphism groups of analytic quotients have proved to be objects worth of study, especially in situations when automorphisms have simple liftings (see e.g., [**18**]). In §3.6 we give a complete description of $\operatorname{Aut}(\mathcal{P}(\mathbb{N})/\mathcal{I}_f)$ for summable ideals $\mathcal{I}_f$, and prove that it always has a subgroup isomorphic to the free group with continuum many generators. We will construct a summable ideal $\mathcal{I}_f$ such that every inner epimorphism of $\mathcal{P}(\mathbb{N})/\mathcal{I}_f$ is automatically an isomorphism and prove some other curious properties of these automorphism groups.

We also investigate which quotients over analytic P-ideals are homogeneous under OCA and MA, and prove several results suggesting that $\mathcal{P}(\mathbb{N})/\operatorname{Fin}$ can be the only one (§3.7). This stands in sharp contrast to the fact that under CH all the quotients over the F_σ-ideals and all the quotients over the EU-ideals are homogeneous (see [**65**]). More consequences of the OCA lifting theorem and its applications to topology can be found in Chapter 4.

The proof of the OCA lifting theorem spans from §3.10 through §3.13, but it is presented as a sequence of self-contained results, some of which are interesting in their own right. In §3.10 we characterize hereditary nonmeager subsets of $\mathcal{P}(\mathbb{N})$, extending the well-known characterization of nonmeager ideals by Jalali-Naini and Talagrand ([**58**], [**121**]). Many of of the results from §3.11 are really results about the complete normed group $\langle \mathcal{P}(\mathbb{N})/\mathcal{I}, \Delta, \|\cdot\|_\varphi \rangle$ (see §1.2) and the stability and completeness properties of partial approximate homomorphisms between such groups. These objects deserve to be studied in their own right. In the proof of Theorem 3.8.1 and Lemma 3.13.5 Martin's Axiom is applied to "multiply" an uncountable set homogeneous for a certain open partition. This approach is likely to be of use in the future study of liftings.

The OCA lifting theorem, together with some results of Todorcevic, implies that the homomorphisms between the analytic quotients frequently preserve their gaps. This result is proved in Chapter 5 (Corollary 5.9.2).

A striking consequence of this Chapter's main result is that OCA is, in some sense, an optimal assumption for investigating quotients over analytic P-ideals. This is because, roughly speaking, two such quotients are provably (without using any additional set-theoretic axioms) isomorphic if and only if they are isomorphic under OCA and MA (see §2.1).

Finally, let us point the reader's attention to David Fremlin's exposition of a proof of the OCA lifting theorem in [**47**].

The organization of this chapter. We start by giving examples of homomorphisms with no Baire-measurable liftings in §3.2. In §3.3 the notion of an almost lifting is introduced, and the OCA lifting theorem is stated. The following four sections, §§3.4–3.7, are devoted to proving various consequences of this theorem. In §3.4 (§3.5, respectively) consequences related to the rigidity of analytic (arbitrary, respectively) quotients are investigated. In §3.6 we investigate the automorphism groups of analytic quotients, and in §3.7 their homogeneity.

In §3.8 and §3.9 we prove the instances of OCA lifting theorems for homomorphisms into quotients over Fin and Fin $\times \emptyset$, respectively. The former section also serves as an introduction to the proof of the general theorem, as it introduces some of the ideas. Sections §3.10 and §3.11 prepare the grounds for the proof of this theorem in its general form. A characterization of hereditary subsets of $\mathcal{P}(\mathbb{N})$ which are nonmeager is given in §3.10. In §3.11 the stabilizers (introduced in §1.5) are modified and utilized in study of partial approximate homomorphisms of the complete normed group $\langle \mathcal{P}(\mathbb{N})/\mathcal{I}, \Delta, \|\cdot\|_\varphi \rangle$. A 'local' version of the OCA lifting theorem is proved in §3.12, and the proof is completed in §3.13. In §3.14 we discuss possible directions for a further study and list some open problems.

3.2. Homomorphisms without Baire-measurable liftings

The two examples presented in this section are most likely already well-known. As we have already remarked, all automorphisms of $\mathcal{P}(\mathbb{N})/\operatorname{Fin}$ can be trivial. However, the situation with homomorphisms rather than automorphisms is by far more complex. If there is a maximal nonprincipal ideal $\mathcal{I}$ on $\mathcal{P}(\mathbb{N})$, then $\mathcal{P}(\mathbb{N})/\mathcal{I}$ is the two-element Boolean algebra and therefore it is embeddable into $\mathcal{P}(\mathbb{N})/\operatorname{Fin}$. This homomorphism, Φ_0, is not trivial in the sense of §1. It is, however, in some sense close to being Baire, because it has a lifting which is equal to the union of two continuous (even constant) functions. It is neither $1-1$ nor onto, but we will see that there are nontrivial homomorphisms with either one of these properties and therefore demonstrate the sharpness of Shelah's result by showing that only automorphisms can be expected to be trivial. Recall the ordering $\leq^+_{\mathrm{BE}}$ on analytic ideals, defined in §1.2 by $\mathcal{I} \leq^+_{\mathrm{BE}} \mathcal{J}$ if and only if $\mathcal{I} \leq_{\mathrm{BE}} \mathcal{J} \upharpoonright A$ for some $\mathcal{J}$-positive set A.

Related to this ordering is the notion of an *amalgamation* $\Phi \oplus \Psi$ of two homomorphisms $\Phi \colon \mathcal{P}(\mathbb{N})/\operatorname{Fin} \to \mathcal{P}(\mathbb{N})/\mathcal{I}$ and $\Psi \colon \mathcal{P}(\mathbb{N})/\operatorname{Fin} \to \mathcal{P}(\mathbb{N})/\mathcal{I}$, which is a homomorphism of $\mathcal{P}(\mathbb{N})/\operatorname{Fin}$ into $\mathcal{P}(\mathbb{N} \oplus \mathbb{N})/\mathcal{I} \oplus \mathcal{J}$ such that the following diagram commutes:

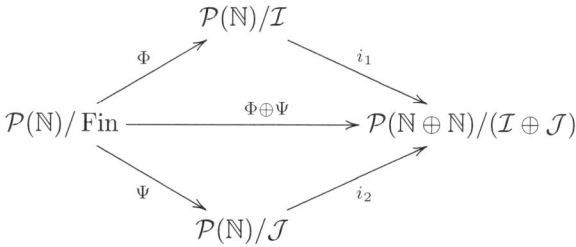

where $i_1 \colon \mathcal{P}(\mathbb{N})/\mathcal{I} \to \mathcal{P}(\mathbb{N} \oplus \mathbb{N})/\mathcal{I} \oplus \mathcal{J}$ is given by its lifting $A \mapsto A \times \{0\}$ and $i_2 \colon \mathcal{P}(\mathbb{N})/\mathcal{J} \to \mathcal{P}(\mathbb{N} \oplus \mathbb{N})/\mathcal{I} \oplus \mathcal{J}$ is given by its lifting $A \mapsto A \times \{1\}$.

EXAMPLE 3.2.1. A monomorphism between two analytic quotients with no Baire-measurable lifting. Let $\Phi_1 \colon \mathcal{P}(\mathbb{N})/\mathcal{I} \to \mathcal{P}(\mathbb{N})/\mathcal{J}_1$ be a monomorphism between analytic quotients which has a Baire lifting and let $\mathcal{J}_2$ be an analytic ideal. Let $\mathcal{K}$ be a maximal ideal extending $\mathcal{I}$ and let $\Phi_2 \colon \mathcal{P}(\mathbb{N})/\mathcal{K} \to \mathcal{P}(\mathbb{N})/\mathcal{J}_2$ be a

monomorphism. Let $\mathcal{J} = \mathcal{J}_1 \oplus \mathcal{J}_2$ and let
$$\Phi = \Phi_1 \oplus \Phi_2 \colon \mathcal{P}(\mathbb{N})/\mathcal{I} \to \mathcal{P}(\mathbb{N} \oplus \mathbb{N})/\mathcal{J}.$$
Then Φ is a monomorphism of $\mathcal{P}(\mathbb{N})/\mathcal{I}$ into $\mathcal{P}(\mathbb{N})/\mathcal{J}$ which has no Baire lifting. If we assume otherwise and let Φ_* be a continuous (see Lemma 1.3.2) lifting of Φ, then we would have
$$\mathcal{K} = \{C \colon \Phi_*(C) \Delta F(C) \in \mathcal{J}\}$$
hence $\mathcal{K}$ would be a nonprincipal maximal analytic ideal, which is impossible by a classical result of Sierpinski (or by, say, Theorem 3.10.1).

Note that the method of the above example implies that one cannot hope that the Baire-embeddability and embeddability of analytic quotients coincide. However, the embeddability of quotients over analytic ideals is closely related with the simply definable ordering $\leq_{\mathrm{BE}}^+$ studied in Chapter 1:

$\mathcal{I} \leq_{\mathrm{BE}}^+ \mathcal{J}$ if and only if $\mathcal{P}(\mathbb{N})/\mathcal{I}$ is Baire-embeddable into $\mathcal{P}(A)/\mathcal{J}$ for some $\mathcal{J}$-positive set A.

PROPOSITION 3.2.2. *If $\mathcal{I}$ and $\mathcal{J}$ are analytic ideals, then the quotient algebra $\mathcal{P}(\mathbb{N})/\mathcal{I}$ can be embedded into $\mathcal{P}(\mathbb{N})/\mathcal{J}$ whenever $\mathcal{I} \leq_{\mathrm{BE}}^+ \mathcal{J}$.*

PROOF. Define an embedding like in Example 3.2.1. □

This, for example, implies that every quotient over a proper summable ideal (see §1.12) is embeddable into any other such quotient; recall that this is false in the case of Baire-embeddability (Theorem 1.12.1).

EXAMPLE 3.2.3. An epimorphism between two analytic quotients without a Baire-measurable lifting. Let $\{\mathcal{U}_n\}$ be a sequence of ultrafilters on $\mathbb{N}$. Define $F_\mathcal{U} \colon \mathcal{P}(\mathbb{N}) \to \mathcal{P}(\mathbb{N})$ by
$$F_\mathcal{U}(A) = \{n \colon A \in \mathcal{U}_n\}.$$
This mapping is clearly an endomorphism of $\mathcal{P}(\mathbb{N})$. If the sequence $\{\mathcal{U}_n\}$ is *discrete*, i.e., for some sequence $\{A_n\}$ of pairwise disjoint subsets of $\mathbb{N}$ we have $A_n \in \mathcal{U}_n$ for every n, then $F_\mathcal{U}$ is an epimorphism. To see this, pick an arbitrary $B \subseteq \mathbb{N}$. Then there is $C \subseteq \mathbb{N}$ such that $A_n \subseteq^* C$ if $n \in B$ and $A_n \perp C$ if $n \notin B$. This proves that $F_\mathcal{U}$ is an epimorphism of $\mathcal{P}(\mathbb{N})$, and therefore a lifting of an epimorphism $\Phi \colon \mathcal{P}(\mathbb{N})/\operatorname{Fin} \to \mathcal{P}(\mathbb{N})/\operatorname{Fin}$. Now assume that moreover each $\mathcal{U}_n$ is a nonprincipal ultrafilter. Then Φ does not have a Baire lifting: Assuming otherwise implies it has a continuous lifting as well (Lemma 1.3.2), therefore $\ker(\Phi)$ is an F_σ ideal. But $\ker(\Phi)$ is nonmeager because it includes an intersection of countably many maximal nonprincipal ideals (see e.g., Theorem 3.10.1 (b)), so it does not have a property of Baire—a contradiction.

Maps of the form $F_\mathcal{U}$ are fairly general. For example, if each $\mathcal{U}_n$ is a principal ultrafilter, then the map $F_\mathcal{U}$ is of the form Φ_h (where $h(n) = \bigcap \mathcal{U}_n$). We do not know whether there are homomorphisms of $\mathcal{P}(\mathbb{N})/\operatorname{Fin}$ other than those constructed in Example 3.2.3 (see Question 3.14.2).

3.3. Almost liftings and lifting theorems

In this section we formulate main results of this Chapter and give some of their applications. Proofs of main theorems will be postponed to latter sections. The following notion of largeness will be of crucial importance in the proof of the main result of this Chapter. Recall that two sets of integers are *almost disjoint* if their intersection is finite.

DEFINITION 3.3.1. An ideal $\mathcal{I}$ is *ccc over* Fin if there is no uncountable family of $\mathcal{I}$-positive, pairwise disjoint modulo Fin, sets.

An ideal which is ccc over Fin is never included in an analytic ideal. Moreover, it can never have the property of Baire, as the following simple Lemma shows.

LEMMA 3.3.2. *Let $\mathcal{J}$ be an ideal on $\mathbb{N}$ which includes* Fin.
(a) *If $\mathcal{P}(\mathbb{N})/\mathcal{J}$ is ccc then $\mathcal{J}$ is ccc over* Fin.
(b) *If $\mathcal{J}$ is ccc over* Fin *and $\mathcal{I} \leq_{\mathrm{RB}} \mathcal{J}$, then $\mathcal{I}$ is ccc over* Fin.
(c) *If $\mathcal{J}$ is ccc over* Fin, *then $\mathcal{J}$ is nonmeager and therefore it does not have the Property of Baire.*
(d) *If $\mathcal{J}$ is ccc over* Fin *and $A \in \mathcal{J}^+$, then $\mathcal{J} \restriction A$ is ccc over* Fin.
(e) *An intersection of two ccc over* Fin *ideals is itself ccc over* Fin.
(f) *If $\mathcal{I}$ is a P-ideal, then it is ccc over* Fin *if and only if the quotient $\mathcal{P}(\mathbb{N})/\mathcal{I}$ has ccc.*

PROOF. (a) Follows from the definitions, since $\mathcal{J} \supseteq$ Fin.

(b) Let $h \colon \mathbb{N} \to \mathbb{N}$ be a Rudin–Blass-reduction of $\mathcal{I}$ to $\mathcal{J}$. If A_ξ ($\xi < \omega_1$) are $\mathcal{I}$-positive sets which are pairwise almost disjoint modulo Fin, then $h^{-1}(A_\xi)$ ($\xi < \omega_1$) are $\mathcal{J}$-positive sets which are pairwise disjoint modulo Fin.

(c) If $\mathcal{J}$ is meager, then by [**58**], [**121**] (see also Theorem 3.10.1) we have Fin $\leq_{\mathrm{RB}} \mathcal{J}$, and therefore $\mathcal{J}$ is not ccc over Fin by (b).

(d) and (e) are obvious.

(f) is proved by an easy recursive construction; see [**60**]. □

Note that "$\mathcal{I}$ is ccc over Fin" is, in general, weaker than "$\mathcal{P}(\mathbb{N})/\mathcal{I}$ is ccc." (As the ideal from Example 3.2.3 shows.) Moreover, David Fremlin has observed that while a ccc over Fin ideal has to be nonmeager (by (c) above), there exist a ccc over Fin ideals of Lebesgue measure zero (see [**47**, page 17]). Also, there are nonmeager ideals that are not ccc over Fin. The following definition is crucial in this Chapter.

DEFINITION 3.3.3. A mapping $F \colon \mathcal{P}(\mathbb{N}) \to \mathcal{P}(\mathbb{N})$ is an *almost lifting* of a homomorphism $\Phi \colon \mathcal{P}(\mathbb{N})/\operatorname{Fin} \to \mathcal{P}(\mathbb{N})/\mathcal{I}$ if it serves as a lifting of Φ on a ccc over Fin ideal, namely if the set (Φ_* is some lifting of Φ)
$$\{A \subseteq \mathbb{N} : F(A) \Delta \Phi_*(A) \in \mathcal{I}\}$$
includes some ccc over Fin ideal.

The following lemma connects the notion of an almost lifting with the notion of amalgamation of homomorphisms introduced in §3.2, and it is a simple consequence of Lemma 3.11.6 which will be proved later on.

LEMMA 3.3.4. *A homomorphism $\Phi \colon \mathcal{P}(\mathbb{N})/\operatorname{Fin} \to \mathcal{P}(\mathbb{N})/\mathcal{I}$ has a Baire-measurable almost lifting if and only if $\Phi = \Phi_1 \oplus \Phi_2$ so that Φ_2 has a continuous lifting and $\ker(\Phi_1)$ is ccc over* Fin.

PROOF. The reverse implication is obvious. To prove the direct implication, assume there is a Baire-measurable mapping F which is a lifting of Φ on a ccc over Fin ideal $\mathcal{J}$. Then Lemma 3.3.2 (c) implies that $\mathcal{J}$ is nonmeager, and Lemma 3.11.6 implies that $\Phi = \Phi_1 \oplus \Phi_2$ as required. □

An important consequence of Lemma 3.3.4 is the fact that the Radon–Nikodym Property of an analytic P-ideal applies not only to Baire-measurable liftings, but also to Baire-measurable almost liftings.

The main results of this Chapter are the following two lifting theorems, showing that every homomorphism into a quotient over an analytic P-ideal or a countably generated ideal is an amalgamation obtained in a manner similar to Example 3.2.1.

THEOREM 3.3.5 (OCA lifting theorem). *Assume OCA and MA and that $\mathcal{I}$ is an analytic P-ideal. If $\Phi\colon \mathcal{P}(\mathbb{N})/\operatorname{Fin} \to \mathcal{P}(\mathbb{N})/\mathcal{I}$ is a homomorphism, then Φ has a continuous almost lifting. Equivalently, Φ can be decomposed as $\Phi_1 \oplus \Phi_2$ so that Φ_1 has a continuous lifting, while the kernel of Φ_2 is ccc over* Fin.

Recall that an ideal $\mathcal{I}$ is *countably generated* if there are sets A_n ($n \in \mathbb{N}$) such that $\mathcal{I} = \{B : B \supseteq A_n \text{ for some } n\}$. Note that Fin and Fin $\times \emptyset$ are the only nontrivial examples of countably generated ideals (up to the isomorphism).

THEOREM 3.3.6. *Assume OCA and MA. If $\mathcal{I}$ is a countably generated ideal and $\Phi\colon \mathcal{P}(\mathbb{N})/\operatorname{Fin} \to \mathcal{P}(\mathbb{N}^2)/\mathcal{I}$ is a homomorphism, then Φ has a continuous almost lifting. Equivalently, Φ can be decomposed as $\Phi_1 \oplus \Phi_2$ so that Φ_1 has a continuous lifting, while the kernel of Φ_2 is ccc over* Fin.

The proofs of these two theorems occupy §§3.11–3.13 and §3.9 (the reader may also want to consult the exposition in [**47**]). We shall first investigate some of their consequences.

3.4. Applications: rigidity of analytic quotients

This section is devoted to applications of the above theorems, which will be proved in the remaining sections of this Chapter. We shall first treat the problem of isomorphisms of two quotients over analytic ideals.

THEOREM 3.4.1. *Assume OCA and MA, and that $\mathcal{I}, \mathcal{J}$ are analytic ideals and at least one of them is either a P-ideal or countably generated. Then every isomorphism $\Phi\colon \mathcal{P}(\mathbb{N})/\mathcal{I} \to \mathcal{P}(\mathbb{N})/\mathcal{J}$ has a Baire-measurable lifting. In particular, quotients over $\mathcal{I}$ and $\mathcal{J}$ are isomorphic if and only if they are Baire-isomorphic.*

PROOF. Assume that $\mathcal{J}$ is a P-ideal or a countably generated ideal, and let $\Phi\colon \mathcal{P}(\mathbb{N})/\mathcal{I} \to \mathcal{P}(\mathbb{N})/\mathcal{J}$ be an isomorphism. Decompose $\Phi = \Phi_1 \oplus \Phi_2$ so that Φ_1 is Baire and $\ker(\Phi_2)$ is ccc over Fin. Since $\ker(\Phi_2)$ is an analytic ideal which is ccc over Fin, it is improper. Therefore Φ_1 coincides with Φ, so the latter has a Baire-measurable lifting, as required. $\square$

Let us recall the basic question of this monograph: How does a change of the ideal $\mathcal{I}$ effect the change of its quotient? The following gives an answer for a rather rich class of analytic ideals (see the definition of when two ideals are isomorphic in §1.9).

COROLLARY 3.4.2. *Assume OCA and MA. If $\mathcal{I}$ and $\mathcal{J}$ are analytic ideals and at least one of them is a non-pathological analytic P-ideal or a countably generated ideal, then their quotients are isomorphic if and only if $\mathcal{I}$ and $\mathcal{J}$ are.*

PROOF. If the quotients are isomorphic, then by Theorem 3.4.1 they are Baire-isomorphic. By Proposition 1.10.4 the ideals are isomorphic as well. $\square$

Recall that Erdös and Monk (see [**20**]) have constructed an isomorphism between $\mathcal{P}(\mathbb{N})/\operatorname{Fin}$ and the summable ideal $\mathcal{P}(\mathbb{N})/\mathcal{I}_{1/n}$ (see §1.12) using the Continuum Hypothesis. Later on, Just and Krawczyk ([**65**]) observed that all quotients over F_σ ideals are *countably saturated*. Therefore, under CH, Cantor's back-and-forth argument is applicable to construct an isomorphism between any two F_σ, and in particular summable, quotients (see e.g., [**12**]). The following shows that the situation under OCA is different, and it also answers a question of Koppelberg ([**79**]).

COROLLARY 3.4.3. *OCA and MA imply that ideals $\mathcal{I}_{1/n}$ and $\mathcal{I}_{1/\sqrt{n}}$ have nonisomorphic quotients.*

PROOF. This follows by the Radon–Nikodym property of summable ideals, Corollary 3.4.2 and Proposition 1.12.8. □

The following gives an alternative solution to a problem of Erdös and Ulam about density ideals solved in [**65**] and [**62**] (see §1.13 for more details).

COROLLARY 3.4.4. *OCA and MA imply that ideals $\mathcal{Z}_0$ and $\mathcal{Z}_{\log}$ have nonisomorphic quotients.*

PROOF. This follows from Proposition 1.13.13 and Theorem 3.4.1. □

COROLLARY 3.4.5. *OCA and MA imply that the ideals Fin, $\operatorname{Fin} \times \emptyset$, $\emptyset \times \operatorname{Fin}$, $\mathcal{I}_{1/n}$, $\mathcal{I}_{1/\sqrt{n}}$, $\mathcal{Z}_0$ and $\mathcal{Z}_{\log}$ have pairwise nonisomorphic quotients.*

PROOF. By Corollary 3.4.2 it suffices to prove that these ideals are pairwise nonisomorphic (in the sense of Definition 1.2.7). Neither Fin not $\operatorname{Fin} \times \emptyset$ can be isomorphic to any of the other ideals, since Fin and $\operatorname{Fin} \times \emptyset$ are countably generated and the others are not (since a countably generated ideal has to be generated by a single set over Fin). The ideal $\emptyset \times \operatorname{Fin}$ is the only one of the remaining ideals whose orthogonal is isomorphic to $\operatorname{Fin} \times \emptyset$. It follows from Proposition 1.13.14 that a dense summable ideal $\mathcal{I}_f$ is never isomorphic to a density ideal, therefore neither of $\mathcal{I}_{1/n}$ and $\mathcal{I}_{1/\sqrt{n}}$ is isomorphic to either $\mathcal{Z}_0$ or $\mathcal{Z}_{\log}$. (This can also be proved by observing that the complexity of a summale ideal is always F_σ, while nontrivial density ideals are complete $F_{\sigma\delta}$ subsets of $\mathcal{P}(\mathbb{N})$.) The proof is concluded by applying Corollary 3.4.3 and Corollary 3.4.4. □

As we have pointed out earlier, the question of triviality of automorphisms of $\mathcal{P}(\mathbb{N})/\operatorname{Fin}$ was the starting point of the line of study of analytic quotients to which this work belongs. Recall that an automorphism of $\mathcal{P}(\mathbb{N})/\mathcal{I}$ is *trivial* if it has a completely additive lifting Φ_h for a $1-1$ partial function h. By applying Theorem 3.3.5 (or Theorem 3.3.6 and Proposition 1.4.6 (b)), we get the following.

COROLLARY 3.4.6. *OCA and MA imply that if $\mathcal{I}$ is either a non-pathological analytic P-ideal or a countably generated ideal, then all automorphisms of $\mathcal{P}(\mathbb{N})/\mathcal{I}$ are trivial.* □

Therefore if $\mathcal{I}$ is a non-pathological ideal then under OCA the automorphism group of $\mathcal{P}(\mathbb{N})/\mathcal{I}$ reduces to its subgroup of trivial automorphisms. This will be used in §3.6, which is devoted to study of automorphism groups of analytic quotients.

Let us now exploit more power of Theorem 3.3.5 and Theorem 3.3.6 by considering the impact of OCA on the embeddability relation on analytic quotients. Let $\mathcal{I}$ and $\mathcal{J}$ be ideals of sets of integers. The symbol

$$\mathcal{P}(\mathbb{N})/\mathcal{I} \hookrightarrow \mathcal{P}(\mathbb{N})/\mathcal{J}$$

denotes the fact that $\mathcal{P}(\mathbb{N})/\mathcal{I}$ is isomorphic to a subalgebra of $\mathcal{P}(\mathbb{N})/\mathcal{J}$ (a fact frequently rendered as $\mathcal{P}(\mathbb{N})/\mathcal{I}$ is *embeddable* into $\mathcal{P}(\mathbb{N})/\mathcal{J}$). Let us define another preordering on analytic ideals.

(O5) $\mathcal{I} \leq_{\mathrm{EM}} \mathcal{J}$ if and only if $\mathcal{P}(\mathbb{N})/\mathcal{I} \hookrightarrow \mathcal{P}(\mathbb{N})/\mathcal{J}$.

This preordering, unlike those defined in §1.2, is not *absolute*. For example, CH implies $\mathcal{I}_{1/n} \leq_{\mathrm{EM}} \mathrm{Fin}$, yet we shall see that this is false under OCA and MA. Also note that Example 3.2.1 shows that we cannot hope that $\mathcal{I} \leq_{\mathrm{EM}} \mathcal{J}$ is equivalent to $\mathcal{I} \leq_{\mathrm{BE}} \mathcal{J}$ because $\mathcal{I} \leq_{\mathrm{BE}}^{+} \mathcal{J}$ (see §1.2) already implies $\mathcal{I} \leq_{\mathrm{EM}} \mathcal{J}$ (Proposition 1.3.1). In particular, every proper summable quotient is embeddable into every other proper summable quotient (see Claim 1 in the proof of Lemma 1.12.3). However, we shall see that, in the OCA-context, $\leq_{\mathrm{EM}}$ indeed coincides with $\leq_{\mathrm{BE}}^{+}$, at least in the realm of analytic P-ideals. The first result of this form is due to Just ([63], [62]), and it says that OCA and MA, together imply that if $\mathcal{I}$ is an analytic ideal then $\mathcal{I} \leq_{\mathrm{EM}} \mathrm{Fin}$ if and only if $\mathcal{I}$ is generated by a single set over Fin. (Later Todorcevic ([130]) proved that OCA alone suffices for this conclusion.)

COROLLARY 3.4.7. *OCA and MA imply that if $\mathcal{I}$ and $\mathcal{J}$ are analytic P-ideals then $\mathcal{I} \leq_{\mathrm{EM}} \mathcal{J}$ if and only if $\mathcal{I} \leq_{\mathrm{BE}}^{+} \mathcal{J}$.*

PROOF. In Proposition 3.2.2 we have proved that $\mathcal{I} \leq_{\mathrm{BE}}^{+} \mathcal{J}$ implies $\mathcal{I} \leq_{\mathrm{EM}} \mathcal{J}$. Let us therefore assume $\mathcal{I} \leq_{\mathrm{EM}} \mathcal{J}$, and let $\Phi \colon \mathcal{P}(\mathbb{N})/\mathcal{I} \to \mathcal{P}(\mathbb{N})/\mathcal{J}$ be a monomorphism. By Theorem 3.3.5, we can decompose Φ as $\Phi_1 \oplus \Phi_2$, where Φ_1 has a Baire-measurable lifting, Φ_{1*}. The set $A = \Phi_{1*}(\mathbb{N})$ is not in $\mathcal{J}$ because $\ker(\Phi) \neq \ker(\Phi_2)$. We claim that $\ker(\Phi) = \ker(\Phi_1)$. Otherwise, since $\ker(\Phi) \subseteq \ker(\Phi_1)$, we have $B \in \ker(\Phi) \setminus \ker(\Phi_2)$. Therefore the ideal $\mathcal{I} \restriction B$ is ccc over Fin, which is impossible because it is analytic. So we have $\mathcal{I} = \ker(\Phi) = \ker(\Phi_1)$ and $\mathcal{I} \leq_{\mathrm{BE}} \mathcal{J} \restriction A$, as required. □

As we have pointed out before, the orderings $\leq_{\mathrm{EM}}$ and $\leq_{\mathrm{BE}}$ differ on the summable ideals. However, they behave similarly on some ideals close to Fin. For example, since $\mathrm{Fin} \restriction A$ is isomorphic to Fin for every positive set A, we have $\mathcal{I} \leq_{\mathrm{RB}} \mathrm{Fin}$ if and only if $\mathcal{I} \leq_{\mathrm{BE}}^{+} \mathrm{Fin}$. Similarly $\mathcal{I} \leq_{\mathrm{RB}} \mathrm{Fin} \times \emptyset$ if and only if $\mathcal{I} \leq_{\mathrm{BE}}^{+} \mathrm{Fin} \times \emptyset$ and $\mathcal{I} \leq_{\mathrm{RB}} \emptyset \times \mathrm{Fin}$ if and only if $\mathcal{I} \leq_{\mathrm{BE}}^{+} \emptyset \times \mathrm{Fin}$ (see also [75]).

3.5. Applications: quotients embeddable into analytic quotients

Now we shall see what can be said about arbitrary ideals on $\mathbb{N}$ whose quotients can be embedded as a subalgebra of some analytic quotient. The following is a slight refinement of Corollary 3.4.7.

COROLLARY 3.5.1. *Assume OCA and MA, that $\mathcal{J}$ is either an analytic P-ideal or a countably generated ideal and that $\mathcal{I}$ is an arbitrary ideal on $\mathbb{N}$ such that $\mathcal{P}(\mathbb{N})/\mathcal{I} \hookrightarrow \mathcal{P}(\mathbb{N})/\mathcal{J}$. Then $\mathcal{I}$ is of the form $\mathcal{I}_0$ or $\mathcal{I}_0 \oplus \mathcal{I}_1$ for some analytic ideal $\mathcal{I}_0$ such that $\mathcal{I}_0 \leq_{\mathrm{BE}}^{+} \mathcal{J}$ and an ideal $\mathcal{I}_1$ which is ccc over Fin.* □

For an ideal $\mathcal{I}$, its *orthogonal*, $\mathcal{I}^{\perp}$, is defined as

$$\mathcal{I}^{\perp} = \{b : b \text{ is almost disjoint from all } a \in \mathcal{I}\}.$$

We say that $\mathcal{I}$ and $\mathcal{I}^{\perp}$ are *separated* (or equivalently, they do not form a *gap*) if there is a $c \subseteq \mathbb{N}$ such that c almost includes all members of $\mathcal{I}$ and is almost disjoint from all members of $\mathcal{I}^{\perp}$. The following result was suggested by Todorcevic, and it is related to his results about analytic gaps ([126]).

3.5. APPLICATIONS: QUOTIENTS EMBEDDABLE INTO ANALYTIC QUOTIENTS

COROLLARY 3.5.2. *Assume OCA and MA and that $\mathcal{I}$ is an arbitrary ideal on $\mathbb{N}$ and that $\mathcal{P}(\mathbb{N})/\mathcal{I} \hookrightarrow \mathcal{P}(\mathbb{N})/\operatorname{Fin}$. Then $\mathcal{I}$ can be separated from its orthogonal, $\mathcal{I}^\perp$.*

PROOF. Let $\Phi \colon \mathcal{P}(\mathbb{N})/\mathcal{I} \to \mathcal{P}(\mathbb{N})/\operatorname{Fin}$ be a monomorphism. By Theorem 3.3.5 and the Radon–Nikodym property of Fin, there is a finite-to-one function $h \colon \mathbb{N} \to \mathbb{N}$ such that Φ_h is a lifting of Φ on a nonmeager ideal $\mathcal{I}_\Phi$. Every nonmeager ideal is dense, so $\mathcal{I}^\perp = (\mathcal{I} \cap \mathcal{I}_\Phi)^\perp$. Now note that $(\mathcal{I} \cap \mathcal{I}_\Phi)^\perp$ is an ideal generated by $\mathbb{N} \setminus h''\mathbb{N}$ and Fin, and therefore the set $h''\mathbb{N}$ separates $\mathcal{I}$ from $\mathcal{I}^\perp$. □

Recall that two families $\mathcal{A}, \mathcal{B}$ of subsets of $\mathbb{N}$ are *countably separated* if there are sets C_n ($n \in \mathbb{N}$) such that for every pair $A \in \mathcal{A}$, $B \in \mathcal{B}$ there is an n such that $A \subseteq^* C_n$ and $C_n \cap B$ is finite.

COROLLARY 3.5.3. *Assume OCA and MA, that $\mathcal{I}$ is an arbitrary ideal on the integers such that $\mathcal{P}(\mathbb{N})/\mathcal{I} \hookrightarrow \mathcal{P}(\mathbb{N})/\mathcal{J}$, and that $\mathcal{J}$ is either an analytic P-ideal or a countably generated ideal. Then $\mathcal{I}$ can be countably separated from its orthogonal.*

PROOF. By Corollary 3.5.1, $\mathcal{I} = \mathcal{I}_0 \oplus \mathcal{I}_1$ where $\mathcal{I}_1$ is nonmeager, and therefore dense, and $\mathcal{I}_1$ is an analytic P-ideal or $\operatorname{Fin} \times \emptyset$. Therefore by Lemma 1.2.9, $\mathcal{I}$ is countably separated from $\mathcal{I}^\perp$. □

COROLLARY 3.5.4. *OCA and MA imply that there are quotients $\mathcal{P}(\mathbb{N})/\mathcal{I}$ that are not embeddable into any quotient over an analytic P-ideal or $\operatorname{Fin} \times \emptyset$.*

PROOF. Fix a Hausdorff gap, i.e., two $\subseteq^*$-increasing ω_1-chains of sets of integers which cannot be separated by a single subset of $\mathbb{N}$. Note that this implies these two chains cannot be countably separated. Let $\mathcal{I}$ be an ideal generated by one of these two chains. Then $\mathcal{I}$ cannot be countably separated from its orthogonal, and therefore, by Corollary 3.5.2, $\mathcal{P}(\mathbb{N})/\mathcal{I}$ cannot be embedded into any quotient over an analytic P-ideal. □

Another application of Corollary 3.5.1 is its topological reformulation. All results about the structure of homomorphisms of quotients $\mathcal{P}(\mathbb{N})/\mathcal{I}$ have their topological duals which talk about the continuous maps of closed subsets of the Čech–Stone remainder $\mathbb{N}^*$ of the integers (equivalently, the Stone space of $\mathcal{P}(\mathbb{N})/\operatorname{Fin}$; see [**97**]). Recall that $A \subseteq \mathbb{N}^*$ is a *P-set* if for every sequence of open neighborhoods $\{U_n\}$ of A set $\bigcap_n U_n$ includes an open neighborhood of A. In the case when A is closed, this reduces to fact that the ideal on $\mathbb{N}$ corresponding to A is a P-ideal. Just ([**60**], [**64**]) proved that OCA and MA together imply that no nowhere dense P-subset of $\mathbb{N}^*$ is homeomorphic to $\mathbb{N}^*$ itself, answering a question of van Mill ([**97**, p. 537]). We shall now give a slight strengthening of this result. Note that Example 3.2.3 shows that the assumption of A being a P-set cannot be dropped from the following Corollary.

COROLLARY 3.5.5. *Assume OCA and MA. Then*

(a) *If a subset of $\mathbb{N}^*$ is a continuous image of $\mathbb{N}^*$, then it is equal to the disjoint union of a clopen set and a nowhere dense set.*
(b) *Every P-subset of $\mathbb{N}^*$ homeomorphic to $\mathbb{N}^*$ must be clopen.*

PROOF. (a) This is just the topological dual of Corollary 3.5.1 (see [**97**]).
(b) We shall prove the dual statement: If $\mathcal{I}$ is a P-ideal and $\mathcal{P}(\mathbb{N})/\mathcal{I}$ embeds into $\mathcal{P}(\mathbb{N})/\operatorname{Fin}$, then $\mathcal{I} = \operatorname{Fin} \oplus \mathcal{I}_1$ so that the quotient $\mathcal{P}(\mathbb{N})/\mathcal{I}_1$ is ccc. By Corollary 3.5.1, $\mathcal{I} = \mathcal{I}_0 \oplus \mathcal{I}_1$ where $\mathcal{I}_0 \leq_{\operatorname{BE}} \operatorname{Fin}$ and $\mathcal{I}_1$ is ccc over Fin, so $\mathcal{I}_0$ is isomorphic to

Fin. Since a P-ideal is ccc over Fin if and only if the corresponding quotient is ccc (see e.g., [60]), this ends the proof. □

An extension of Corollary 3.5.5 and more applications of Theorem 3.3.6 to topology can be found in Chapter 4.

3.6. Automorphism groups

As remarked earlier, the fact that the automorphism group $\mathrm{Aut}(\mathcal{P}(\mathbb{N})/\mathcal{I})$ of an analytic quotient under OCA reduces to its subgroup of all trivial automorphisms can be interesting in its own right. For example, van Douwen ([18]) has noticed that the group $\mathrm{Aut}(\mathcal{P}(\mathbb{N})/\,\mathrm{Fin})$ is not simple, under the assumption that all automorphisms of $\mathcal{P}(\mathbb{N})/\,\mathrm{Fin}$ are trivial. This gives an example of a homogeneous (see below) Boolean algebra whose automorphism group is not simple. It is still unknown whether a Boolean algebra with this properties can be constructed without using any additional set-theoretic assumptions (see Question 1 of [20]). This section is devoted to the study of automorphism groups of analytic quotients, with the particular emphasis on summable ideals. All results of this section talk about arbitrary isomorphisms between analytic quotients and their automorphism groups under OCA and MA, so we shall not explicitly state these assumptions. Alternatively, these results can be regarded as results about the subgroup of $\mathrm{Aut}(\mathcal{P}(\mathbb{N})/\mathcal{I})$ consisting of those automorphisms which have continuous liftings. By Theorem 3.4.1 (see also Theorem 2.1.1 in §3.14), there is no difference between these two groups. Let us first prove that an automorphism group of a quotient over a summable ideal is rather large. Recall the notation $\mu_f(A) = \sum_{i \in A} f(i)$.

PROPOSITION 3.6.1. *If $\mathcal{I}_f$ is a summable ideal, then $\mathrm{Aut}(\mathcal{P}(\mathbb{N})/\mathcal{I}_f)$ has a subgroup isomorphic to the free Abelian group with continuum many generators.*

PROOF. Let us first assume $\mathcal{I}_f$ is a proper summable ideal. Then we can moreover assume that f is nonincreasing. Find a sequence $1 = n_1 < n_2 < \ldots$ such that for all i
  (1) $\mu_f([n_i, n_{i+1})) \geq 1$, and
  (2) $\mu_f(\{n_{i+1} - 1 : i \in \mathbb{N}\}) < \infty$.
For $B \subseteq \mathbb{N}$ let $h_B \colon \mathbb{N} \to \mathbb{N}$ be defined by

$$h_B(k) = \begin{cases} k+1, & \text{if } k \in [n_i, n_{i+1}) \text{ for } i \in B, \\ k, & \text{if } k \in [n_i, n_{i+1}) \text{ for } i \notin B. \end{cases}$$

CLAIM 1. *Φ_{h_B} is a lifting of an automorphism Φ_B^* of $\mathcal{P}(\mathbb{N})/\mathcal{I}_f$.*

PROOF. For $\varepsilon > 0$ let

$$A_\varepsilon = \left\{ n : \frac{f(n)}{f(n-1)} < \varepsilon \right\}.$$

If $\mu_f(A_\varepsilon) = \infty$ for every $\varepsilon > 0$, then we can find a sequence of finite sets of integers s_i such that for all i
  (3) $s_i \subseteq A_{1/i}$,
  (4) $\max s_i < \min s_{i+1}$, and
  (5) $\mu_f(s_i) > 1$.

Let $B = \bigcup_i s_i$, and enumerate B increasingly as $\{k_j\}$. Then by (3), (4) and the monotonicity of f we have $\lim_j f(k_j)/f(k_{j-1}) = 0$ which implies $\mu_f(B) < \infty$, contradicting (5). Therefore we can find an $\varepsilon > 0$ such that $\mu_f(A_\varepsilon) < \infty$. Then for all $k \in \mathbb{N} \setminus A_\varepsilon$ we have $0 < \varepsilon \leq f(k)/f(k-1) \leq 1$. Since $h_B^{-1}(k) \subseteq \{k, k-1\}$ for all k, this implies that for $k \in A_\varepsilon$ we have

$$1 \leq \frac{f(h_B^{-1}(k))}{f(k)} \leq 1 + \frac{1}{\varepsilon}.$$

Therefore Lemma 1.12.2 implies that Φ_B^* is a homomorphism of $\mathcal{P}(\mathbb{N})/\mathcal{I}_f$. Moreover, h_B is $1-1$ on the set $\mathbb{N} \setminus \{n_{i+1} - 1 : i \in \mathbb{N}\}$ and by (2) this set is in the dual filter of $\mathcal{I}_f$. Therefore Φ_B^* is an automorphism of $\mathcal{P}(\mathbb{N})/\mathcal{I}_f$. □

Let F be the free Abelian group with generators $\{g_\xi : \xi \in \mathbb{R}\}$, and let $\mathcal{A} = \{B_\xi : \xi \in \mathbb{R}\}$ be a family of infinite almost disjoint subsets of $\mathbb{N}$. Note that the subgroup G of $\mathrm{Aut}(\mathcal{P}(\mathbb{N})/\mathcal{I}_f)$ generated by Φ_B ($B \in \mathcal{A}$) is Abelian, since for $B, C \in \mathcal{A}$ we have

$$\Phi_B \circ \Phi_C = \Phi_{B \cup C}.$$

This is because $B \cap C$ is finite, and therefore in $\mathcal{I}_f$. Let $\Omega \colon F \to G$ be a homomorphism uniquely defined by

$$\Omega(g_\xi) = \Phi_{B_\xi}$$

for $\xi \in \mathbb{R}$. We claim Ω is an isomorphism. Since it is an epimorphism, it will suffice to prove that it is a monomorphism. To see this, we pick $a = g_1^{l_1} g_2^{l_2} \ldots g_n^{l_n}$ in G different from unity and prove that $\Phi = \Omega(a)$ is not the identity on $\mathcal{P}(\mathbb{N})/\mathcal{I}_f$. By the definitions, for $h = h_n^{l_n} \circ \ldots h_2^{l_2} \circ h_1^{l_1}$, mapping Φ_h is a lifting of Φ. Since $a \neq 1_F$, we can assume $l_1 \neq 0$ and $g_i \neq g_1$ for $i = 2, \ldots, n$ if $n > 1$. This implies that h has no fixed points in the set $C = \bigcup_{j \in B_1} [n_j, n_{j+1})$. Therefore, by Lemma 9.1 of [14], we can write $C = C_0 \cup C_1 \cup C_2$ so that $h''C_i$ and C_i are disjoint for $i = 0, 1, 2$. By (1), one of these sets is not in $\mathcal{I}_f$. Since this set is moved by Φ, Φ is not the identity. This proves that Ω is an isomorphism.

If $\mathcal{I}_f$ is not a proper summable ideal, then $\mathrm{Aut}(\mathcal{P}(\mathbb{N})/\mathcal{I}_f)$ has a subgroup isomorphic to $\mathrm{Aut}(\mathcal{P}(\mathbb{N})/\mathrm{Fin})$, and it is easy to modify the above construction to prove that this group has a subgroup isomorphic to the free Abelian group with continuum many generators. □

If in the proof of Proposition 3.6.1 instead of a family of pairwise almost disjoint subsets of $\mathbb{N}$ we use a family of independent subsets of $\mathbb{N}$, then by essentially the same proof we have the following:

PROPOSITION 3.6.2. *If $\mathcal{I}_f$ is a summable ideal, then $\mathrm{Aut}(\mathcal{P}(\mathbb{N})/\mathcal{I}_f)$ has a subgroup isomorphic to the free group with continuum many generators.* □

For a function $f \colon \mathbb{N} \to \mathbb{R}^+$ such that $\mu_f(\mathbb{N}) = \infty$, define the following two subgroups of the infinite symmetric group S_∞.

$G_f = \{\ \pi \in S_\infty :$ the set $A_{f,\pi}[p, q] = \{n : p \leq f(\pi(n))/f(n) \leq q\}$ is in the dual filter $\mathcal{I}_f^*$ of $\mathcal{I}_f$ for some $0 < p \leq q < \infty\}$.

$H_f = \{\ \pi \in S_\infty :$ the set $\{n : f(\pi(n)) = f(n)\}$ is in $\mathcal{I}_f^*\}$.

Obviously H_f is a normal subgroup of G_f.

THEOREM 3.6.3. *Assume OCA and MA and let $\mathcal{I}_f$ be a summable ideal other than Fin. Then $\mathrm{Aut}(\mathcal{P}(\mathbb{N})/\mathcal{I}_f)$ is isomorphic to the quotient G_f/H_f.*

PROOF. We define a transformation
$$\Lambda \colon S_\infty \to \operatorname{Aut}(\mathcal{P}(\mathbb{N}))$$
by letting $\Lambda(\pi)$ be $\Phi_{\pi^{-1}}$ (the homomorphism with the lifting $A \mapsto \pi''A$). This transformation is obviously an isomorphism between groups S_∞ and $\operatorname{Aut}(\mathcal{P}(\mathbb{N}))$. By Lemma 1.12.2, $\Lambda(\pi) = \Phi_{\pi^{-1}}$ is in $\operatorname{Aut}(\mathcal{P}(\mathbb{N})/\mathcal{I}_f)$ if and only if π is in G_f. Moreover, every automorphism of $\mathcal{P}(\mathbb{N})/\mathcal{I}_f$ has a lifting of the form $\Lambda(\pi)$ for some π in G_f. To see this, let $\Phi \colon \mathcal{P}(\mathbb{N})/\mathcal{I}_f \to \mathcal{P}(\mathbb{N})/\mathcal{I}_f$ be an automorphism. By the Radon–Nikodym property of summable ideals and proof of Proposition 1.10.4, Φ has a completely additive lifting Φ_h for a $1-1$ partial mapping $h \colon \mathbb{N} \to \mathbb{N}$. Since $\mathcal{I}$ is not Fin, we can assume that the domain and the range of h are cofinite and therefore extend h^{-1} to a permutation π of the positive integers. Lemma 1.12.2 implies that π is in G_f. Therefore Λ maps G_f onto $\operatorname{Aut}(\mathcal{P}(\mathbb{N})/\mathcal{I}_f)$, and it will suffice to prove that $\ker(\Lambda) = H_f$. It is obvious that $H_f \subseteq \ker \Lambda$, so let us prove the reverse inclusion. Assume that $\pi \notin H_f$ is such that $\Lambda(\pi)$ is a lifting of the identity of $\operatorname{Aut}(\mathcal{P}(\mathbb{N})/\mathcal{I}_f)$. Therefore the set $A = \{n : \pi(n) \neq n\}$ is not in $\mathcal{I}_f$. By [14, Lemma 9.1] we can split A into sets A_0, A_1 and A_2 such that $\pi''A_i$ is disjoint from A_i for $i = 0, 1, 2$. One of these sets is not in $\mathcal{I}_f$, and this set is moved by $\Lambda(\pi)$, contradicting the assumption that $\Lambda(\pi)$ is the identity of $\operatorname{Aut}(\mathcal{P}(\mathbb{N})/\mathcal{I}_f)$. □

THEOREM 3.6.4. *Assume OCA and MA and let $\mathcal{I}$ be an analytic P-ideal. Then $\operatorname{Aut}(\mathcal{P}(\mathbb{N})/\mathcal{I}_f)$ is isomorphic to a quotient of a Borel subgroup of S_∞ over its Borel ideal.*

PROOF. If there is an infinite set in $\mathcal{I}$, then Λ defined in proof of Theorem 3.6.3 maps some subgroup $G_\mathcal{I}$ of S_∞ onto $\operatorname{Aut}(\mathcal{P}(\mathbb{N})/\mathcal{I})$, and its restriction to $G_\mathcal{I}$ is a lifting of the desired isomorphism by the proof of Theorem 3.6.3. In the case when $\mathcal{I} = \operatorname{Fin}$, the ideal $\mathbb{N} \oplus \operatorname{Fin}$ is isomorphic to Fin and the above proof applies to it. □

Now we shall see that Boolean algebras of the form $\mathcal{P}(\mathbb{N})/\mathcal{I}_f$ can have rather curious properties.

THEOREM 3.6.5. *Assume OCA and MA. There is a summable ideal $\mathcal{I}_f$ such that every epimorphism $\Phi \colon \mathcal{P}(\mathbb{N})/\mathcal{I}_f \to \mathcal{P}(\mathbb{N})/\mathcal{I}_f$ is automatically an isomorphism.*

Before we prove Theorem 3.6.5, we shall prove another property of the ideal $\mathcal{I}_f$ which satisfies its conclusion.

PROPOSITION 3.6.6. *Assume OCA and MA. There is a summable ideal $\mathcal{I}_f$ such that $\operatorname{Aut}(\mathcal{P}(\mathbb{N})/\mathcal{I}_f)$ is isomorphic to some quotient of the group $\prod_{n=1}^\infty S_{n!}$.*

PROOF. Let f be a nonincreasing function uniquely determined by the following conditions: (i) $\operatorname{range}(f) = \{1/n! : n \in \mathbb{N}\}$, (ii) the set
$$I_n = \left\{i : f(i) = \frac{1}{n!}\right\}$$
is an interval of $\mathbb{N}$, and (iii) $\mu_f(I_n) = 1$ for all n. Let Λ be as defined in the proof of Theorem 3.6.3. It will suffice to prove that every automorphism Φ of $\mathcal{P}(\mathbb{N})/\mathcal{I}_f$ has a lifting of the form $\Lambda(\pi)$ for some $\pi \in \prod_{n=1}^\infty S_{I_n}$. So assume $\Phi \in \operatorname{Aut}(\mathcal{P}(\mathbb{N})/\mathcal{I}_f)$

and let τ be such that $\Lambda(\tau)$ is a lifting of Φ. Let p, q be such that $A_{f,\tau}[p,q] \in \mathcal{I}_f^*$, pick $\bar{k} > \max(q, 1/p)$ and $\bar{n}$ such that

$$f(\bar{n} - 1) = \frac{1}{\bar{k}!} \quad \text{and} \quad f(\bar{n}) = \frac{1}{(\bar{k}+1)!}.$$

Then for all i, j which are bigger than $\bar{k}$ and distinct we have $\pi'' I_i \cap I_j = \emptyset$. Therefore the restriction of τ to $A_{f,\tau} \setminus [1, \bar{n}]$ can be extended to an element π of $\prod_{n=1}^{\infty} S_{I_n}$. This π is as required. □

PROOF OF THEOREM 3.6.5. We shall prove that the ideal defined in proof of Proposition 3.6.6 satisfies the requirements. Assume that $\Phi \colon \mathcal{P}(\mathbb{N})/\mathcal{I}_f \to \mathcal{P}(\mathbb{N})/\mathcal{I}_f$ is an epimorphism. By Theorem 3.3.5 it has a completely additive lifting Φ_h, and by proof of Proposition 1.10.4 applied to the ideals $\mathcal{J} = \mathcal{I}_f$ and $\mathcal{I} = \ker(\Phi)$ we can assume h is $1-1$, and therefore that Φ has a lifting of the form $\Lambda(\pi)$ for some $\pi \in S_\infty$. Since $\ker(\Phi) \supseteq \mathcal{I}_f$, there is a $q < \infty$ such that the set

$$A(\cdot, q] = \left\{ i : \frac{f(\pi(i))}{f(i)} \leq q \right\}$$

is in the dual filter $\mathcal{I}_f^*$ of $\mathcal{I}_f$. This was proved as Case 1 in the proof of Lemma 1.12.2. By the same lemma, to prove that Φ is an isomorphism it will suffice to find a $p > 0$ such that

$$A[p, \cdot) = \left\{ i : \frac{f(\pi(i))}{f(i)} \geq p \right\}$$

is in $\mathcal{I}_f^*$. We claim that $p = 1/q$ works. So let us assume this is not true, and the set $A[p, \cdot)$ is not in $\mathcal{I}_f$, namely $\mu_f(A[p, \cdot)) = \infty$. Find $\bar{n} \geq q/(q-2)$ large enough so that for some k we have

$$f(\bar{n} - 1) = \frac{1}{k!}, \qquad f(\bar{n}) = \frac{1}{(k+1)!}$$

and

$$\mu_f(\pi'' A(q, \cdot) \setminus [1, \bar{n})) < 1.$$

Let $t \subseteq A(\cdot, q] \setminus [1, \bar{n}]$ be such that $\mu_f(t) > \bar{n}$, and find $\bar{m} > \bar{n}$ so that $t \subseteq [\bar{n}, \bar{m}]$ and for some k' we have

$$\frac{1}{k'!} = f(\bar{m}) > f(\bar{m} + 1) = \frac{1}{(k'+1)!}.$$

Define $s \subseteq [\bar{n}, \bar{m}]$ to be

$$s = \{i \in [\bar{n}, \bar{m}] : f(i) \neq f(\pi^{-1}(i))\},$$

and note that for every $l \in [k, k']$ we have $|I_l \cap s| \geq |I_l \cap t|$ because π is a permutation. In particular, we have $\mu_f(s) \geq \mu_f(t)$. Also, since for $i \in t$ we have $f(\pi(i)) \leq f(i)/k$, we have

$$\mu_f(\pi'' t) \leq \frac{1}{k} \mu_f(t) < \frac{1}{q} \mu_f(t).$$

Similarly, since $f(\pi(i)) < f(i)/k$ for $i \in [1, \bar{n}]$ such that $f(i) > \bar{n}$, we have

$$\mu_f((\pi''[1, \bar{n}]) \cap [\bar{n}, \infty)) \leq \frac{\bar{n}}{k}.$$

Therefore we have

$$\mu_f(s \setminus (\pi''[1, \bar{n}] \cup t)) > \mu_f(t) - \frac{1}{q} \mu_f(t) - \frac{\bar{n}}{q} > \bar{n} \frac{q-2}{q} > 1.$$

But $i \in s \setminus (\pi''[1,\bar{n}] \cup t)$ implies $f(\pi^{-1}(i)) > i$. Since $f(i) \leq 1/(k+1)!$ for all $i \in s$, we have $s \subseteq \pi''A(q,\cdot)$ and therefore $\mu_f((\pi''A(q,\cdot)) \setminus [1,\bar{n}]) > 1$, contradicting the choice of $\bar{n}$. $\square$

Boolean algebras $\mathcal{B}$ such that every epimorphism $\Phi \colon \mathcal{P}(\mathbb{N})/\mathcal{I}_f \to \mathcal{P}(\mathbb{N})/\mathcal{I}_f$ is automatically an isomorphism are sometimes called *Hopfian* (see [20]). Therefore Theorem 3.6.5 says that a quotient over a summable ideal can sometimes be a Hopfian Boolean algebra. (Note that this cannot happen under CH, since $\mathcal{P}(\mathbb{N})/\operatorname{Fin}$ is not Hopfian and all quotients over F_σ ideals are isomorphic under CH.) Not all quotients over summable ideals share this property. The conclusion of Theorem 3.6.5 is equivalent to saying that $\mathcal{P}(\mathbb{N})/\mathcal{I}_f$ is not isomorphic to $\mathcal{P}(\mathbb{N})/\mathcal{J}$ for every analytic ideal $\mathcal{J} \supsetneq \mathcal{I}$, and for $\mathcal{I}_{1/n}$ there are many such ideals $\mathcal{J}$. For example, the ideal generated by $2\mathbb{N}$ over $\mathcal{I}_{1/n}$. Recall that a Boolean algebra $\mathcal{B}$ is *dual Hopfian* (see [20]) if every monomorphism $\Phi \colon \mathcal{P}(\mathbb{N})/\mathcal{I}_f \to \mathcal{P}(\mathbb{N})/\mathcal{I}_f$ is automatically an automorphism. So the following statement (which should be compared to Lemma 1.12.5) says that $\mathcal{P}(\mathbb{N})/\mathcal{I}_f$ is never dual Hopfian.

PROPOSITION 3.6.7. *If $\mathcal{I}_f$ is a summable ideal, then there is always a monomorphism $\Phi \colon \mathcal{P}(\mathbb{N})/\mathcal{I}_f \to \mathcal{P}(\mathbb{N})/\mathcal{I}_f$ which is not an automorphism.*

PROOF. It suffices to prove this for proper summable ideals, so we can assume f is nonincreasing. Define an increasing sequence $1 = n_1 < n_2 < n_3 < \ldots$ such that
$$2f(i) \leq \mu_f([n_i, n_{i+1})) \leq 3f(i).$$
This is possible because we have $n_i \geq i$ and therefore $f(n_i) \leq f(i)$ for all i. Note that this inequality also implies that the size of each interval $[n_i, n_{i+1})$ is at least two. Let h be a mapping which collapses $[n_i, n_{i+1})$ to i. Then Φ_h is a lifting of a monomorphism $\Phi \colon \mathcal{P}(\mathbb{N})/\mathcal{I}_f \to \mathcal{P}(\mathbb{N})/\mathcal{I}_f$. If it was an isomorphism, then there would be a set $A \in \mathcal{I}_f^*$ such that the restriction of h to A is $1-1$. Since there is no such set, Φ is not an isomorphism. $\square$

Recall that $\mathcal{Z}_0$ is the ideal of sets of zero density (see §1.13).

QUESTION 3.6.8. *Is the group $\operatorname{Aut}(\mathcal{P}(\mathbb{N})/\mathcal{Z}_0)$ always simple?*

In an earlier version of this monograph it was claimed that $\operatorname{Aut}(\mathcal{P}(\mathbb{N})/\mathcal{Z}_0)$ is never simple. David Fremlin has pointed out to an error in the proof, and moreover to the proposition below whose proof is included with his kind permission.

PROPOSITION 3.6.9. *Assume CH. Then $\operatorname{Aut}(\mathcal{P}(\mathbb{N})/\mathcal{Z}_0)$ is not simple.*

PROOF. Continuum Hypothesis implies that the algebra $\mathcal{P}(\mathbb{N})/\mathcal{Z}_0$ is homogeneous (Corollary 1.13.7). CH also implies that $\mathcal{P}(\mathbb{N})/\mathcal{Z}_0$ is isomorphic to its countable power, $(\mathcal{P}(\mathbb{N})/\mathcal{Z}_0)^\mathbb{N}$ (Corollary 1.13.8). By a result of P. Štěpánek and M. Rubin ([143, Corollary 5.9a]), if a homogeneous Boolean algebra $\mathcal{B}$ is isomorphic to its countable power, then its automorphism group is simple. $\square$

Let us end this section with a question an (unlikely) positive answer to which would simplify some of the above proofs.

QUESTION 3.6.10. *Does $\mathcal{I} \leq_{\mathrm{RB}} \mathcal{J}$ necessarily imply that $\operatorname{Aut}(\mathcal{P}(\mathbb{N})/\mathcal{I})$ is a subgroup of $\operatorname{Aut}(\mathcal{P}(\mathbb{N})/\mathcal{J})$?*

3.7. Homogeneity of analytic quotients

Boolean algebra $\mathcal{B}$ is *homogeneous* if for all $a, b \in \mathcal{B}$ distinct from $0_\mathcal{B}$ and $1_\mathcal{B}$ there is an automorphism of $\mathcal{B}$ sending a to b. It is *weakly homogeneous* if for all $a, b \in \mathcal{B}$ distinct from $0_\mathcal{B}$ there is an automorphism Φ of $\mathcal{B}$ such that $\Phi(a) \cap b \neq 0_\mathcal{B}$. By a result of Just and Krawczyk ([65, Theorem 1]; see also [20, Question 48]), Continuum Hypothesis implies that every quotient over an F_σ ideal is isomorphic to $\mathcal{P}(\mathbb{N})/\operatorname{Fin}$, and therefore homogeneous. Similarly, CH implies that all quotients over Erdös–Ulam ideals are homogeneous (Corollary 1.13.7). Once again, the impact of OCA is in the opposite.

PROPOSITION 3.7.1. *Assume OCA and MA. If $\mathcal{I}_f$ is a summable ideal other than* Fin, *then the algebra $\mathcal{P}(\mathbb{N})/\mathcal{I}$ is not weakly homogeneous.*

PROOF. Since $\mathcal{I}_f$ is not isomorphic to Fin, there is a positive set C such that $\mathcal{I}_f \restriction C$ is a proper summable ideal and therefore it will suffice to prove the statement assuming $\mathcal{I}_f$ is a proper summable ideal. Let s_i be disjoint finite sets of integers such that for all $i < k$ we have

$$1 \leq \mu_f(s_i) \leq 2 \quad \text{and} \quad \min_{j \in s_i} f(j) > i \cdot \max_{j \in s_k} f(j).$$

Let $A = \bigcup_i s_{2i}$ and $B = \bigcup_i s_{2i+1}$. Then it is easy to see that Lemma 1.12.2 implies $\mathcal{I}_f \restriction A'$ is not Baire-isomorphic (and therefore not isomorphic) to $\mathcal{I}_f \restriction B'$, whenever $A' \subseteq A$ and $B' \subseteq B$ are $\mathcal{I}_f$-positive sets. $\square$

By a proof similar to that of Proposition 3.7.1, Theorems 1.13.3 and 1.13.12 can be used to prove the following (see §1.13 for definitions).

PROPOSITION 3.7.2. *Assume OCA and MA. If $\mathcal{Z}_\mu$ is a density ideal not isomorphic to* Fin *or to $\emptyset \times$ Fin, then the algebra $\mathcal{P}(\mathbb{N})/\mathcal{Z}_\mu$ is not weakly homogeneous.* $\square$

There is an analytic P-ideal different from Fin whose quotient is weakly homogeneous; the ideal $\emptyset \times$ Fin verifies this assertion. One can say more (recall that an ideal $\mathcal{I}$ has the *Fréchet property* if $(\mathcal{I}^\perp)^\perp = \mathcal{I}$).

LEMMA 3.7.3. *If $\mathcal{I}$ has the Fréchet property, then its quotient is weakly homogeneous.*

PROOF. If $\mathcal{I}$ has the Fréchet property, then every positive set A has a subset B such that $\mathcal{I} \restriction B$ is isomorphic to Fin. Therefore if C, D are $\mathcal{I}$-positive sets, then we can easily define a bijection $h \colon \mathbb{N} \to \mathbb{N}$ (we can assume that $\mathcal{I} \neq \operatorname{Fin}$) which sends $\mathcal{I}$ into itself and such that $h''C \cap D$ is infinite. Then Φ_h is an automorphism of the quotient algebra as required. $\square$

We do not know whether there are analytic P-ideals other than Fin and $\emptyset \times$ Fin whose quotients are weakly homogeneous (Corollary 1.2.11 suggests that a variation of the above proof of Proposition 3.7.1 may apply to give a negative answer), so let us turn to the homogeneous ones.

PROPOSITION 3.7.4. *Assume OCA and MA. If $\mathcal{I}$ is a non-pathological analytic P-ideal different from* Fin, *then the algebra $\mathcal{P}(\mathbb{N})/\mathcal{I}$ is not homogeneous.*

PROOF. Let $\mathcal{I} = \operatorname{Exh}(\varphi)$ be a non-pathological P-ideal whose quotient is homogeneous, where φ is a lower semicontinuous submeasure guaranteed by Solecki's theorem. The ideal $\mathcal{I}$ is dense, since otherwise the ideal $\mathcal{I} \upharpoonright A$ would be isomorphic to Fin for some positive set A, contradicting the homogeneity and the assumption that $\mathcal{I}$ itself is not isomorphic to Fin. Therefore we can assume that $\mathcal{I}$ is dense, or equivalently that

(1) $\lim_i \varphi(\{i\}) = 0$.

Now construct sequences u_i, v_i ($i \in \mathbb{N}$) and $f \colon \mathbb{N} \to \mathbb{N}$ so that for all i

(2) $u_1 < u_2 < \ldots$ are finite subsets of $\mathbb{N}$,
(3) $v_1 < v_2 < \ldots$ are finite subsets of $\mathbb{N}$,
(4) $|\varphi(u_i) - 1| < 2^{-i-1}$,
(5) $|\varphi(v_i) - 1| < 2^{-i-1}$,
(6) $f(i) = \sum_{j=1}^{i} |u_j|$,
(7) $\varphi(\{k\}) < 1/(2f(i))$, for all $k \in v_i$,
(8) $\varphi(\{l\}) < 1/(2^i|v_i|)$ for all $l \in u_{i+1}$.

Knowing that (1) holds, this construction is straightforward, so let us assume that u_i, v_i and f are as above. Let us write

(9) $U_i = \bigcup_{j=1}^{i} u_i$.

CLAIM 1. *For every i and every $1-1$ map $h \colon v_i \to \mathbb{N}$ there is $s \subseteq v_i$ such that $h''s \cap U_i = \emptyset$ and $\varphi(s) \geq 1/4$.*

PROOF. Let $t = h^{-1}(U_i)$. Since h is $1-1$, (7) implies
$$\varphi(t) \leq \frac{1}{2f(i)} |U_i| = \frac{1}{2},$$
and therefore $s = v_i \setminus t$ is as required. □

CLAIM 2. *For every $h \colon v_i \to \bigcup_{j=1}^{\infty} u_j$ we have $\varphi(h''v_i \setminus U_i) < 2^{-i}$.*

PROOF. We can assume that $h''v_i \cap U_i = \emptyset$. By (8),
$$\varphi(h''v_i) \leq \sum_{k \in v_i} \varphi(\{h(k)\}) < \frac{1}{2^i|v_i|}|v_i| = 2^{-i},$$
and this completes the proof. □

Let $A = \bigcup_i u_i$ and $B = \bigcup_i v_i$. By (4) and (5), both sets are positive. Since $\mathcal{P}(\mathbb{N})/\mathcal{I}$ is homogeneous, there is an isomorphism $\Phi \colon \mathcal{P}(B)/\mathcal{I} \to \mathcal{P}(A)/\mathcal{I}$. By Corollary 3.4.2, there is a bijection $h \colon A \to B$ which defines an isomorphism between $\mathcal{I} \upharpoonright B$ and $\mathcal{I} \upharpoonright A$. (Recall that, although in general we cannot assume that h is a bijection, in this case we can since both $\mathcal{I} \upharpoonright A$ and $\mathcal{I} \upharpoonright B$ are dense.) By Claim 1 and Claim 2, for every i there is $s_i \subseteq v_i$ such that $\varphi(s_i) \geq 1/4$ and $\varphi(h''s_i) < 2^{-i}$. Therefore $\bigcup_i s_i$ is a positive set which Φ_h sends to a set in $\mathcal{I}$, contradicting our assumption on h. □

CONJECTURE 3.7.5. *Proposition 3.7.4 remains true even when the assumption that $\mathcal{I}$ is non-pathological is dropped.*

This would follow from OCA lifting theorem, an (expected) positive answer to Question 1.14.3 and Corollary 1.2.11.

In [65, Question C] Just and Krawczyk asked whether all homogeneous quotients over ideals which are $F_{\sigma\delta}$ but not F_σ are pairwise isomorphic. This question may turn out to have a positive answer trivially. By the above, Fin seems to be the only $F_{\sigma\delta}$ P-ideal whose quotient is homogeneous, and it turns out, at least when the ideal has the Radon–Nikodym property, that the quotient is homogeneous only when the ideal itself is homogeneous (under a natural definition). The only homogeneous $F_{\sigma\delta}$ ideal which is not F_σ known to me is Fin × Fin. This ideal coincides with the ordinal ideal $\mathcal{I}_{\omega^2}$ (see §1.14), and it is clear that all these ideals $\mathcal{I}_\alpha$ are homogeneous. The topological ordinal ideals $\mathcal{W}_\alpha$ (for α multiplicatively indecomposable) of Weiss (see §1.14) are also homogeneous.

QUESTION 3.7.6. *Are there any other homogeneous analytic ideals?*

3.8. Almost liftings of embeddings into $\mathcal{P}(\mathbb{N})/\mathrm{Fin}$

This section includes a proof of a special case of Theorem 3.3.5, when the range of the homomorphism is $\mathcal{P}(\mathbb{N})/\mathrm{Fin}$. Our reasons for proving this special case in a separate section are twofold: (i) Its proof shows some of the ideas involved in the (quite long) proof of Theorem 3.3.5 in a more transparent way, and (ii) This is the only instance of Theorem 3.3.5 used in topological applications given in Chapter 4; these applications rely on Theorem 3.3.6, which is proved using Theorem 3.8.1.

THEOREM 3.8.1. *Assume OCA and MA. If $\Phi \colon \mathcal{P}(\mathbb{N})/\mathrm{Fin} \to \mathcal{P}(\mathbb{N})/\mathrm{Fin}$ is a homomorphism, then Φ has a continuous almost lifting. Equivalently, Φ can be decomposed as $\Phi_1 \oplus \Phi_2$ so that Φ_1 has a continuous lifting, while the kernel of Φ_2 is ccc over* Fin.

Before we start the proof, let us note the more useful reformulation of this result:

COROLLARY 3.8.2. *Assume OCA and MA. If $\Phi \colon \mathcal{P}(\mathbb{N})/\mathrm{Fin} \to \mathcal{P}(\mathbb{N})/\mathrm{Fin}$ is a homomorphism, then Φ has a completely additive almost lifting. More precisely, there is $h \colon A \to \mathbb{N}$ for some $A \subseteq \mathbb{N}$ such that Φ_h is an almost lifting of Φ.*

PROOF. Apply the Radon–Nikodym Property of Fin to the homomorphism Φ_1 obtained by Theorem 3.8.1. □

PROOF OF THEOREM 3.8.1. We will use the following local version of Theorem 3.8.1 taken from [64, Theorem 11] (in Proposition 3.12.1 we prove a more general statement).

LEMMA 3.8.3. *The ideal*
$$\mathcal{J}_{\mathrm{cont}} = \{A : \Phi \upharpoonright \mathcal{P}(A) \text{ has a continuous lifting}\}$$
is ccc over Fin. □

By Theorem 1.6.1, the ideal $\mathcal{J}_{\mathrm{cont}}$ equals the set of all A for which $\Phi \upharpoonright \mathcal{P}(A)$ has a completely additive lifting Φ_{h_A} for some finite-to-one $h_A \colon \mathbb{N} \to \mathbb{N}$. Let us consider another ideal associated with Φ:
$$\mathcal{J}_1 = \{A : \Phi \upharpoonright \mathcal{P}(A) \text{ has a completely additive almost lifting}\}.$$
Then clearly $\mathcal{J}_1$ includes $\mathcal{J}_{\mathrm{cont}}$ and therefore it is ccc over Fin as well. Recall that a family $\mathcal{F}$ of partial functions on the integers is *coherent modulo* Fin if for all $f, g \in \mathcal{F}$ there are at most finitely many $n \in \mathrm{dom}(f) \cap \mathrm{dom}(g)$ such that $f(n) \neq g(n)$. For $A \in \mathcal{J}_1$ let $g_A \colon A \to \mathrm{Fin}$ be defined by
$$g_A(n) = h_A^{-1}(n).$$

LEMMA 3.8.4. *The family $\{g_A : A \in \mathcal{J}_1\}$ is coherent modulo* Fin.

PROOF. Assume the contrary, that for some $A, B \in \mathcal{J}_1$ the set of all $n \in A \cap B$ such that $g_A(n) \neq g_B(n)$ is infinite. Assume $g_A(n) \setminus g_B(n) \neq \emptyset$ for infinitely many n. Let Φ_* be a lifting of Φ. Since both $\{C : \Phi_{h_A}(C) =^* \Phi_*(C)\}$ and $\{C : \Phi_{h_B}(C) =^* \Phi_*(C)\}$ include a ccc over Fin (and therefore dense) ideal on $A \cap B$, we can find an infinite $C \subseteq A \cap B$ such that $g_A(n) \setminus g_B(n) \neq \emptyset$ for all $n \in C$ and
$$\Phi_{h_A}(C) =^* \Phi_*(C) =^* \Phi_{h_B}(C).$$
Since $g_A(n), g_B(n)$ are finite for all n, we can furthermore assume that $g_A(n) \cap g_B(m) = \emptyset$ for all distinct n and m in C. Then $\Phi_{h_A}(C) \setminus \Phi_{h_B}(C)$ is infinite, contradicting the above formula. The case when $g_B(n) \setminus g_A(n) \neq \emptyset$ for infinitely many n is treated in an analogous way. This completes the proof. □

For $D \subseteq \mathbb{N}$ define a partition $[\mathcal{J}_1]^2 = L_0(D) \cup L_1(D)$ by letting $\{A, B\}$ in $L_0(D)$ if and only if there exists $n \in A \cap B \cap D$ such that
$$g_A(n) \neq g_B(n).$$
By identifying A with the graph of g_A, we see that OCA applies to this partition.

LEMMA 3.8.5. *The following are equivalent*

(D1) *the ideal $\mathcal{J}_1$ is σ-$L_1(D)$-homogeneous,*
(D2) *the set D is in $\mathcal{J}_1$,*

for every $D \subseteq \mathbb{N}$.

PROOF. Assume $D \in \mathcal{J}_1$ with $h_D : \mathbb{N} \to D$ as a witness. For $n \in \mathbb{N}$ let
$$\mathcal{X}_n = \{A \in \mathcal{J}_1 : g_A(m) = g_B(m) \text{ for all } m \in (A \cap B) \setminus [1, n]\}.$$
Then each $\mathcal{X}_n$ can be covered by countably many $L_1(D)$-homogeneous sets, according to what is the behavior of g_A on $[1, n]$.

Assume now $\mathcal{J}_1$ is σ-$L_1(D)$-homogeneous. We can also assume that $D = \mathbb{N}$, and therefore that $\mathcal{J}_1$ is σ-$L_1(\mathbb{N})$-homogeneous. Then there is a nonmeager $L_1(\mathbb{N})$-homogeneous set $\mathcal{X} \subseteq \mathcal{J}_1$. Note that if $A \subseteq B$ then the map h_A witnesses that $B \in \mathcal{J}_{\text{cont}}$, and therefore we can assume that $\mathcal{X}$ is *hereditary*, i.e., closed under taking subsets of its elements.

CLAIM 1. *There are at most finitely many integers m such that $h_A(m) \neq h_B(m)$ for some $A, B \in \mathcal{X}$.*

PROOF. Suppose otherwise, let $\{m_i\}$ be a sequence such that n_i^0 and n_i^1 are in $\{h_A(m_i) : A \in \mathcal{X}\}$ for some $n_i^0 \neq n_i^1$ and all i. By $L_1(\mathbb{N})$-homogeneity of $\mathcal{X}$, sets $\{n_i^0, n_i^1\}$ ($i \in \mathbb{N}$) are pairwise disjoint. Since $\mathcal{X}$ is nonmeager and closed under taking subsets of its element, there is an $A \in \mathcal{X}$ which includes infinitely many of $\{n_i^0, n_i^1\}$ ($i \in \mathbb{N}$) (this follows by Theorem 3.10.1 (a) which will be proved later). But by the homogeneity we have both $h_A(m) = n_i^0$ and $h_A(m) = n_i^1$, a contradiction. □

By Claim 1 and making finite changes to some of h_A for $A \in \mathcal{X}$, we can assume that $h = \bigcup_{A \in \mathcal{X}} h_A$ is indeed a function.

CLAIM 2. *For every $B \in \mathcal{J}_{\text{cont}}$ the set $\{m : h_B(m) \neq h(m)\}$ is finite.*

PROOF. If this set was infinite, then since X is nonmeager there would be an $A \in \mathcal{X}$ having an infinite intersection with this set. But then the pair A, B would contradict the conclusion of Lemma 3.8.4. □

This implies that the set $\{n : h^{-1}(n) \neq h_A^{-1}(n)\}$ is finite for all $A \in \mathcal{J}_{\mathrm{cont}}$, and therefore $\mathcal{I}$ includes $\mathcal{J}_{\mathrm{cont}}$. □

LEMMA 3.8.6. *The ideal $\mathcal{J}_1$ is a P-ideal.*

PROOF. Assume $\mathcal{J}_1$ is not a P-ideal, so for some sequence $A_1 \subseteq A_2 \subseteq A_3 \subseteq \ldots$ of sets in $\mathcal{J}_1$ there is no $A_\infty \in \mathcal{J}_1$ which almost includes all A_n's. For convenience, we assume $\mathcal{J}_{\mathrm{cont}}, \mathcal{J}_1$ are ideals on $\mathbb{N} \times \mathbb{N}$ and that $A_k = [1,k] \times \mathbb{N}$. For an increasing function $f \colon \mathbb{N} \to \mathbb{N}$ consider the set
$$\Gamma_f = \{\langle k,m\rangle : m \geq f(k)\}.$$
By the choice of $\{A_k\}$, the set $\mathbb{N}^2 \setminus \Gamma_f$ is $\mathcal{J}_1$-positive. We claim there is a $\bar{f} \colon \mathbb{N} \to \mathbb{N}$ such that for all $f \geq^* \bar{f}$ the set $\Gamma_f \setminus \Gamma_{\bar{f}}$ is in $\mathcal{J}_{\mathrm{cont}}$. Otherwise, we can recursively find $\{f_\xi\}_{\xi < \omega_1}$ such that $\Gamma_{f_{\xi+1}} \setminus \Gamma_{f_\xi} \notin \mathcal{J}_{\mathrm{cont}}$, contradicting the fact that $\mathcal{J}_{\mathrm{cont}}$ is ccc over Fin (Lemma 3.8.3). Since $\Gamma_{\bar{f}} \cap A_n$ is finite for all n, we can assume that $\bar{f}$ is identically equal to zero. Then $\mathcal{F} = \{g_{\Gamma_f} : f \in \mathbb{N}^\mathbb{N}\}$ is a coherent family of functions. By OCA and Theorem 2.2.7, there is a function $\bar{g} \colon \mathbb{N}^2 \to \mathrm{Fin}$ which trivializes $\mathcal{F}$.

CLAIM 3. *For every $A \in \mathcal{J}_1$ there is a $k(A) \in \mathbb{N}$ such that*
$$\{\langle i,j\rangle \in A : \bar{g}(\langle i,j\rangle) \neq g_A(\langle i,j\rangle)\} \subseteq [1, k(A)] \times \mathbb{N}.$$

PROOF. If this was false, there would be an infinite $C = \{\langle m_i, n_i\rangle : i \in \mathbb{N}\} \subseteq A$ in $\mathcal{J}_{\mathrm{cont}}$ such the sequence $\{m_i\}$ is strictly increasing and for all i we have
$$\bar{g}(\langle m_i, n_i\rangle) \neq g_A(\langle m_i, n_i\rangle).$$
Since $\mathcal{Y}$ is cofinal in $\mathbb{N}^\mathbb{N}$, there is an $f \in \mathcal{Y}$ such that $\langle m_i, n_i\rangle \in \Gamma_f$ for all i. But this implies that g_{Γ_f} and g_A differ on infinitely many places, contradicting the fact that the family $\mathcal{F}$ is coherent. □

By Claim 3, for every $k \in \mathbb{N}$ the set $\mathcal{H}_k = \{A \in \mathcal{J}_1 : k(A) = k\}$ is $L_1((\mathbb{N}^2) \setminus A_k)$-homogeneous. For $A \in \mathcal{H}_k$ we can replace function g_A with
$$\langle i,j\rangle \mapsto \begin{cases} g_{A_k}(\langle i,j\rangle), & \text{if } i \leq k, \\ g_A(\langle i,j\rangle), & \text{otherwise.} \end{cases}$$
After this adjustment, each $\mathcal{H}_k$ becomes σ-$L_1(\mathbb{N}^2)$-homogeneous, and therefore Lemma 3.8.5 implies that $\mathbb{N}^2 \in \mathcal{J}_1$, contradicting our assumption. □

Now that we have proved $\mathcal{J}_1$ is a P-ideal, we can assume $\mathcal{J}_{\mathrm{cont}}, \mathcal{J}_1$ are ideals on $\mathbb{N}$. Assume $\mathcal{J}_1$ is not σ-$L_1(\mathbb{N})$-homogeneous, and fix an uncountable $L_0(\mathbb{N})$-homogeneous $\mathcal{X} \subseteq \mathcal{J}_1$. Since $\mathcal{J}_1$ is a P-ideal, by increasing some of the elements of $\mathcal{X}$ we can assume $\mathcal{X}$ forms a chain of type ω_1 in the ordering $\subseteq^*$ of almost inclusion. Let $\mathcal{P}$ be the poset of all $\langle s, k, F\rangle$, where

($\mathcal{P}$1) $k \in \mathbb{N}$ and s is a subset of k,
($\mathcal{P}$2) F is a finite $L_0(s)$-homogeneous subset of $\mathcal{X}$.

Define the ordering on $\mathcal{P}$ by letting $p \leq q$ if (recall that $p = \langle s^p, k^p, F^p\rangle$)

($\mathcal{P}$3) $s^p \cap k^q = s^q$ and $F^p \supseteq F^q$.

We will prove that $\mathcal{P}$ has a strong form of a countable chain condition that will assure ccc-ness of a certain amalgamation $\mathcal{P}_{\omega_1}$ of uncountably many copies of $\mathcal{P}$. A pair of uncountable subsets X, Y of $\mathcal{P}$ such that every $p \in X$ is incompatible with every $q \in Y$ is called an *uncountable rectangle of incompatible conditions*. One similarly defines the notion of an *uncountable rectangle of compatible conditions*.

CLAIM 4. *There are no $L_1([n, \infty))$-homogeneous rectangles with both sides uncountable, for any $n \in \mathbb{N}$.*

PROOF. Suppose the contrary, and let $Y \times Z$ be an uncountable $L_1([n, \infty))$-homogeneous rectangle; we will prove that this implies there is an uncountable $L_1([n, \infty))$-homogeneous subset of X. Let

$$h = \bigcup_{A \in Y} h_A.$$

Since Z is cofinal in $\langle X, \subseteq^* \rangle$, the set of all $n \in A$ for which $h^{-1}(n) \neq h_A^{-1}(n)$ is finite for all $A \in X$. This implies that for some $\bar{k}$ the set of all $A \in X$ for which $h_A^{-1}(n) = h^{-1}(n)$ for all $n > \bar{k}$ in A is uncountable. This set is $L_1([\bar{k}, \infty))$-homogeneous, and therefore a finite union of $L_1(\mathbb{N})$-homogeneous sets; but X is $L_0(\mathbb{N})$-homogeneous, a contradiction. □

A *working part* of $p \in \mathcal{P}$ is $\langle s^p, k^p \rangle$. Working parts of p, q are *compatible* if $s^p \cap [1, k^q] = s^q \cap [1, k^p]$.

CLAIM 5. *If X, Y are uncountable subsets of $\mathcal{P}$ such that for all $p, q \in X \cup Y$ their working parts are compatible, then there is an uncountable rectangle $X' \subseteq X$, $Y' \subseteq Y$ of compatible conditions.*

PROOF. Let $\bar{k}$ be such that for uncountably many $p \in X$ and uncountably many $q \in Y$ we have $k^p, k^q \leq \bar{k}$. Let $X = \{p_{0\xi}\}$ and $Y = \{p_{1\xi}\}$. By going to an uncountable subset, we can assume that the size of each $F_{0\xi} = F^{p_{0\xi}}$ (each $F_{1\xi} = F^{p_{1\xi}}$) is equal to some fixed l (m, respectively), and let

$$F_{0\xi} = \{A_{0\xi}^1, \ldots, A_{0\xi}^l\} \quad \text{and} \quad F_{1\xi} = \{A_{1\xi}^1, \ldots, A_{1\xi}^m\}.$$

This enumeration is chosen so that $A_{i\xi}^j \supseteq^* A_{i\xi}^{j+1}$ for all ξ, i, and j, and (again by going to an uncountable subset) we can assume that there is a $\bar{k}_1 \geq \bar{k}$ such that for uncountably many ξ and all i, j we have

$$A_{i\xi}^{j+1} \setminus A_{i\xi}^j \subseteq [1, \bar{k}_1] \quad \text{and} \quad g_{A_{i\xi}^{j+1}}(n) = g_{A_{i\xi}^j}(n) \quad \text{for all } n \in A_{i\xi}^{j+1} \setminus [1, \bar{k}_1].$$

To a condition $p_\xi \in X$ we can associate the $2l$-tuple

$$\langle A_{0\xi}^1, \ldots, A_{0\xi}^l, g_{A_{0\xi}^1}, \ldots, g_{A_{0\xi}^l} \rangle$$

in $\mathcal{P}(\mathbb{N})^l \times (\text{Fin}^{\mathbb{N}})^l$. Since this is a separable metric space, we can assume (by going to an uncountable subset of X) that the set of these $2l$-tuples is $\aleph_1$-*dense in itself*, i.e., that every open set which contains one of these $2l$-tuples contains uncountably many of them. We can also assume that the set of $2m$-tuples associated to elements of Y in an analogous way is an $\aleph_1$-dense in itself subset of $\mathcal{P}(\mathbb{N})^m \times (\text{Fin}^{\mathbb{N}})^m$. Finally, by Claim 4 there are ξ, η such that

$$\{A_{0\xi}^1, A_{1\eta}^1\} \in L_0([\bar{k}_1, \infty)).$$

Since the partition $L_0([\bar{k}_1, \infty))$ is open, there are uncountable sets of ξ's and η's for which this happens. By the choice of $\bar{k}_1$, for all these ξ, η and all i, j we also have

$$\{A_{0\xi}^i, A_{1\xi}^j\} \in L_0([\bar{k}_1, \infty)).$$

Since the working parts of these conditions are compatible, the conditions are compatible themselves. □

The poset $\mathcal{P}_{\omega_1}$ is defined as an amalgamation of $\aleph_1$ many copies of $\mathcal{P}$ as follows: a typical condition is

$$p = \langle I, k, s(\xi)(\xi \in I), F(\xi)(\xi \in I) \rangle,$$

where I is a finite subset of ω_1 and $p(\xi) = \langle k, s(\xi), F(\xi) \rangle$ is in $\mathcal{P}$ for all $\xi \in I$. An ordering is defined by letting $p \leq q$ if

($\mathcal{P}4$) $I^p \supseteq I^q$ and $p(\xi) \leq_{\mathcal{P}} q(\xi)$ for all $\xi \in I^q$, and
($\mathcal{P}5$) the sets $s^p(\xi) \setminus \{1, \ldots, k\}$ ($\xi \in I^q$) are pairwise disjoint.

CLAIM 6. *The poset $\mathcal{P}_{\omega_1}$ is ccc.*

PROOF. Let p_α ($\alpha < \omega_1$) be an uncountable subset of $\mathcal{P}_{\omega_1}$. We can assume sets $I^\alpha = I^{p_\alpha}$ from their first coordinates form a Δ-system with root $\bar{I}$. It follows from the definitions that p and q are compatible in $\mathcal{P}_{\omega_1}$ if and only if the conditions

$$\langle I, s^p_\xi, k^p_\xi, F^p_\xi : \xi \in I \rangle \quad \text{and} \quad \langle I, s^q_\xi, k^q_\xi, F^q_\xi : \xi \in I \rangle, \qquad \text{where } I = I^p \cap I^q,$$

are compatible. Therefore we can assume that

$$I^\alpha = \bar{I} = \{\xi_1, \ldots, \xi_l\}$$

for all p_α. Uniformizing further, we can assume that for every fixed $i \in \{1, \ldots, l\}$ the working parts of all conditions $p^\alpha(\xi_i)$ ($\alpha < \omega_1$) are equal, say

$$s^\alpha_{\xi_i} = \bar{s}_i \text{ and } k^\alpha_{\xi_i} = \bar{k}_i, \text{ for all } \alpha \text{ and } i \in \{1, \ldots, l\}.$$

We can, moreover, assume that sets $F^\alpha_{\xi_i} = F^{p_\alpha}_{\xi_i}$ from the third coordinates of $p^\alpha(\xi_i)$'s form a Δ-system with root F_{ξ_i}. Now observe that by the definition of $\mathcal{P}$ the conditions $p, q \in \mathcal{P}$ are compatible if and only if the conditions

$$\langle s^p, k^p, F^p \setminus F \rangle \quad \text{and} \quad \langle s^q, k^q, F^q \setminus F \rangle, \qquad \text{where } F = F^p \cap F^q,$$

are compatible. Therefore we can assume that the sets $F^\alpha_{\xi_i}$ ($\alpha < \omega_1$) are pairwise disjoint for every fixed $i \in \{1, \ldots, l\}$. We can apply Claim 4 to $p_\alpha(\xi_1)$ ($\alpha < \omega_1$) and get subsets X_1, Y_1 of ω_1 such that $p_\alpha(\xi_1)$ ($\alpha \in X_1$) and $p_\beta(\xi_1)$ ($\beta \in Y_1$) form an uncountable rectangle of compatible conditions. Let $\bar{k}'_1 > \bar{k}_1$ and $\bar{s}'_1 \subseteq [1, \bar{k}'_1]$ be such that

$$\langle \bar{s}'_1, \bar{k}'_1, F^\alpha_{\xi_1} \cup F^\beta_{\xi_1} \rangle$$

is a joint extension of $p_\alpha(\xi_1)$ and $p_\beta(\xi_1)$ for all $\alpha \in X_1$ and $\beta \in Y_1$. Now we can extend every $p_\alpha(\xi_2)$ ($\alpha \in X_1 \cup Y_1$) to

$$p'_\alpha(\xi_2) = \langle \bar{s}_2, \bar{k}'_2, F^\alpha(\xi_2) \rangle$$

so that $\bar{k}'_2 > \bar{k}_1$. Another application of Claim 4 gives sets $X_2 \subseteq X_1$ and $Y_2 \subseteq Y_1$ such that $p'_\alpha(\xi_2)$ ($\alpha \in X_1$) and $p'_\beta(\xi_2)$ ($\beta \in Y_2$) is an uncountable rectangle of compatible conditions. Note that, if

$$\langle \bar{s}'_2, \bar{k}''_2, F^\alpha_{\xi_2} \cup F^\beta_{\xi_2} \rangle$$

is a joint extension of $p'_\alpha(\xi_2)$ and $p'_\beta(\xi)$, then by our choice of $\bar{k}'_2$ the condition ($\mathcal{P}7$) is satisfied between $\bar{s}'_1$ and $\bar{s}'_2$. By continuing this construction for $i = 3, \ldots, l$, we get an uncountable rectangle of compatible conditions in $\mathcal{P}_{\omega_1}$. □

Let G be a sufficiently generic filter of $\mathcal{P}_{\omega_1}$ and for $\xi < \omega_1$ define
$$D_\xi = \bigcup_{p \in G} s_\xi^p \quad \text{and} \quad \mathcal{X}_\xi = \bigcup_{p \in G} F_\xi^p.$$

Then the sets D_ξ are pairwise almost disjoint and $\mathcal{X}_\xi$ is $L_0(D_\xi)$-homogeneous for all ξ. By ccc-ness of $\mathcal{P}_{\omega_1}$ there must be a condition p in it which forces there are uncountably many ξ's with uncountable $\mathcal{X}_\xi$'s. Then we can choose a suitable family of $\aleph_1$ many dense open subsets of $\mathcal{P}_{\omega_1}$ so that if G is a filter of $\mathcal{P}_{\omega_1}$ containing p and intersecting all these dense open sets, then for uncountably many ξ the set $\mathcal{X}_\xi$ is uncountable. By Lemma 3.8.5, for such ξ the set D_ξ is not in $\mathcal{J}_1$ and therefore not in $\mathcal{J}_{\text{cont}}$. But this is in contradiction with Lemma 3.8.3, and this concludes the proof of Theorem 3.8.1. $\square$

3.9. Almost liftings of embeddings into $\mathcal{P}(\mathbb{N}^2)/\operatorname{Fin} \times \emptyset$

This section is devoted to a proof of Theorem 3.3.6, which will be of crucial importance in topological applications given in Chapter 4. We shall first prove its special case for the ideal $\operatorname{Fin} \times \emptyset$, but we will need a lemma first.

LEMMA 3.9.1. *Assume $\mathfrak{b} > \aleph_1$. If $\Phi \colon \mathcal{P}(\mathbb{N})/\operatorname{Fin} \to \mathcal{P}(\mathbb{N}^2)/\operatorname{Fin} \times \emptyset$ is a homomorphism, and $h \colon \mathbb{N}^2 \to \mathbb{N}$ is such that $\Phi_{h \restriction \Gamma_f}$ is an almost lifting of Φ^{Γ_f} for every $f \in \mathbb{N}^{\mathbb{N}}$ then Φ_h is an almost lifting of Φ.*

PROOF. Let $\mathcal{J} = \{A : \Phi_*(A) =^{\operatorname{Fin} \times \emptyset} \Phi_h(A)\}$. Assume $\mathcal{J}$ is not ccc over Fin, and let $\mathcal{A}$ be a family of $\aleph_1$ many pairwise almost disjoint, $\mathcal{J}$-positive sets. For $A \in \mathcal{A}$ let f_A in $\mathbb{N}^{\mathbb{N}}$ be such that the set
$$(\Phi_*(A) \Delta \Phi_h(A)) \cap \Gamma_{f_A} = \Phi_*(A) \Delta \Phi_{h \restriction \Gamma_{f_A}}(A)$$
is $\mathcal{J}$-positive (equivalently, infinite). By Theorem 2.2.6 find g in $\mathbb{N}^{\mathbb{N}}$ which eventually dominates all f_A ($A \in \mathcal{A}$). But $\Phi_{h \restriction \Gamma_g}$ is, by the assumption, an almost lifting of Φ^{Γ_g}, and therefore for all but countably many $A \in \mathcal{A}$ the set
$$(\Phi_*(A) \Delta \Phi_h(A)) \cap \Gamma_{g_A} \supseteq (\Phi_*(A) \Delta \Phi_h(A)) \cap \Gamma_{f_A}$$
is finite; contradiction. $\square$

THEOREM 3.9.2. *Assume OCA and MA. If $\Phi \colon \mathcal{P}(\mathbb{N})/\operatorname{Fin} \to \mathcal{P}(\mathbb{N}^2)/\operatorname{Fin} \times \emptyset$ is a homomorphism, then Φ can be decomposed as $\Phi_1 \oplus \Phi_2$ so that Φ_1 has a continuous lifting, while the kernel of Φ_2 is ccc over Fin. Equivalently, a continuous lifting of Φ_1 serves as a lifting of Φ on an ideal which is ccc over Fin.*

PROOF. This proof is similar to the proof that $\operatorname{Fin} \times \emptyset$ has the Radon–Nikodym property (Theorem 1.5.2). Assume $\Phi \colon \mathcal{P}(\mathbb{N})/\operatorname{Fin} \to \mathcal{P}(\mathbb{N}^2)/\operatorname{Fin} \times \emptyset$ is a homomorphism. For $f \colon \mathbb{N} \to \mathbb{N}$ let $\Phi_f \colon \mathcal{P}(\mathbb{N}) \to \mathcal{P}(\Gamma_f)$ be defined by
$$\Phi_f(A) = \Phi_*(A) \cap \Gamma_f.$$
Then Φ_f is a lifting of a Baire homomorphism $\Phi^{\Gamma_f} \colon \mathcal{P}(\mathbb{N})/\operatorname{Fin} \to \mathcal{P}(\Gamma_f)/\operatorname{Fin}$. Since the restriction of $\operatorname{Fin} \times \emptyset$ to the set Γ_f is equal to the ideal Fin on Γ_f, by Corollary 3.8.2 there is an $E_f \subseteq \Gamma_f$ and a finite-to-one $h_g^0 \colon E_f \to \mathbb{N}$ such that the ideal
$$\mathcal{J}_f = \{A : \Phi_f(A) =^* \Phi_{h_f^0}(A)\}$$
is ccc over Fin. To simplify the notation, let us define
$$h_f \colon \Gamma_f \to \mathbb{N} \cup \{\emptyset\}$$

by letting h_f coincide with h_f^0 on E_f and be equal to $\emptyset$ otherwise. Note that Φ_{h_f} is still a completely additive almost lifting of Φ^{Γ_f}, and that $\{h_f \colon f \in \mathbb{N}^{\mathbb{N}}\}$ is a coherent family indexed by $\mathbb{N}^{\mathbb{N}}$ (by Lemma 3.8.4, or rather by its proof).

By Theorem 2.2.7, pick a function $h \colon \mathbb{N}^2 \to \mathbb{N} \cup \{\emptyset\}$ which trivializes this family, i.e., such that
$$h \restriction \Gamma_f =^* h_f$$
for all f. Then clearly $\Phi_{h \restriction \Gamma_f}$ is a completely additive almost lifting of Φ^{Γ_f} for every f, and by Lemma 3.9.1 Φ_h is a completely additive almost lifting of Φ. $\square$

PROOF OF THEOREM 3.3.6. By Proposition 1.2.8, every countably generated ideal is isomorphic (see Definition 1.2.7) to either Fin or Fin $\times \emptyset$. Therefore the result follows from Theorem 3.9.2 and Theorem 3.8.1. $\square$

3.10. Nonmeager hereditary sets

In this short section we study a largeness property of subsets of $\mathcal{P}(\mathbb{N})$ which will play a key role in the proof of the OCA lifting theorem (see §3.11). Recall that a set of reals is *meager* if it can be covered by a countable union of nowhere dense sets. A set of reals is *comeager* if its complement is meager. If $\mathcal{X}$ is a family of subsets of $\mathbb{N}$ by $\hat{\mathcal{X}}$ we denote the *downwards closure* (or, the *hereditary closure*) of $\mathcal{X}$, i.e., $\hat{\mathcal{X}} = \bigcup_{a \in X} \mathcal{P}(a)$, and we say that $\mathcal{X}$ is *hereditary* if $\mathcal{X} = \hat{\mathcal{X}}$.

THEOREM 3.10.1. (a) *If $\mathcal{X}$ is a hereditary subset of $\mathcal{P}(\mathbb{N})$, then it is non-meager if and only if for every sequence s_i of disjoint finite sets of integers there is an infinite $a \subseteq \mathbb{N}$ such that $\bigcup_{i \in a} s_i \in \mathcal{X}$.*
(b) *The family of nonmeager hereditary subsets of $\mathcal{P}(\mathbb{N})$ is closed under taking finite intersections of its elements.*
(c) *The family of nonmeager hereditary subsets of $\mathcal{P}(\mathbb{N})$ which are closed under finite changes of their elements is closed under taking countable intersections.*

PROOF. (a) This proof is similar to those in Jalali–Naini ([58]) and Talagrand ([121]), where a slightly weaker statement (Corollary 3.10.2 below) was proved. For $m \in \mathbb{N}$ and $s \subseteq [1, m]$ define
$$[s; m] = \{a \subseteq \mathbb{N} : a \cap [1, m) = s\}.$$
Assume first $\mathcal{X}$ is meager, therefore it is covered by a countable union $\bigcup_{m=1}^{\infty} F_m$ of closed nowhere dense sets. Recursively find $1 \leq n_1 < n_2 < \ldots$ and $u_i \subseteq [n_i, n_{i+1})$ so that
$$[t \cup u_i; n_{i+1}] \cap \bigcup_{m \leq i} F_m = \emptyset$$
for all $t \subseteq [1, n_i)$. This is possible because each F_m is nowhere dense. Then any infinite union of $s_i = [n_i, n_{i+1})$ has a subset which is equal to an infinite union of the u_i's and therefore avoids all F_m's. On the other hand, if there is a sequence s_i such that no infinite union of its elements lies in $\mathcal{X}$, then for all i the set
$$U_i = \{A : s_n \subseteq A \text{ for some } n \geq i\}.$$
is a dense open subset of $\mathcal{P}(\mathbb{N})$, and $\bigcap_{i=1}^{\infty} U_i$ is a dense G_δ set. But this is the set of all subsets A of $\mathbb{N}$ which include infinitely many of the s_i's, and $\mathcal{X}$ is included in the complement of $\bigcap_{i=1}^{\infty} U_i$.

(b) Let $\mathcal{H}_1, \ldots, \mathcal{H}_k$ be nonmeager and hereditary. Since the set $\bigcap_{i=1}^k \mathcal{H}_i$ is clearly hereditary, we need only to prove that it is nonmeager. Let s_i ($i \in \mathbb{N}$) be a sequence of pairwise disjoint subsets of $\mathbb{N}$. By (a), we can find infinite sets $A_1 \supseteq A_2 \supseteq \cdots \supseteq A_k$ such that
$$\bigcup_{j \in A_i} s_j \in \mathcal{H}_i$$
for $i = 1, \ldots, k$. In particular, $\bigcup_{j \in A_k} s_j$ is in $\bigcap_{i=1}^k \mathcal{H}_i$, and therefore by (a) this set is nonmeager.

(c) Assume $\bigcap_i \mathcal{X}_i$ is meager, and let $\{s_k\}$ be as guaranteed by (a). Using (a), recursively pick infinite sets $\mathbb{N} = A_1 \supseteq A_2 \supseteq \ldots$ such that $\bigcup_{k \in A_i} s_k \in \mathcal{X}_i$ and $n_i = \min A_{i+1} > \min A_i$ for all i. Then $\bigcup_i s_{n_i}$ is in $\bigcap_i \mathcal{X}_i$—a contradiction. $\square$

COROLLARY 3.10.2 (Jalali–Naini, Talagrand). *An ideal $\mathcal{I}$ which includes* Fin *has the property of Baire if and only if it is meager if and only if* Fin $\leq_{\mathrm{RB}} \mathcal{I}$. $\square$

Let us note that the above proof also gives the following well-known characterization of comeager subsets of $\mathcal{P}(\mathbb{N})$, which we state for future reference.

LEMMA 3.10.3. *A subset $\mathcal{X}$ of $\mathcal{P}(\mathbb{N})$ is comeager if and only if there is a sequence $0 = n_0 < n_1 < \ldots$ of naturals and $s_i \subseteq [n_i, n_{i+1})$ (for $i \geq 0$) such that $\mathcal{X}$ includes the set $\{A \subseteq \mathbb{N} : A \cap [n_i, n_{i+1}) = s_i$ for infinitely many $i\}$.* $\square$

Nonmeager, hereditary sets closed under finite changes of their elements are also called *groupwise dense*, and the minimal cardinality of the family of groupwise dense sets whose intersection is not groupwise dense is called *groupwise density* (note that Theorem 3.10.1 (c) says that this cardinal is uncountable; of course, this fact is well-known). These notions were introduced in [**9**] in order to provide a succinct combinatorial formulation of a statement with strong implications to rather diverse mathematical structures (see [**8**]).

3.11. Approximate homomorphisms; more on stabilizers

In this section we present two results which do not require any additional set-theoretic axioms and which will be used in the proof of the OCA lifting theorem, Theorem 3.3.5. One of them is Theorem 3.11.3, a technical statement used to uniformize a sequence of Borel-measurable partial liftings of a given homomorphism. Another one is Lemma 3.11.6, which makes the notion of an "almost lifting" more useful, by making the results of Chapter 1 (in particular, the Radon-Nikodym Property) applicable to almost liftings. The proofs make use of the stabilizers (see §1.5 and Definition 3.11.5) in a situation where we do not necessarily have a Baire-measurable lifting.

Let $\mathcal{I}$ be an analytic P-ideal and let φ be a lower semicontinuous submeasure (see §1.2) supported by $\mathbb{N}$ such that (recall that $\|A\|_\varphi = \limsup_k \varphi(A \setminus [1, k))$) we have
$$\mathcal{I} = \mathrm{Exh}(\varphi) = \{A : \|A\|_\varphi = 0\}.$$
The existence of such φ is guaranteed by Solecki's theorem (Theorem 1.2.5). Possibly by replacing φ with $\min(1, \varphi)$ we can assume φ, and therefore $\|A\|_\varphi$, is finite. Note that this does not change the ideal. Recall that
$$d_\varphi(A, B) = \|A \Delta B\|_\varphi$$

is a complete metric on $\mathcal{P}(\mathbb{N})/\mathcal{I}$ (Lemma 1.3.3). Some results of this and the following section could be regarded as results about the complete normed group $\langle \mathcal{P}(\mathbb{N})/\mathcal{I}, \|\cdot\|_\varphi \rangle$. In fact, we shall be interested in the space of all homomorphisms $\Phi\colon \mathcal{P}(\mathbb{N})/\operatorname{Fin} \to \mathcal{P}(\mathbb{N})/\mathcal{I}$ equipped with the natural uniform metric induced by d_φ.

DEFINITION 3.11.1. Assume φ is a lower semicontinuous submeasure, $\varepsilon > 0$, $\mathcal{X} \subseteq \mathcal{P}(\mathbb{N})$, and $F\colon \mathcal{P}(\mathbb{N}) \to \mathcal{P}(\mathbb{N})$. A a map $\Gamma\colon \mathcal{P}(\mathbb{N}) \to \mathcal{P}(\mathbb{N})$ is an ε-*approximation of* F *on* $\mathcal{X}$ if
$$\|\Gamma(A) \Delta F(A)\|_\varphi \leq \varepsilon$$
for all $A \in \mathcal{X}$. We say that a homomorphism $\Phi\colon \mathcal{P}(\mathbb{N})/\operatorname{Fin} \to \mathcal{P}(\mathbb{N})/\operatorname{Exh}(\varphi)$ is ε-*tame* on $\mathcal{H}$ if its lifting Φ_* has a Baire ε-approximation on $\mathcal{H}$.

Note that the ε-tameness of Φ depends on the choice of φ. It can, however, be shown that the truth of "Φ is ε-tame for all $\varepsilon > 0$" does not depend on the choice of φ (see [**47**]).

We shall use the following classical theorem (see, e.g., [**74**, 18.A]).

THEOREM 3.11.2 (Jankov, von Neumann). *If* $R \subseteq \mathcal{P}(\mathbb{N}) \times \mathcal{P}(\mathbb{N})$ *is analytic and* $\mathcal{X} = \{a : \langle a, b \rangle \in R \text{ for some } b\}$, *then there is a function* $f\colon \mathcal{X} \to \mathcal{P}(\mathbb{N})$ *such that the graph of* f *is included in* R *and the* f-*preimage of every open subset of* $\mathcal{P}(\mathbb{N})$ *belongs to the* σ-*algebra generated by the analytic subsets of* $\mathcal{P}(\mathbb{N})$. $\square$

The following theorem gives the conditions which imply that Φ satisfies a slightly weaker statement than the conclusion of Theorem 3.3.5, but without assuming any additional set-theoretic axioms.

THEOREM 3.11.3. *Assume* $\mathcal{I} = \operatorname{Exh}(\varphi)$ *is an analytic P-ideal and*
$$\Phi\colon \mathcal{P}(\mathbb{N})/\operatorname{Fin} \to \mathcal{P}(\mathbb{N})/\mathcal{I}$$
is a homomorphism. If there is a nonmeager ideal $\mathcal{J}$ *which for every* $\varepsilon > 0$ *can be covered by countably many hereditary sets such that* Φ *is* ε-*tame on each one of them, then* Φ *can be decomposed as* $\Phi_1 \oplus \Phi_2$ *so that* Φ_1 *has a Baire-measurable lifting and* $\ker(\Phi_2)$ *includes* $\mathcal{J}$, *in particular it is nonmeager.*

Before we give a proof of this theorem, let us note that the requirement that the sets on which Φ is ε-tame be hereditary may seem to be artificial. However, dropping this requirement would make Theorem 3.11.3 false: If Φ is a homomorphism given in Example 3.2.1, then $\mathcal{P}(\mathbb{N})$ can be split into two sets ($\mathcal{U}$ and its dual, $\mathcal{U}^*$) such that Φ has a continuous lifting on each one of these two sets, but nevertheless Φ has no Baire lifting. This also shows a difference between Theorem 3.11.3 and Theorem 2 of [**142**], which states that if an *automorphism* of $\mathcal{P}(\mathbb{N})/\operatorname{Fin}$ has a σ-Borel lifting, then it is trivial.

LEMMA 3.11.4. *Assume* $\mathcal{I} = \operatorname{Exh}(\varphi)$, Φ, *and* $\mathcal{J}$ *are like in Theorem 3.11.3, but that* $\mathcal{J}$ *can only be covered by countably many hereditary sets such that* Φ *is* ε-*tame on each one of them for some fixed* $\varepsilon > 0$. *Then* Φ_* *has a continuous* 28ε-*approximation on* $\mathcal{J}$.

PROOF. Let Θ_i ($i \in \mathbb{N}$) be a sequence of Baire-measurable functions from $\mathcal{P}(\mathbb{N})$ to $\mathcal{P}(\mathbb{N})$, and for each i let $\mathcal{H}_i$ be a hereditary set on which Θ_i is an ε-approximation of Φ_*, so that $\mathcal{J} = \bigcup_{i=1}^\infty \mathcal{H}_i$.

Recall that for $m \in \mathbb{N}$ and $s \subseteq [1, m]$ we write
$$[s; m] = \{a \subseteq \mathbb{N} : a \cap [1, m] = s\}.$$

CLAIM 1. *We can assume that functions Θ_i are continuous 2ε-approximations of Φ_* on $\mathcal{H}_i$.*

PROOF. To this end, pick a dense G_δ set $G \subseteq \mathcal{P}(\mathbb{N})$ such that $\Theta_i \upharpoonright G$ is continuous for all i and that $G \cap \mathcal{H}_i$ is empty whenever $\mathcal{H}_i$ is meager. Now find an increasing sequence $\{n_i\}$ and $s_i \subseteq [n_i, n_{i+1})$ for all i so that G includes

$$\{a \subseteq \mathbb{N} : a \cap [n_i, n_{i+1}) = s_i \text{ for infinitely many } i \in \mathbb{N}\}$$

(see Lemma 3.10.3). Since $\mathcal{J}$ is nonmeager, we have $\bigcup_{k \in C} s_k \in \mathcal{J}$ for some infinite $C = \{k(1), k(2), \dots\}$. For $\varepsilon = 0, 1$ let

$$a_\varepsilon = \bigcup_{j=0}^{\infty} [n_{k(2j+\varepsilon)}, n_{k(2j+\varepsilon+1)}) \quad \text{and} \quad c_\varepsilon = \bigcup_{j=0}^{\infty} s_{k(2j+\varepsilon)}.$$

For $i, j \in \mathbb{N}$ define a function Θ_{ij} by

$$\Theta_{ij0}(a) = \Theta_i((a \cap a_0) \cup c_1) \cap \Phi_*(a_0),$$
$$\Theta_{ij1}(a) = \Theta_j((a \cap a_1) \cup c_0) \setminus \Phi_*(a_0),$$
$$\Theta_{ij}(a) = \Theta_{ij0}(a) \cup \Theta_{ij1}(a).$$

Therefore Θ_{ij} is an amalgamation of Θ_{ij0} and Θ_{ij1}. Moreover, Θ_{ij} is continuous, and for every $a \in \mathcal{J}$ we have i, j such that a belongs to the set

$$\mathcal{H}_{ij} = \{a : (a \cap a_0) \cup c_1 \in \mathcal{H}_i \quad \text{and} \quad (a \cap a_1) \cup c_0 \in \mathcal{H}_j\}.$$

(This is because both c_0 and c_1 belong to $\mathcal{J}$.) Each $\mathcal{H}_{ij}$ is clearly hereditary. For $a \in \mathcal{H}_{ij}$ we have:

$$\Theta_{ij}(a) \Delta \Phi_*(a) = ((\Theta_i((a \cap a_0) \cup c_1) \cap \Phi_*(a_0))$$
$$\cup (\Theta_j((a \cap a_1) \cup c_0) \setminus \Phi_*(a_0))) \Delta \Phi_*(a)$$
$$=^{\mathcal{I}} ((\Theta_i((a \cap a_0) \cup c_1) \Delta \Phi_*(a)) \cap \Phi_*(a_0))$$
$$\cup ((\Theta_j((a \cap a_1) \cup c_0) \Delta \Phi_*(a)) \setminus \Phi_*(a_0))$$
$$=^{\mathcal{I}} ((\Theta_i((a \cap a_0) \cup c_1) \Delta \Phi_*((a \cap a_0) \cup c_1)) \cap \Phi_*(a_0))$$
$$\cup ((\Theta_j((a \cap a_1) \cup c_0) \Delta \Phi_*((a \cap a_1) \cup c_0)) \setminus \Phi_*(a_0))$$
$$=^{\mathcal{I}} (\Theta_i((a \cap a_0) \cup c_1) \Delta \Phi_*((a \cap a_0) \cup c_1))$$
$$\cup (\Theta_j((a \cap a_1) \cup c_0) \Delta \Phi_*((a \cap a_1) \cup c_0)),$$

and therefore $\|\Theta_{ij}(a) \Delta \Phi_*(a)\|_\varphi \leq 2\varepsilon$. Hence each Θ_{ij} is a continuous 2ε-approximation to Φ_* on $\mathcal{H}_{ij}$. By reindexing $\mathcal{H}_{ij}$ ($i, j \in \mathbb{N}$) we can assume that each Θ_i is a continuous 2ε-approximation of Φ_* on $\mathcal{H}_i$ and that each $\mathcal{H}_i$ is hereditary. □

Recall that a set is *everywhere nonmeager* if its intersection with any nonempty open set is nonmeager.

CLAIM 2. *We can furthermore assume each $\mathcal{H}_i$ is either meager or everywhere nonmeager.*

PROOF. Assume $\mathcal{H}_i$ is nonmeager. Let $[u]$ be a basic clopen subset of $\mathcal{P}(\mathbb{N})$ such that $[v] \cap \mathcal{H}_i$ is nonmeager for every $[v] \subseteq [u]$. Since $\mathcal{H}_i$ is hereditary, we can assume that $[u] = [\emptyset; p]$ for some $p \in \mathbb{N}$. Let

$$\Theta'_i(a) = \Theta_i(a \setminus p)$$

Then clearly the set
$$\mathcal{H}'_i = \{a : a \setminus p \in \mathcal{H}_i\}$$
is everywhere nonmeager, hereditary, includes $\mathcal{H}_i$, and $\|\Theta'_i(a)\Delta\Phi_*(a))\|_\varphi \leq 2\varepsilon$, for all $a \in \mathcal{H}'_i$.

By replacing Θ_i with Θ'_i and $\mathcal{H}_i$ with $\mathcal{H}'_i$, we get the desired claim. $\square$

We need to redefine the notion of stabilizer given in Definition 1.5.1.

DEFINITION 3.11.5. Assume $\Omega \colon \mathcal{P}(\mathbb{N}) \to \mathcal{P}(\mathbb{N})$. For $n < n'$ and $\varepsilon > 0$ we say that $c \subseteq [n, n')$ is an (n, n')-ε-stabilizer for Ω if there exists $k \in (n, n')$ such that for all $s, t \subseteq [1, n)$ and all $X, Y \subseteq [n', \infty)$:
(S'1) $\min(\Omega(s \cup c \cup X)\Delta\Omega(s \cup c \cup Y)) \geq k$,
(S'2) $\varphi((\Omega(s \cup c \cup X)\Delta\Omega(t \cup c \cup X) \setminus [1, k))) \leq \varepsilon$.

CLAIM 3. *Assume that $\Omega \colon \mathcal{P}(\mathbb{N}) \to \mathcal{P}(\mathbb{N})$ is continuous and there is an everywhere nonmeager, hereditary set $\mathcal{H}$ such that for all $A, B \in \mathcal{H}$ satisfying $A\Delta B \in$ Fin we have*
$$\|\Omega(A)\Delta\Omega(B)\|_\varphi \leq \delta.$$
Then for every n and all large enough n' there is an (n, n')-δ-stabilizer for Ω.

PROOF. Assume this fails for some n. By the uniform continuity of Ω for all large enough $n' > n$ the condition (S'1) is satisfied for all $c \subseteq [n, n')$. Let us fix a $\bar{k}$ such that this is true for all $n' > \bar{k}$. Note that (S'2) fails for all $n' > \bar{k}$. By the continuity of Ω and lower semicontinuity of φ the set X which witnesses that (S'2) fails can always be chosen to be finite. We shall choose a sequence $n = n_1 < \bar{k} = k_1 < n_2 < k_2 < n_3 < \ldots$, and $w_i \subseteq [n_i, n_{i+1})$ so that for every $l \in \mathbb{N}$ and $w \subseteq [1, l]$ the parameters
$$c = \bigcup_{i \in w} w_i, \qquad k = k_l, \qquad n' = n_{l+1}$$
satisfy (S'1) for all u, X and Y, but $X = w_{l+1}$ witnesses that (S'2) fails for some $s, t \subseteq [1, n)$. To construct w_{l+1}, we need a few auxiliary objects. Let a_j ($j = 1, \ldots, 2^l$) enumerate all subsets of $[1, l-1]$. We shall find:
(1) $n_l = m_1 < m_2 < \cdots < m_{2^l + 1}$,
(2) $X_j \subseteq [m_j, m_{j+1})$ for $j \in \{1, \ldots, 2^l\}$,

so that $w_l = \bigcup_{j=1}^{2^l} X_j$ is as required. The construction proceeds as follows. Since $\bigcup_{i \in a_1} w_i$ is not an (n, n_{l+1})-δ-stabilizer with k_l as a witness, we can find $s_1, t_1 \subseteq [1, n)$ and $X_1 \subseteq [n_l, \infty)$ witnessing that (S'2) fails. By the uniform continuity of Ω we can assume X_1 is finite. Pick an arbitrary $m_2 > \max X_1$. Since $s = \bigcup_{i \in a_2} w_i \cup X_1$ is not an (n, m_2)-δ-stabilizer with k_l as a witness, we can find $s_2, t_2 \subseteq [1, n)$ and a finite $X_2 \subseteq [m_2, \infty)$ witnessing that (S'2) fails. In this manner we construct $w_l = \bigcup_{j=1}^{2^l} X_1$ such that for all $j < 2^l$ set X_{l+1} witnesses that $\bigcup_{i \in a_j} w_i \cup \bigcup_{p=1}^{j} X_p$ is not an (n, m_{j+1})-δ-stabilizer with k_l as a witness. It is easy to see that w_l is as required.

The recursive construction of sequences $\{n_i\}$, $\{k_i\}$ and $\{w_i\}$ is now obvious—if we have $n_1 < k_1 < \cdots < n_l$, then find $k_l > n_l$ and w_l in the manner described above and pick n_{l+1} to be larger than $\max(w_l)$. Assume these sequences are chosen. For $u \subseteq [1, n)$ let
$$\mathcal{H}_u = \{B \subseteq [n, \infty) : u \cup B \in \mathcal{H}\}.$$

Then by (b) of Theorem 3.10.1 the set $\bigcap_{u \subseteq [1,n)} \mathcal{H}_u$ is nonmeager and we can find an infinite $C \subseteq \mathbb{N}$ such that $B = \bigcup_{j \in C} w_j$ is in this set. We may assume that some fixed $\bar{u}, \bar{v} \subseteq [1, n)$ appear as witnesses that (S'2) fails for $s = B \cap [n_0, n_p)$, $k = n_p$ and $x = w_p$ for infinitely many $p \in C$. Therefore

$$\|\Omega(\bar{u} \cup B) \Delta \Omega(\bar{v} \cup B)\|_\varphi > \delta,$$

contradicting the assumption on Ω and $\mathcal{H}$, for both $\bar{u} \cup B$ and $\bar{v} \cup B$ are in $\mathcal{H}$. □

Continuing the proof of Lemma 3.11.4, note that if Θ_i is continuous and a 2ε-approximation to a homomorphism on some set $\mathcal{H}_i$ which is hereditary and everywhere nonmeager, then by the triangle inequality for $\|\cdot\|_\varphi$, the function Θ_i satisfies the assumptions of Claim 3 with $\delta = 4\varepsilon$. Therefore we can construct $m_1 < k_1 < m_2 < k_2 < \ldots$ and $s_i \subseteq [m_i, m_{i+1})$ so that for all i:

(3) s_i is an (m_i, m_{i+1})-4ε-stabilizer for Θ_j with the witness k_i for all $j \leq i$.

By Theorem 3.10.1 (a), find a $b \in \mathcal{J}$ which is equal to some infinite union $\bigcup_{i \in c} s_i$, say $c = \{l(j)\}$ and $l(1) < l(2) < l(3) < \ldots$. Since $\mathcal{J} \restriction (\mathbb{N} \setminus b)$ is a nonmeager ideal, we can find an $\bar{m}$ such that

$$\{a \subseteq \mathbb{N} \setminus b : b \cup a \in \mathcal{H}_{\bar{m}}\}$$

is a nonmeager subset of $\mathcal{P}(\mathbb{N} \setminus b)$. We can assume $\bar{m} < l(1)$. Note that

(5) $s_{l(i)}$ is an $(m_{l(i)}, m_{l(i+1)})$-δ-stabilizer for Θ_j for all $j \leq i$.

Now we let $\Upsilon \colon \mathcal{P}(\mathbb{N}) \to \mathcal{P}(\mathbb{N})$ be the stabilization of $\Theta_{\bar{m}}$ by $\{m_{l(j)}\}$ and $\{s_{l(j)}\}$ ($j \in \mathbb{N}$), namely let

$$a_0 = \bigcup_{j=0}^\infty [m_{l(2j)}, m_{l(2j+1)}), \qquad c_0 = \bigcup_{j=0}^\infty s_{l(2j)},$$

$$a_1 = \bigcup_{j=0}^\infty [m_{l(2j+1)}, m_{l(2j+2)}), \qquad c_1 = \bigcup_{j=0}^\infty s_{l(2j+1)},$$

and let

$$\Upsilon_0(a) = \Theta_{\bar{m}}((a \cap a_0) \cup c_1) \cap \Phi_*(a_0),$$
$$\Upsilon_1(a) = \Theta_{\bar{m}}((a \cap a_1) \cup c_0) \setminus \Phi_*(a_0),$$
$$\Upsilon(a) = \Upsilon_0(a) \cup \Upsilon_1(a).$$

We claim that the continuous function Υ is a 28ε-approximation of Φ_* on $\mathcal{J}$. Assume not, then either Υ_0 is not a 14ε-approximation of Φ_* on $\mathcal{J} \restriction a_0$ or Υ_1 is not a 14ε-approximation of Φ_* on $\mathcal{J} \restriction a_1$. Assume the first possibility applies, and let $d \in \mathcal{J} \restriction a_0$ be such that

$$\|\Phi_*(d) \Delta \Upsilon_0(d)\|_\varphi = \|\Phi_*(d) \Delta (\Theta_{\bar{m}}(d \cup c_1) \cap \Phi_*(a_0))\|_\varphi > 14\varepsilon.$$

Denote the set $d \cup c_1$ by d'. Since $\Theta_{\bar{m}}(d') \Delta \Phi_*(d') \supseteq^{\mathcal{I}} \Phi_*(d) \Delta (\Theta_{\bar{m}}(d \cup c_1) \cap \Phi_*(a_0))$, we have

(6) $\|\Theta_{\bar{m}}(d') \Delta \Phi_*(d')\|_\varphi > 14\varepsilon.$

Let $\bar{n}$ be such that

(7) $\|\Phi_*(d') \Delta \Theta_{\bar{n}}(d')\|_\varphi \leq 2\varepsilon$, in particular $d' \in \mathcal{H}_{\bar{n}}$.

Note that for every j there is a large enough $j' > j$ such that

(8) $\varphi((\Theta_{\bar{m}}(d') \Delta \Phi_*(d')) \cap [j, j')) \geq 14\varepsilon.$

Therefore we can find a subsequence $\{l'(i)\}$ of $\{l(2i+1)\}$ such that

(9) $\varphi((\Theta_{\bar{m}}(d')\Delta\Phi_*(d')) \cap [k_{l'(i)}, k_{l'(i+1)-1})) \geq 14\varepsilon$ for all i.

Let us define

(10) $I_i = [m_{l'(i)}, m_{l'(i+1)})$,

(11) $\varphi_i(X) = \varphi(X \cap [k_{l'(i)}, k_{l'(i+1)-1}))$,

so that (9) becomes

(12) $\varphi_i(\Theta_{\bar{m}}(d')\Delta\Phi_*(d')) \geq 14\varepsilon$, for all i.

A careful reader may recognize that a version of Claim 4 below was the main idea behind the use of stabilizers when we were transforming an arbitrary continuous lifting into asymptotically additive one in Theorem 1.5.2.

CLAIM 4. *If $a \cap I_i = b \cap I_i$, $j \leq i$ and a, b and $c = (a \cap \max I_i) \cup (b \setminus \max I_i)$ are in $\mathcal{H}_j$, then*
$$\varphi_i(\Theta_j(a)\Delta\Theta_j(b)) \leq 4\varepsilon.$$

PROOF. Note that
$$(\Theta_j(a)\Delta\Theta_j(c)) \cap [k_{l'(i)}, k_{l'(i+1)-1}] = \emptyset,$$
$$\varphi_i(\Theta_j(c)\Delta\Theta_j(b)) \leq 4\varepsilon,$$

by (S'1) and (S'2), respectively, since $(a\Delta c) \cap \max I_i = \emptyset$ and $c\Delta b \subseteq [1, \min I_i)$. Recalling that $\varphi_i(X) = \varphi(X \cap [k_{l'(i)}, k_{l'(i+1)-1}))$, we have
$$\varphi_i(\Theta_j(a)\Delta\Theta_j(b)) \leq \varphi_i(\Theta_j(a)\Delta\Theta_j(c)) + \varphi_i(\Theta_j(c)\Delta\Theta_j(b)) \leq 4\varepsilon,$$
completing the proof. □

By Theorem 3.10.1 (b) and (a), find an infinite $C \subseteq \mathbb{N}$ such that

(13) $d'' = \bigcup_{i \in C}(d' \cap I_i) \in \mathcal{H}_{\bar{m}}$.

(Also note that $d'' \subseteq d' \in \mathcal{H}_{\bar{n}}$, and therefore $d'' \in \mathcal{H}_{\bar{n}}$ by hereditarity.) Let i be a large enough element of C. Then since by everywhere nonmeagerness and hereditarity all finite sets are in $\mathcal{H}_{\bar{m}}$, we have

(14) $d'', d' \cap [1, m_{l'(i)}), d'' \cap [1, m_{l'(i)})$ are in $\mathcal{H}_{\bar{m}}$

and since $d' \in \mathcal{H}_{\bar{n}}$ we also have

(15) $d'' \subseteq d' \in \mathcal{H}_{\bar{n}}$,

and therefore (we need i to be "large enough" in (18), (19) and (21) below):

(16) $\varphi_i(\Theta_{\bar{m}}(d')\Delta\Theta_{\bar{m}}(d' \cap [1, m_{l'(i)+1})) = 0$ (since this set is, by (S'1), empty),

(17) $\varphi_i(\Theta_{\bar{m}}(d' \cap [1, m_{l'(i)+1}))\Delta\Theta_{\bar{m}}(d'')) \leq 4\varepsilon$ (by $d' \cap I_i = d'' \cap I_i$, (14) this follows from Claim 4),

(18) $\varphi_i(\Theta_{\bar{m}}(d'')\Delta\Phi_*(d'')) \leq 2\varepsilon$ (since $d'' \in \mathcal{H}_{\bar{m}}$, $\|\cdot\|_\varphi$ of this set is $\leq 2\varepsilon$),

(19) $\varphi_i(\Phi_*(d'')\Delta\Theta_{\bar{n}}(d'')) \leq 2\varepsilon$ (since $d'' \in \mathcal{H}_{\bar{n}}$, $\|\cdot\|_\varphi$ of this set is $\leq 2\varepsilon$),

(20) $\varphi_i(\Theta_{\bar{n}}(d'')\Delta\Theta_{\bar{n}}(d')) \leq 4\varepsilon$ (by $d' \cap I_i = d'' \cap I_i$, (15), and Claim 4),

(21) $\varphi_i(\Theta_{\bar{n}}(d')\Delta\Phi_*(d')) \leq 2\varepsilon$ (since $d' \in \mathcal{H}_{\bar{n}}$, $\|\cdot\|_\varphi$ of this set is $\leq 2\varepsilon$),

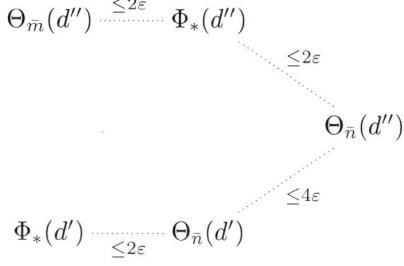

and all this implies $\varphi_i(\Theta_{\bar{m}}(d')\Delta\Phi_*(d')) \leq 14\varepsilon$, contradicting (12). This completes the proof of Lemma 3.11.4. □

PROOF OF THEOREM 3.11.3. Using Lemma 3.11.4, from now on we can assume that for every n there is a continuous $1/n$-approximation Υ_n to Φ_* on $\mathcal{J}$. For $a \subseteq \mathbb{N}$ let

$$\mathcal{Z}_a = \{b \subseteq \mathbb{N} : \text{ for every } n \text{ there is an } m$$
$$\text{for which } \varphi((b\Delta\Upsilon_n(a)) \cap [m,\infty)) \leq 1/n\}.$$

Then $\bigcup_{a \subseteq \mathbb{N}} \{a\} \times \mathcal{Z}_a$ is a Borel subset of $\mathcal{P}(\mathbb{N}) \times \mathcal{P}(\mathbb{N})$, hence by Theorem 3.11.2 there is a Baire-measurable function $\Upsilon \colon \mathcal{P}(\mathbb{N}) \to \mathcal{P}(\mathbb{N})$ such that $\Upsilon(a) \in \mathcal{Z}_a$ for all a satisfying $\mathcal{Z}_a \neq \emptyset$. Since Υ_n is a $1/n$-approximation to Φ_* on $\mathcal{J}$ for all n, the analytic set $\{a : \mathcal{Z}_a \neq \emptyset\}$ includes $\mathcal{J}$. It follows that $\Upsilon(a)\Delta\Phi_*(a) \in \mathcal{I}$ for all $a \in \mathcal{J}$, so we can say Υ is a *lifting of of Φ on $\mathcal{J}$*.

LEMMA 3.11.6. *If Φ has a Baire lifting Υ on a nonmeager ideal $\mathcal{J}$, then Φ can be decomposed as $\Phi_1 \oplus \Phi_2$, where Φ_1 has a continuous lifting and $\ker \Phi_2 \supseteq \mathcal{J}$.*

PROOF. First of all, like in the proof of Lemma 3.11.4, we can assume Υ is continuous. We shall prove that there is an asymptotically additive mapping Ω (see §1.5) such that $\|\Omega(A)\Delta\Upsilon(A)\|_\varphi = 0$ for all $A \in \mathcal{J}$. This construction is almost identical to the construction of an asymptotically additive lifting in Theorem 1.5.2. By using Claim 3, we find a sequence $1 = n_1 < k_1 < n_2 < k_2 < \ldots$ and $s_i \subseteq [n_i, n_{i+1})$ which is an (n_i, n_{i+1})-2^{-i}-stabilizer for every i with k_i witnessing this fact. Since $\mathcal{J}$ is a nonmeager ideal, we can assume $\bigcup_i s_i \in \mathcal{J}$. Since $\mathcal{I}$ is a P-ideal, we can assume Υ maps finite sets into finite sets. Let

$$A_0 = \bigcup_i [n_{2i}, n_{2i+1}), \qquad C_0 = \bigcup_i s_{2i},$$
$$A_1 = \bigcup_i [n_{2i+1}, n_{2i+2}), \qquad C_1 = \bigcup_i s_{2i+1}.$$

Define functions $f_0, f_1, \bar{\Upsilon} \colon \mathcal{P}(\mathbb{N}) \to \mathcal{P}(\mathbb{N})$ by

$$f_0(X) = \Upsilon((X \cap A_0) \cup C_1) \cap \Phi_*(A_0)$$
$$f_1(X) = \Upsilon((X \cap A_1) \cup C_0) \setminus \Phi_*(A_0)$$
$$\bar{\Upsilon}(X) = f_0(X) \cup f_1(X).$$

Let $H_i \colon \mathcal{P}([n_i, n_{i+1})) \to \text{Fin}$ be defined by

$$H_i(s) = \bar{\Upsilon}(s) \cap [k_{i-1}, k_{i+1}).$$

Then, by the proof of Theorem 1.5.2, we conclude that the mapping $\Omega = \Psi_H$ with parameters $\{n_i\}$ and $\{H_i\}$ satisfies

$$\Omega(B)\Delta\Upsilon(B) \in \mathcal{I} \text{ for all } B \in \mathcal{J},$$

and in particular it is a lifting of Φ on $\mathcal{J}$. Let $A = \Omega(\mathbb{N})$ and (recall that $\Phi^A(B) = \Phi(B) \cap A$):

$$\Phi_1 = \Phi^A, \qquad \Phi_2 = \Phi^{\mathbb{N} \setminus A}.$$

We claim Ω is a lifting of Φ_1. Pick $a, b \subseteq \mathbb{N}$ and assume $(\Omega(a) \cup \Omega(b))\Delta\Omega(a \cup b)$ is not in $\mathcal{I}$, namely

$$\|(\Omega(a) \cup \Omega(b))\Delta\Omega(a \cup b)\|_\varphi > \varepsilon$$

for some $\varepsilon > 0$. Since Ω is of the form Ψ_H, there is a subsequence n'_i of n_i such that
$$\varphi((\Omega(a \cap [n'_i, n'_{i+1})) \cup \Omega(b \cap [n'_i, n'_{i+1})))\Delta\Omega((a \cup b) \cap [n'_i, n'_{i+1}))) > \varepsilon$$
for all i. But we can find an infinite C such that $e = \bigcup_{i \in C}[n'_i, n'_{i+1})$ is in $\mathcal{J}$, and then letting $a' = a \cap e$ and $b' = b \cap e$, it follows that $\Omega(a') \cup \Omega(b') \neq^{\mathcal{I}} \Omega(a' \cup b')$, contradicting the fact that Ω is a lifting of Φ on $\mathcal{J}$. A similar proof shows that $\Omega(B)\Delta(A \setminus \Omega(\mathbb{N} \setminus B)) \in \mathcal{I}$ for all $B \subseteq \mathbb{N}$, and therefore Ω is as required. This implies that Φ_1 has a Baire-measurable lifting Φ_{1*}, and also that
$$\ker(\Phi_2) \supseteq \{a : \Phi_*(a) =^{\mathcal{I}} \Phi_{1*}(a)\} \supseteq \mathcal{J}.$$
Therefore mappings Φ_1 and Φ_2 satisfy the requirements. □

This ends the proof of Theorem 3.11.3. □

3.12. A local version of the OCA lifting theorem

The main result of this section is Proposition 3.12.1, a "local" version of Theorem 3.3.5 which already suffices to prove many of its consequences. We should remark that Proposition 3.12.1 was proved by Just ([**64**]) in the case when $\mathcal{I}$ is an F_σ-ideal, who also proved it is consistent that the conclusion of Theorem 3.10.1 is true for all analytic ideals ([**63**]). We do not know whether this stronger conclusion follows from OCA and MA.

Recall that if we are given a homomorphism $\Phi \colon \mathcal{P}(\mathbb{N})/\operatorname{Fin} \to \mathcal{P}(\mathbb{N})/\mathcal{I}$ with a lifting Φ_*, then for $B \subseteq \mathbb{N}$ by Φ^B we denote a homomorphism of $\mathcal{P}(\mathbb{N})/\operatorname{Fin}$ into $\mathcal{P}(B)/\mathcal{I}$ with lifting
$$\Phi_*^B(c) = \Phi_*(c) \cap B.$$

PROPOSITION 3.12.1. *Assume OCA and MA. If $\mathcal{I}$ is an analytic P-ideal and $\Phi \colon \mathcal{P}(\mathbb{N})/\operatorname{Fin} \to \mathcal{P}(\mathbb{N})/\mathcal{I}$ is a homomorphism given by a lifting $\Phi_* \colon \mathcal{P}(\mathbb{N}) \to \mathcal{P}(\mathbb{N})$, then the ideal*
$$\mathcal{J}_{\operatorname{cont}} = \{a : \Phi^{\Phi_*(a)} \text{ has a continuous lifting}\}$$
is ccc over Fin.

Before we start the proof, we need some definitions. Recall that infinite sets A and B of integers are *almost disjoint* if their intersection is finite, and that A *almost includes* B, $A \supseteq^* B$, if $B \setminus A$ is finite. The following notion will be very useful in setting up an open partition (see [**106**, II.4.1], [**142**]).

DEFINITION 3.12.2. A family $\mathcal{A}$ of almost disjoint sets of integers is *tree-like* if there is an ordering $\prec$ on its domain $\mathcal{D} = \bigcup \mathcal{A}$ such that $\langle \mathcal{D}, \prec \rangle$ is a tree of height ω and each element of $\mathcal{A}$ is included in the unique maximal branch of this tree.

Equivalently, $\mathcal{A}$ is tree-like if there is a $1-1$ mapping f of $\mathbb{N}$ into $\{0,1\}^{<\mathbb{N}}$ (finite sequences of 0's and 1's) such that the image of every $A \in \mathcal{A}$ is included in a single branch of the tree $\{0,1\}^{<\mathbb{N}}$ (the ordering is given by the end-extension), and for different A and B in $\mathcal{A}$ the corresponding branches are different. (In [**142**] and [**64**] such families are called *neat*.)

PROOF OF PROPOSITION 3.12.1. Fix a lower semicontinuous submeasure φ supported by $\mathbb{N}$ such that $\mathcal{I} = \operatorname{Exh}(\varphi)$. Assume that $\mathcal{J}_{\operatorname{cont}}$ is not ccc over Fin and let $\mathcal{B}$ be an uncountable almost disjoint family of $\mathcal{J}_{\operatorname{cont}}$-positive sets.

CLAIM 1. *For every $B \in \mathcal{B}$ there is an $\varepsilon > 0$ such that there are no Borel ε-approximations to Φ_* on $\mathcal{P}(B)$.*

PROOF. Assume the contrary, and let $B \in \mathcal{B}$ be such that there is a Borel 2^{-n}-approximation Θ^n to Φ_* on $\mathcal{P}(B)$ for all n. For $a \subseteq B$ let

$$\mathcal{Y}_a = \{b \subseteq \mathbb{N} : \|b \Delta \Theta^n(a)\|_\varphi \leq 2^{-n} \text{ for every } n\}.$$

Then $\mathcal{Y}_a$ is equal to $\{b : b =^{\mathcal{I}} \Phi_*(a)\}$ and nonempty because $\Phi_*(a) \in \mathcal{Y}_a$. Moreover $\bigcup_{a \subseteq B} \{a\} \times \mathcal{Y}_a$ is Borel, and by Jankov, von Neumann uniformization theorem (Theorem 3.11.2) there is a Baire measurable function $\Theta \colon \mathcal{P}(B) \to \mathcal{P}(\mathbb{N})$ such that $\Theta(a) \in \mathcal{Y}_a$ for all $a \subseteq B$, and therefore Θ is a Baire lifting of a homomorphism Φ on $\mathcal{P}(B)$. By Lemma 1.3.2 there is a continuous lifting of $\Phi \upharpoonright \mathcal{P}(B)$ and therefore there is a continuous lifting of $\Phi^{\Phi_*(B)}$, contradicting our original assumption that B is $\mathcal{J}_{\text{cont}}$-positive. □

Thus we can assume from now on that there is an $\bar{n}$ such that for no $B \in \mathcal{B}$ there is a Borel $2^{-\bar{n}+6}$-approximation to Φ_* on $\mathcal{P}(B)$. By MA, we can find an almost disjoint family $\mathcal{A}'$ of size $\aleph_1$ such that every element of $\mathcal{A}'$ almost includes infinitely many members of $\mathcal{B}$. Use MA again to find an uncountable $\mathcal{A} \subseteq \mathcal{A}'$ and a partition $A = A_0 \cup A_1$ of each $A \in \mathcal{A}$ such that $\mathcal{A}_i = \{A_i : A \in \mathcal{A}\}$ is a tree-like almost disjoint family for $i = 0, 1$ (see [**142**, Lemma 2.3]). Let us concentrate at $\mathcal{A}_0$ for a moment. Let $\hat{\mathcal{A}}_0 = \bigcup_{a \in \mathcal{A}_0} [a]^\omega$, and let B_b be the unique element of $\mathcal{A}_0$ including some $b \in \hat{\mathcal{A}}_0$. Consider the set

$$\mathcal{X} = \{\langle a, b \rangle : b \subseteq a \in \hat{\mathcal{A}}_0\}.$$

Now we shall apply OCA to an open partition of $[\mathcal{X}]^2$ which is a variation of a partition of Velickovic ([**142**, p. 6]; see also [**64**, p. 286]). Define $[\mathcal{X}]^2 = K_0^{\bar{n}} \cup K_1^{\bar{n}}$ by letting $\{\langle a, b \rangle, \langle a', b' \rangle\}$ in $K_0^{\bar{n}}$ if

- (K1) $B_a \neq B_{a'}$
- (K2) $b' \cap a = b \cap a'$
- (K3) $\varphi((\Phi_*(b') \cap \Phi_*(a)) \Delta (\Phi_*(b) \cap \Phi_*(a'))) > 2^{-\bar{n}}$.

Since the family $\mathcal{A}_0$ is tree-like, this is an open partition. [To see this, fix a bijection $f \colon \mathbb{N} \to \{0,1\}^{<\mathbb{N}}$ such that $f''a$ is a chain for every $a \in \mathcal{A}_0$ and identify every $b \in \hat{\mathcal{A}}_0$ with the triple $\langle b, \Phi_*(b), f''B_b \rangle$. The condition (K3) is open by the lower semicontinuity of φ.]

CLAIM 2. *There are no uncountable $K_0^{\bar{n}}$-homogeneous sets.*

PROOF. Assume the contrary and let $\mathcal{H}$ be such a set. Let $c = \bigcup_{\langle a, b \rangle \in \mathcal{H}} b$. Then by (K2) we have $c \cap a = b$ for every $\langle a, b \rangle \in \mathcal{H}$, and therefore $\Phi_*(c) \cap \Phi_*(a) =^{\mathcal{I}} \Phi_*(b)$. So we can find $\bar{s}, \bar{t} \subseteq [1, \bar{m})$ for some $\bar{m}$ and an uncountable $\bar{\mathcal{H}} \subseteq \mathcal{H}$ so that for all $\langle a, b \rangle \in \bar{\mathcal{H}}$:

- (1) $\varphi((\Phi_*(c) \cap \Phi_*(a)) \Delta \Phi_*(b)) \cap [\bar{m}, \infty)) \leq 2^{-\bar{n}-1}$,
- (2) $\Phi_*(b) \cap [1, \bar{m}) = \bar{s}$,
- (3) $\Phi_*(a) \cap [1, \bar{m}) = \bar{t}$.

Then for $\langle a,b\rangle$ and $\langle a',b'\rangle \in \bar{\mathcal{H}}$ we have
$$\begin{aligned}d &= (\Phi_*(b) \cap \Phi_*(a'))\Delta(\Phi_*(b') \cap \Phi_*(a))\\ &\subseteq (\Phi_*(b) \cap \Phi_*(a'))\Delta(\Phi_*(c) \cap \Phi_*(a) \cap \Phi_*(a'))\\ &\quad\cup (\Phi_*(b') \cap \Phi_*(a))\Delta(\Phi_*(c) \cap \Phi_*(a) \cap \Phi_*(a'))\\ &\subseteq (\Phi_*(b)\Delta(\Phi_*(c) \cap \Phi_*(a)))\\ &\quad\cup (\Phi_*(b')\Delta(\Phi_*(c) \cap \Phi_*(a'))).\end{aligned}$$

Thus $\varphi(d \cap [\bar{m},\infty)) \leq 2^{-\bar{n}-1} + 2^{-\bar{n}-1} = 2^{-\bar{n}}$ but (2)–(3) imply $d \cap [\bar{m},\infty) = d$, and (K3) implies $\varphi(d) > 2^{-\bar{n}}$, a contradiction. □

Therefore OCA implies that $\mathcal{X}$ can be covered by a countable union $\bigcup_{i=1}^{\infty} \mathcal{H}_i^{\bar{n}}$ of $K_1^{\bar{n}}$-homogeneous sets. Let $\mathcal{D}_i^{\bar{n}}$ be a countable subset of $\mathcal{H}_i^{\bar{n}}$ such that the set
$$\{\langle a,b,\Phi_*(a),\Phi_*(b)\rangle : \langle a,b\rangle \in \mathcal{D}_i^{\bar{n}}\}$$
is dense in $\{\langle a,b,\Phi_*(a),\Phi_*(b)\rangle : \langle a,b\rangle \in \mathcal{H}_i^{\bar{n}}\}$. Fix $B \in \mathcal{A}_0$ which is different from all B_a ($\langle a,b\rangle \in \mathcal{D}_i^{\bar{n}}$ for some b and $i \in \mathbb{N}$). Recall that a function is said to be σ-Borel if its graph can be covered by the union of countably many graphs of total Borel functions.

CLAIM 3. *There exists a σ-Borel $2^{-\bar{n}+2}$-approximation $\Gamma_B^{\bar{n}} : \mathcal{P}(B) \to \mathcal{P}(\mathbb{N})$ to Φ_* on $\mathcal{P}(B)$.*

PROOF. Fix $\bar{m} \in \mathbb{N}$. For pairs $\langle a,b\rangle$ and $\langle a',b'\rangle$ in $\mathcal{P}(\mathbb{N}) \times \mathcal{P}(\mathbb{N})$ and $k \in \mathbb{N}$ we say that $\langle a,b\rangle$ and $\langle a',b'\rangle$ *agree below* k, and write $\langle a,b\rangle \upharpoonright k = \langle a',b'\rangle \upharpoonright k$, if $\min(a\Delta a') > k$ and $\min(b\Delta b') > k$. This terminology and notation extend to n-tuples in an obvious way. For $m \in \mathbb{N}$ and $\langle d,e\rangle \in \mathcal{D}_{\bar{m}}^{\bar{n}}$ let
$$m^+(d,e) = \min\{\max(d' \cap B) : \langle d',e'\rangle \in \mathcal{D}_{\bar{m}}^{\bar{n}}$$
$$\text{and}\quad \langle d,e,\Phi_*(d),\Phi_*(e)\rangle \upharpoonright m = \langle d',e',\Phi_*(d'),\Phi_*(e')\rangle \upharpoonright m\},$$
$$m^+ = \max\{m^+(d,e) : \langle d,e\rangle \in \mathcal{D}_{\bar{m}}^{\bar{n}}\}.$$
Then both $m^+(c,d)$ and m^+ are well-defined, since

(1) $m^+(d,e) \leq \max(d\Delta B)$, and this value is finite by the choice of B and d,
(2) $m^+(d,e)$ depends only on $\langle d,e\rangle \upharpoonright m$, and therefore m^+ is equal to a maximum of a finite set.

Now recursively define a sequence $\{m_j = m_j(B,\bar{n},\bar{m})\}$ by picking $m_1 = 1$ and
$$m_{j+1} = m_j^+ + 1.$$
Let
$$B_0 = B \cap \bigcup_{j=0}^{\infty} [m_{2j}, m_{2j+1}), \qquad B_1 = B \cap \bigcup_{j=0}^{\infty} [m_{2j+1}, m_{2j+2}).$$

We now prove there is a σ-Borel $2^{-\bar{n}+1}$-approximation $\Gamma_0 : \mathcal{P}(B_0) \to \mathcal{P}(\mathbb{N})$ to Φ_*. Towards this end, we shall first define a set $\mathcal{Y}_a$ of "candidates" for $\Gamma_0(a)$. For $a \subseteq B_0$ let $T_a = T_a(B_0,\bar{n},\bar{m})$ be the set of all pairs $\langle s,k\rangle$ such that

$(T_a 1)$ $k \in \mathbb{N}$ and $s \subseteq [1,k] \cap \Phi_*(B_0)$,

and there exists $\langle d,e\rangle \in \mathcal{D}_{\bar{m}}^{\bar{n}}$ such that:

$(T_a 2)$ $e \cap B_0 = a \cap d$,
$(T_a 3)$ $\Delta(\Phi_*(B_0), \Phi_*(d)) \geq k$, and
$(T_a 4)$ $\varphi((\Phi_*(e)\Delta s) \cap \Phi_*(B_0) \cap [1,k)) \leq 2^{-\bar{n}}$.

Note that if $\langle s, k \rangle \in T_a$ then for some $\langle d, e \rangle \in \mathcal{D}_{\bar{m}}^{\bar{n}}$ such that $\Delta(\Phi_*(B_0), \Phi_*(d)) \geq k$ the pair $\{\langle B_0, a \rangle, \langle d, e \rangle\}$ satisfies (K1), (K2) and s approximates $\Phi_*(e) \cap \Phi_*(B_0)$ up to k. Consider T_a as a tree ordered by the end-extension: $\langle s, k \rangle \leq \langle t, l \rangle$ if $k \leq l$ and $t \cap [1, k) = s$. Note that T_a is *downwards closed*: if $\langle s, k \rangle \in T_a$ then $\langle s \cap [1, k'), k' \rangle$ is in T_a for all $k' < k$. Let

$$\mathcal{Y}_a = \mathcal{Y}_a(B_0, \bar{n}, \bar{m}) = \{b : \langle b \cap [1, k), k \rangle \in T_a \text{ for all } k \in \mathbb{N}\}.$$

Since T_a is a downwards closed subset of a binary tree, the set $\mathcal{Y}_a$ is closed for all a and the function $a \mapsto \mathcal{Y}_a$ is a Borel mapping of $\mathcal{P}(B_0)$ into the exponential space of $\mathcal{P}(\mathbb{N})$.

SUBCLAIM. *Assume $\langle B_0, a \rangle$ belongs to $\mathcal{H}_{\bar{m}}^{\bar{n}}$. Then*
(a) $\mathcal{Y}_a = \mathcal{Y}_a(B_0, \bar{n}, \bar{m})$ *is nonempty, and*
(b) $\varphi(b\Delta(\Phi_*(a) \cap \Phi_*(B_0))) \leq 2^{-\bar{n}-1}$ *for all $b \in \mathcal{Y}_a$.*

PROOF. (a) Since $\mathcal{D}_{\bar{m}}^{\bar{n}}$ is dense in $\mathcal{H}_{\bar{m}}^{\bar{n}}$ (in a rather strong way—see the paragraph before Claim 3) and $\mathcal{H}_{\bar{m}}^{\bar{n}}$ is a $K_1^{\bar{n}}$-homogeneous set, for every m_{2j} there is a $\langle d, e \rangle \in \mathcal{D}_{\bar{m}}^{\bar{n}}$ such that

(*) $\langle d, e, \Phi_*(d), \Phi_*(e) \rangle \upharpoonright m_{2j+1} = \langle B_0, a, \Phi_*(B_0), \Phi_*(a) \rangle \upharpoonright m_{2j+1}.$

Then by the choice of m_{2j+2} there is a $\langle d, e \rangle$ in $\mathcal{D}_{\bar{m}}^{\bar{n}}$ satisfying (*) and such that, moreover, $d \cap B_0 \subseteq [1, m_{2j+2})$. Since B_0 is disjoint from $[m_{2j+1}, m_{2j+2})$, we automatically have

$$d \cap B_0 \subseteq [1, m_{2j+1}).$$

This implies $e \cap B_0 \subseteq [1, m_{2j+1})$ and $d \cap a \subseteq [1, m_{2j+1})$ and therefore $e \cap B_0 = a \cap d$, also $\varphi(\Phi_*(e)\Delta(\Phi_*(a) \cap \Phi_*(B_0))) = 0$, and the conditions for putting $\langle \Phi_*(a) \cap \Phi_*(B) \cap [1, m_{2j}), m_{2j} \rangle$ in T_a are satisfied. Since m_{2j} was arbitrarily large, we have $\Phi_*(a) \cap \Phi_*(B_0) \in \mathcal{Y}_a$, completing the proof of (a).

(b) Assume $b \in \mathcal{Y}_a$, fix $k \in \mathbb{N}$, and let $\langle d, e \rangle \in \mathcal{D}_{\bar{m}}^{\bar{n}}$ be a witness that $\langle b, k \rangle \in T_a$, i.e., such that
(1) $e \cap B_0 = a \cap d$,
(2) $\Delta(\Phi_*(B_0), \Phi_*(d)) \geq k$, and
(3) $\varphi((\Phi_*(e)\Delta b) \cap \Phi_*(B_0) \cap [1, k)) \leq 2^{-\bar{n}}$.

Note that

$$F = (\Phi_*(e) \cap \Phi_*(B_0))\Delta(\Phi_*(a) \cap \Phi_*(d))$$
$$\supseteq ((\Phi_*(e) \cap \Phi_*(B_0))\Delta(\Phi_*(a) \cap \Phi_*(d))) \cap [1, k)$$
$$= ((\Phi_*(e) \cap \Phi_*(B_0))\Delta(\Phi_*(a) \cap \Phi_*(B_0))) \cap [1, k)$$
$$= (\Phi_*(e)\Delta\Phi_*(a)) \cap \Phi_*(B_0) \cap [1, k),$$

and we have $\Phi_*(a)\Delta b \subseteq F \cup (\Phi_*(e)\Delta b)$. Moreover, by (K3) we have $\varphi(F) \leq 2^{-\bar{n}}$, and therefore

$$\varphi((b\Delta\Phi_*(a)) \cap \Phi_*(B_0) \cap [1, k)) \leq 2^{-\bar{n}} + 2^{-\bar{n}} = 2^{-\bar{n}+1}.$$

The submeasure φ is lower semicontinuous and the inequality is true for all k, so we have $\varphi(b\Delta(\Phi_*(a) \cap \Phi_*(B_0))) \leq 2^{-\bar{n}+1}$. □

We have already noted that the mapping $a \mapsto \mathcal{Y}_a$ is Borel (as a mapping of $\mathcal{P}(B_0)$ into the exponential space of $\mathcal{P}(\mathbb{N})$), and that each T_a is a binary tree. It follows that $\{a : \mathcal{Y}_a \neq \emptyset\}$ is a Borel set, and therefore the mapping $\Xi_{0,\bar{m}}$ defined on $\mathcal{P}(B_0)$ by (by $\min \mathcal{Y}_a$ we denote the lexicographically minimal element of $\mathcal{Y}_a$, corresponding to the leftmost branch of T_a):

$$\Xi_{0,\bar{m}}(a) = \begin{cases} \min \mathcal{Y}_a, & \text{if } \mathcal{Y}_a \neq \emptyset, \\ \emptyset, & \text{if } \mathcal{Y}_a = \emptyset, \end{cases}$$

is Borel. By Subclaim (b), $\Xi_{0,\bar{m}}$ is also a $2^{-\bar{n}+1}$-approximation to Φ_* on $\mathcal{H}_{\bar{m}}^{\bar{n}} \cap \mathcal{P}(B_0)$. By letting $\bar{m}$ run over all positive integers, we find such $\Xi_{0,m}$ for all $m \in \mathbb{N}$. Therefore the mapping $\Xi_0 \colon \mathcal{P}(B_0) \to \mathcal{P}(\mathbb{N})$ defined by

$$\Xi_0(a) = \Xi_{0,m}(a), \qquad \text{whenever } \langle B_0, a \rangle \in \mathcal{H}_m^{\bar{n}} \setminus \bigcup_{k<m} \mathcal{H}_k^{\bar{n}}$$

is a σ-Borel $2^{-\bar{n}+1}$-approximation to Φ_* on $\mathcal{P}(B_0)$. In a similar fashion we define mapping $\Xi_1 \colon \mathcal{P}(B_1) \to \mathcal{P}(\mathbb{N})$ which is a σ-Borel $2^{-\bar{n}+1}$-approximation to Φ_* on $\mathcal{P}(B_1)$ and finally define $\Xi = \Xi_B^{\bar{n}}$ by

$$\Xi(a) = \Xi_0(a \cap B_0) \cup \Xi_1(a \cap B_1), \qquad \text{for } a \subseteq B.$$

Then Ξ is a σ-Borel $2^{-\bar{n}+2}$-approximation to Φ_* on $\mathcal{P}(B)$. $\square$

We are now in a position to conclude that for all but countably many $B \in \mathcal{A}_0$ there is a σ-Borel $2^{-\bar{n}+2}$-approximation to Φ_* on $\mathcal{P}(B)$. Similarly, for all but countably many $B \in \mathcal{A}_1$ there is a σ-Borel $2^{-\bar{n}+2}$-approximation to Φ_* on $\mathcal{P}(B)$. It follows that for all but countably many $B \in \mathcal{A}$ there is a σ-Borel $2^{-\bar{n}+3}$-approximation to Φ_* on $\mathcal{P}(B)$. Now we have to relate the existence of a σ-Borel approximation to Φ_* with the existence of a Borel approximation to Φ_*. Regarding this, we should note that in Theorem 2 of [142] it was proved that, in the case of automorphisms of $\mathcal{P}(\mathbb{N})/\operatorname{Fin}$, the existence of a σ-Borel lifting implies the existence of a Borel lifting. In our case, when we have a homomorphism instead of an automorphism, this is no longer true (see Example 3.2.1). Let

$\mathcal{J}_\sigma^n = \{a \subseteq \mathbb{N} : \Phi_* \text{ can be } 2^{-n}\text{-approximated on } \mathcal{P}(a) \text{ by a } \sigma\text{-Borel function}\}$,
$\mathcal{J}_{\text{cont}}^n = \{a \subseteq \mathbb{N} : \Phi_* \text{ can be } 2^{-n}\text{-approximated on } \mathcal{P}(a) \text{ by a continuous function}\}$.

It is straightforward to check that these are ideals on $\mathbb{N}$ which include Fin. We shall need the following diagonalization lemma which relates $\mathcal{J}_\sigma^n$ and $\mathcal{J}_{\text{cont}}^n$ (compare this with [64, Lemma 6a] and [142, pages 4–5]).

LEMMA 3.12.3. *For any given integer n there is no infinite sequence of pairwise disjoint $\mathcal{J}_{\text{cont}}^{n-3}$-positive sets whose union is in $\mathcal{J}_\sigma^n$.*

PROOF. Assume the contrary, and let a_k ($k \in \mathbb{N}$) be pairwise disjoint $\mathcal{J}_{\text{cont}}^{n-3}$-positive sets whose union a is in $\mathcal{J}_\sigma^n$. Let $\Theta_m \colon \mathcal{P}(a) \to \mathcal{P}(\mathbb{N})$ ($m \in \mathbb{N}$) be Borel functions which 2^{-n}-approximate Φ_* on $\mathcal{P}(a)$. We shall recursively find sequences a'_j, x_j such that for every $j \in \mathbb{N}$:

(A1) $x_j \subseteq a'_j =^* a_j$,
(A2) $\|(\Theta_j(\bigcup_i x_i) \cap \Phi_*(a'_j)) \Delta \Phi_*(x_j)\|_\varphi > 2^{-n}$.

If we succeed in finding such sequences, let j be such that

$$\left\|\Theta_j\left(\bigcup_i x_i\right) \Delta \Phi_*\left(\bigcup_i x_i\right)\right\|_\varphi \leq 2^{-n}.$$

Since $a'_j \cap \bigcup_i x_i = x_j$, we have

$$\left\|\left(\Theta_j\left(\bigcup_i x_i\right) \cap \Phi_*(a'_j)\right) \Delta \Phi_*(x_j)\right\|_\varphi$$
$$\leq \left\|\left(\Theta_j\left(\bigcup_i x_i\right) \Delta \Phi_*\left(\bigcup_i x_i\right)\right) \cap \Phi_*(a'_j)\right\|_\varphi \leq 2^{-n},$$

contradicting (A2). Therefore it will suffice to construct a'_j and x_j satisfying (A1) and (A2). In order to make the recursive construction possible and to assure (A2), we shall need some auxiliary objects satisfying the following for all $j < l$ in $\mathbb{N}$:

(A3) $c_j = a \setminus \bigcup_{i=1}^{j} a'_i$.
(A4) $\mathcal{Y}_j = \{y \subseteq c_j : \|(\Theta_j(\bigcup_{i=1}^{j} x_i \cup y) \cap \Phi_*(a'_j))\Delta\Phi_*(x_j)\|_\varphi > 2^{-n}\}$ is relatively comeager in $\mathcal{P}(c_j)$.
(A5) $U_{j1} \supseteq U_{j2} \supseteq \ldots$ are dense open subsets of $\mathcal{P}(c_j)$ such that $\bigcap_i U_{ji} \subseteq \mathcal{Y}_j$.
(A6) $\mathcal{Y}_{j,l} = \{y \subseteq c_l : \bigcup_{i=j+1}^{l} x_i \cup y \in \mathcal{Y}_j\}$ is relatively comeager in $\mathcal{P}(c_j)$.
(A7) $\{y \subseteq c_j : y \cap \bigcup_{i=j+1}^{l} a'_i = \bigcup_{i=j+1}^{l} x_i\}$ is included in U_{jl}.

Assume these objects are chosen so that (A1) and (A3)–(A7) are satisfied and let $x = \bigcup_i x_i$. For $m \in \mathbb{N}$, (A7) implies that $\bigcup_{i=m+1}^{\infty} x_i \in \bigcap_{l=1}^{\infty} U_{ml} \in \mathcal{Y}_m$, so (A4) implies (A2). Therefore it will suffice to assure (A1) and (A3)–(A7).

We shall now describe the recursive construction of these objects. Assume x_k, a'_k, and $\{U_{ki}\}_{i=1}^{\infty}$ are chosen for $k = 1, \ldots, m-1$ to satisfy (A1) and (A3)–(A7) for $j < l \leq m-1$. Let $\bar{a}_m = a_m \setminus \bigcup_{i=1}^{m-1} a'_i$, and note that by (A1) we have $\bar{a}_m =^* a_m$. Also, let $\bar{c}_m = c_{m-1} \setminus \bar{a}_m$, and for $x \subseteq \bar{a}_m$ define

$$H_m(x) = \left\{ y \subseteq \bar{c}_m : \left\|\left(\Theta_m\left(\bigcup_{i=1}^{m-1} x_i \cup x \cup y\right) \cap \Phi_*(\bar{a}_m)\right) \Delta \Phi_*(x) \right\|_\varphi \leq 2^{-n} \right\}.$$

Since Θ_m is a Borel function, this is a Borel subset of $\mathcal{P}(\bar{c}_m)$ for every $x \subseteq \bar{a}_m$. Now let

$$\mathcal{X}_m = \{x \subseteq \bar{a}_m : H_m(x) \text{ is relatively comeager in } \mathcal{P}(\bar{c}_m)\}$$

and note that if $x_m \notin \mathcal{X}_m$, then $\mathcal{Y}_m$ is relatively comeager in $\mathcal{P}(c_m)$ and (A4) is satisfied. The following claim will make it easier to pick $x_m \notin \mathcal{X}_m$.

CLAIM 4. *The intersection of the complement of $\mathcal{X}_m$ with any nonempty clopen subset of $\mathcal{P}(\bar{a}_m)$ is relatively nonmeager.*

PROOF. Assume the contrary, and let $v \subseteq u \subseteq \bar{a}_m$ be finite and such that $\mathcal{X}_m$ is relatively comeager in the set

$$\{x \subseteq \bar{a}_m : x \cap u = v\}.$$

Then there is a disjoint partition $\bar{a}_m \setminus u = a_m^0 \dot\cup a_m^1$ and $t^i \subseteq a_m^i$ ($i = 0, 1$) so that for every $x \subseteq \bar{a}_m$ both $(x \setminus (a_m^1 \cup u)) \cup v \cup t^1$ and $(x \setminus (a_m^0 \cup u)) \cup v \cup t^0$ are in $\mathcal{X}_m$ (see e.g., [141] or proof of Lemma 1.3.2). Define $\Theta'_m : \mathcal{P}(\bar{a}) \to \mathcal{P}(\mathbb{N})$ by

$$\Theta'_{m0}(x) = \Theta_m((x \setminus (a_m^1 \cup u)) \cup v \cup t^1) \cap \Phi_*(a_m^0),$$
$$\Theta'_{m1}(x) = \Theta_m((x \setminus (a_m^0 \cup u)) \cup v \cup t^0) \setminus \Phi_*(a_m^0),$$
$$\Theta'_m(x) = \Theta_{m0}(x) \cup \Theta_{m1}(x).$$

Note that Θ'_m is a Borel function. For $x \subseteq \bar{a}_m$ and $y \subseteq \mathbb{N}$, define
$$H'_m(x,y) = \left\{ z \subseteq \bar{c}_m \,:\, \left\| \left(\Theta'_m \left(\bigcup_{i=1}^{m-1} x_i \cup x \cup z \right) \cap \Phi_*(\bar{a}_m) \right) \Delta y \right\|_\varphi \le 2^{-n+1} \right\}.$$
We claim that for any $x \subseteq \bar{a}_m$ the set $H'_m(x, \Phi_*(x))$ includes
$$H = H_m((x \setminus (a_m^1 \cup u)) \cup v \cup t^1) \cap H_m((x \setminus (a_m^0 \cup u)) \cup v \cup t^0),$$
and in particular it is comeager. This is so because H is the set of all $z \subseteq \bar{c}_m$ for which
$$\varphi\left(\left(\Theta'_{m\varepsilon} \left(\bigcup_{i=1}^{m-1} x_i \cup x \cup z \right) \cap \Phi_*(\bar{a}_m) \right) \Delta \Phi_*(x) \right) \le 2^{-n}$$
for both $\varepsilon = 0, 1$ and Θ'_m is defined as an amalgamation of Θ'_{m0} and Θ'_{m1}. Since the set H'_m is Borel, the set
$$\mathcal{Z} = \{\langle x, b \rangle \in \mathcal{P}(\bar{a}_m) \times \mathcal{P}(\mathbb{N}) \,:\, H'_m(x,b) \text{ is relatively comeager in } \mathcal{P}(\bar{c}_m)\}$$
is an analytic subset of $\mathcal{P}(\bar{a}_m) \times \mathcal{P}(\mathbb{N})$ (again by [74, 29E]). Note that $\langle x, b \rangle \in \mathcal{Z}$ implies that for some (actually, comeager many) $y \subseteq \bar{c}_m$ we have
$$\|\Phi_*(x) \Delta b\|_\varphi \le \left\| \Phi_*(x) \Delta \left(\Theta'_m \left(\bigcup_{i=1}^{m-1} x_i \cup x \cup y \right) \cap \Phi_*(a'_m) \right) \right\|_\varphi$$
$$+ \left\| \left(\Theta'_m \left(\bigcup_{i=1}^{m-1} x_i \cup x \cup y \right) \cap \Phi_*(a'_m) \right) \Delta b \right\|_\varphi \le 2^{-n+2}$$
and therefore by Jankov, von Neumann theorem (Theorem 3.11.2) applied to $\mathcal{Z}$ there is a Baire 2^{-n+2}-approximation Ξ to Φ_* on $\mathcal{P}(\bar{a}_m)$. Since Ξ is continuous on a dense G_δ subset G of $\mathcal{P}(\bar{a}_m)$, we can modify Ξ to obtain a continuous 2^{-n+3}-approximation to Φ_* on $\mathcal{P}(\bar{a}_m)$ (like in Lemma 1.3.2 or the definition of Θ'_m above). Since $\bar{a}_m =^* a_m$, this gives a continuous 2^{-n+3}-approximation to Φ_* on $\mathcal{P}(a_m)$, contradicting the fact that $\bar{a}_m$ is not in $\mathcal{J}_{\text{cont}}^{n-3}$. This concludes the proof. □

Fix $k < m$ and recall the inductive assumption (A6), that the set $\mathcal{Y}_{k,m-1}$ is relatively comeager in $\mathcal{P}(c_{m-1})$. By the Kuratowski–Ulam theorem (see e.g., [74, 8.K]) applied to the product $\mathcal{P}(c_{m-1}) \cong \mathcal{P}(\bar{a}_m) \times \mathcal{P}(\bar{c}_m)$ the set
$$(*) \ \mathcal{Z}_k = \left\{ x \subseteq \bar{a}_m \,:\, \left\{ y \subseteq \bar{c}_m \,:\, \bigcup_{i=k}^{m-1} x_i \cup x \cup y \in \mathcal{Y}_{k,m-1} \right\} \text{ is comeager in } \mathcal{P}(\bar{c}_m) \right\}$$
is comeager in $\mathcal{P}(\bar{a}_m)$. Therefore $\bigcap_{k=1}^{m-1} \mathcal{Z}_k$ is comeager in $\mathcal{P}(\bar{a}_m)$ as well. By Claim 4 we can find
$$\bar{x}_m \in (\mathcal{P}(\bar{a}_m) \setminus \mathcal{X}_m) \cap \bigcap_{k=1}^{m-1} \mathcal{Z}_k.$$
Since we shall choose $a'_m =^* \bar{a}_m$ and $x_m =^* \bar{x}_m$, this takes care of (A4) (see definitions of $\mathcal{Y}_m$ and $H_m(x)$), and also of (A6). The set $H_m(\bar{x}_m)$ is not comeager in $\mathcal{P}(\bar{c}_m)$, because $\bar{x}_m \notin \mathcal{X}_m$. Also, for every $k < m$ the vertical section
$$\left\{ y \subseteq \bar{c}_m \,:\, \bigcup_{i=k}^{m-1} x_i \cup \bar{x}_m \cup y \in U_{km} \right\}$$

is an open subset of $\mathcal{P}(\bar{c}_m)$, which is moreover dense since it includes

$$\left\{ y \subseteq \bar{c}_m : \bigcup_{i=k}^{m-1} x_i \cup \bar{x}_m \cup y \in \mathcal{Y}_k \right\},$$

and the latter set is dense by the choice of $\bar{x}_m$. Therefore we can find a finite $t \subseteq u \subseteq \bar{c}_m$ such that $H_m(\bar{x}_m)$ is relatively meager on $\{y \subseteq \bar{c}_m : y \cap u = t\}$ and that $\bigcup_{i=k}^{m-1} x_i \cup \bar{x}_m \cup t \cup y \in U_{km}$ for all $y \subseteq \bar{c}_m \setminus u$.

This, together with (*), assures (A6) and (A7) for $l = m$, if we let $a'_m = \bar{a}_m \cup u$, $x_m = \bar{x}_m \cup t$, and pick a sequence $\{U_{mi}\}_{i=1}^{\infty}$ of dense open subsets of $\mathcal{P}(c_m)$ whose intersection is included in $\mathcal{Y}_m$. So (A5) is taken care of. Conditions (A1) and (A3) are obviously satisfied, so this describes the recursive construction. Objects constructed in this way satisfy (A1) and (A3)–(A5), and it was remarked before that this implies (A2). This ends the proof of Lemma 3.12.3. □

Recall that every element of $\mathcal{A}$ has infinitely many disjoint subsets which belong to $\mathcal{B}$ and that all members of $\mathcal{B}$ are $\mathcal{J}_{\text{cont}}^{n-6}$-positive. Recall also the conclusion that all but countably many elements of $\mathcal{A}$ belong to $\mathcal{J}_{\sigma}^{n-3}$. All this is in contradiction with Lemma 3.12.3. This shows that the assumption that $\mathcal{J}_{\text{cont}}$ is not ccc over Fin leads to a contradiction, therefore finishing the proof of Proposition 3.12.1. □

3.13. The proof of the OCA lifting theorem for the analytic P-ideals

We are now in a position to start the proof of the main result of this Chapter, Theorem 3.3.5, so let us first recall its statement.

THEOREM 3.13.1. *Assume OCA and MA and let $\mathcal{I}$ be an analytic P-ideal. If $\Phi \colon \mathcal{P}(\mathbb{N})/\text{Fin} \to \mathcal{P}(\mathbb{N})/\mathcal{I}$ is a homomorphism, then Φ has a continuous almost lifting. Equivalently, Φ can be decomposed as $\Phi_1 \oplus \Phi_2$ so that Φ_1 has a continuous lifting, while the kernel of Φ_2 is ccc over Fin.*

By Proposition 3.12.1, it will suffice to find a single mapping Ω which uniformizes all Ψ_a for $a \in \mathcal{J}_{\text{cont}}$ (for definitions see §3.12). OCA has proved to be extremely useful in similar situations (see [**125**, §8], [**135**, §10], [**142**, Lemma 2.4]).

PROOF OF THEOREM 3.3.5. The key object in this proof is the following ideal extending the ideal $\mathcal{J}_{\text{cont}}$:

$\mathcal{J}_1 = \{\, a \subseteq \mathbb{N} :$ there is a partition $\Phi_*(a) = C_0^a \dot{\cup} C_1^a$ such that $\Phi^{C_0^a}$ has a continuous
  lifting and the kernel of $\Phi^{C_1^a}$ is ccc over Fin$\}$

Hence Theorem 3.3.5 translates as "$\mathcal{J}_1$ is an improper ideal." We claim $\mathcal{J}_1$ is equal to the following ideal

$\mathcal{J}_2 = \{\, a \subseteq \mathbb{N} :$ there exists a continuous lifting $\Psi_c \colon \mathcal{P}(a) \to \mathcal{P}(\Phi_*(a))$ of a homomorphism of $\mathcal{P}(\mathbb{N})/\text{Fin}$ into $\mathcal{P}(\Psi_a(a))/\mathcal{I}$ such that the ideal

$$\mathcal{K}_a = \{b \subseteq a : \Phi_*(c) =^{\mathcal{I}} \Psi_a(c) \text{ for all } c \subseteq b\}$$

on a is ccc over Fin$\}$.

To see that $\mathcal{J}_1 \subseteq \mathcal{J}_2$, take some $a \in \mathcal{J}_1$ and let $\Psi_a = \Phi_*^{C_0^a}$. Then for $b \in \ker(\Phi_*^{C_1^a})$ we have $\Psi_a(b) =^{\mathcal{I}} \Phi_*(b)$. Therefore $\ker(\Phi_*^{C_1^a})$ includes $\mathcal{K}_a$, so it it ccc over Fin. To see that $\mathcal{J}_2 \subseteq \mathcal{J}_1$ (a fact which we will actually not need), assume $a \in \mathcal{J}_2$, and let $C_0^a = \Psi_a(a)$. For $b \subseteq a$ let

$$D_b = (\Phi_*(b) \Delta \Psi_a(b)) \cap C_0^a =^{\mathcal{I}} (\Phi_*(b) \setminus \Psi_a(b)) \cap C_0^a$$

Then $\{c \subseteq a : \Psi_a(c) \cap D_b \in \mathcal{I}\}$ is an analytic ideal and it includes the nonmeager set $\{c \subseteq a : \Phi_*(c)\Delta\Psi_a(c) \in \mathcal{I}\}$. Therefore b belongs to this ideal, and $D_b \in \mathcal{I}$. So $(\Phi_*(b)\Delta\Psi_a(b)) \cap C_0^a$ is in $\mathcal{I}$ for all $b \subseteq a$, and Ψ_a is a Baire lifting of the homomorphism $\Phi_*^{C_0^a}$. On the other hand, if $b \in \mathcal{K}_a$, then $\Phi_*(b) \setminus C_0^a \in \mathcal{I}$, and therefore $\ker(\Phi_*^{C_1^a})$ is ccc over Fin.

CLAIM 1. *For all $a \in \mathcal{J}_2$ we have $\mathcal{K}_a \supseteq \mathcal{J}_{\mathrm{cont}} \upharpoonright a$.*

PROOF. Pick $c \in \mathcal{J}_{\mathrm{cont}} \upharpoonright a$ and consider Ψ_c, a continuous lifting of Φ_* on $\mathcal{P}(c)$. The ideal
$$\{d \subseteq c : \Psi_c(e) =^\mathcal{I} \Psi_a(e) \text{ for all } e \subseteq d\}$$
is a coanalytic subset of $\mathcal{P}(c)$ and it includes a nonmeager ideal $\mathcal{K}_a \upharpoonright c$ of subsets of c so it is improper and therefore $\Psi_a(c) =^\mathcal{I} \Psi_c(c) =^\mathcal{I} \Phi_*(c)$. □

By Lemma 3.11.6 and Theorem 1.5.2 we can assume each Ψ_a for $a \in \mathcal{J}_2$ is asymptotically additive, namely of the form Ψ_H for some strictly increasing sequence of integers $\{n_i = n_i^a\}$, sequence of disjoint finite sets of integers $\{v_i = v_i^a\}$ and functions $\{H_i = H_i^a\}$ from subsets of the interval $[n_i, n_{i+1})$ to subsets of v_i (see Definition 1.5.1). The following coherence property is an important consequence of this representation of functions Ψ_a.

CLAIM 2. *For all $a, b \in \mathcal{J}_2$ and all n there are m, m' such that for all $s \subseteq a \cap b$ we have*
$$\varphi((\Psi_a(s)\Delta\Psi_b(s)) \setminus [1, m)) \leq 2^{-n},$$
$$\varphi((\Psi_a(s \setminus [1, m'))\Delta\Psi_b(s \setminus [1, m')))) \leq 2^{-n}.$$

PROOF. Assume the first inequality fails for some n and for all m. Then, using the fact that Ψ_a and Ψ_b are asymptotically additive, we can pick subsequences $\{m_i^a\}$ of $\{n_i^a\}$ and $\{m_i^b\}$ of $\{n_i^b\}$ such that
$$0 = m_0^a = m_0^b < m_1^a \leq m_1^b < m_2^a \leq m_2^b < \ldots$$
and for every i there is an $s_i \subseteq [m_{2i}^b, m_{2i+1}^a) \cap a \cap b$ such that
$$\varphi((\Psi_a(s_i)\Delta\Psi_b(s_i)) \cap [m_{2i}^a, \infty)) > 2^{-n}.$$
By Proposition 3.12.1 ideal $\mathcal{J}_{\mathrm{cont}}$ is ccc over Fin and therefore nonmeager, so by (a) of Theorem 3.10.1 and we can furthermore assume $s = \bigcup_{i=1}^\infty s_i \in \mathcal{J}_{\mathrm{cont}}$. Therefore Claim 1 implies $\Psi_a(s) =^\mathcal{I} \Phi_*(s) =^\mathcal{I} \Psi_b(s)$. By the choice of $\{s_i\}$ we have
$$\Psi_a\left(\bigcup_{i=1}^\infty s_i\right) = \bigcup_{i=1}^\infty \Psi_a(s_i) \quad \text{and} \quad \Psi_b\left(\bigcup_{i=1}^\infty s_i\right) = \bigcup_{i=1}^\infty \Psi_b(s_i),$$
therefore $\lim_{k\to\infty} \varphi((\Psi_a(s)\Delta\Psi_b(s)) \setminus [1, k]) \geq 2^{-n}$ and $\Psi_a(s)\Delta\Psi_b(s)$ is not in $\mathcal{I}$—a contradiction. To see that m' can be found so that the second inequality is satisfied, find j such that $m < n_j$ and pick m' bigger than $\max(\bigcup_{i \leq j}(H_i^a([n_i, n_{i+1})) \cup H_i^b([n_i, n_{i+1})))$. This m' is obviously as required. □

The following partition is the key tool for applying OCA in order to obtain a single function which uniformizes all Ψ_a ($a \in \mathcal{J}_2$).

DEFINITION 3.13.2. For $D \subseteq \mathbb{N}$ and $n \in \mathbb{N}$ let $[\mathcal{J}_2]^2 = L_0^n(D) \cup L_1^n(D)$ be a partition defined by letting $\{a,b\}$ into $L_0^n(D)$ if there is an $s \subseteq a \cap b \cap D$ such that
$$\varphi(\Psi_a(s)\Delta\Psi_b(s)) > 2^{-n}.$$
We shall write L_0^n, L_1^n instead of $L_0^n(\mathbb{N})$, $L_1^n(\mathbb{N})$.

Since each Ψ_a is continuous and φ is lower semicontinuous, these partitions are open in a natural separable metric topology on $\mathcal{J}_2$ obtained when each $a \in \mathcal{J}_2$ is identified with the pair $\langle a, \Psi_a\rangle$. Observe that the proof of the following statement does not make use of either OCA or MA.

PROPOSITION 3.13.3. *For $D \subseteq \mathbb{N}$ the following are equivalent:*
(L1) *The ideal $\mathcal{J}_2$ is σ-$L_1^n(D)$-homogeneous for all n,*
(L2) *D is in $\mathcal{J}_2$.*

PROOF. Assume (L2) and let Ψ_D be a witness for $D \in \mathcal{J}_2$. By Claim 2, for every $a \in \mathcal{J}_2$ there is an m_a such that for all $s \subseteq [m_a, \infty) \cap D \cap a$ we have
$$\varphi(\Psi_D(s)\Delta\Psi_a(s)) \leq 2^{-n-1}.$$
This makes it easy to cover $\mathcal{J}_2$ by countably many $L_1^n(D)$-homogeneous subsets.

Now assume (L1). Without a loss of generality, we can assume $D = \mathbb{N}$, so there are L_1^n-homogeneous sets $\{\mathcal{H}_i^n\}_{i=1}^{\infty}$ covering $\mathcal{J}_2$. Recall that $\hat{\mathcal{H}}$ stands for the downwards closure under $\subseteq$ of a subset $\mathcal{H}$ of $\mathcal{P}(\mathbb{N})$.

CLAIM 3. *If $\mathcal{H} \subseteq \mathcal{J}_2$ is L_1^n-homogeneous, then there is a Borel-measurable mapping $\Theta\colon \mathcal{P}(\mathbb{N}) \to \mathcal{P}(\mathbb{N})$ which is a 2^{-n+2}-approximation to Φ_* on $\hat{\mathcal{H}} \cap \mathcal{J}_{\mathrm{cont}}$.*

PROOF. Let $\mathcal{X}$ be a countable subset of $\mathcal{H}$ such that the set $\{\langle a, \Psi_a\rangle : a \in \mathcal{X}\}$ is dense in $\{\langle a, \Psi_a\rangle : a \in \mathcal{H}\}$. For $b \in \hat{\mathcal{H}}$ fix some element $x(b)$ of $\mathcal{H}$ such that $b \subseteq x(b)$. Recall that for $d \in \mathcal{X}$ the function Ψ_d is asymptotically additive, and let $\{n_i^d\}$, $\{v_i^d\}$ and $\{H_i^d\}$ be the parameters determining it (see Definition 1.5.1). For $d \in \mathcal{X}$ and $a \subseteq \mathbb{N}$ let $i(d,a)$ be the minimal i such that $a \cap [n_i^d, n_{i+1}^d)$ is not included in d, and let
$$k(d,a) = \min \bigcup_{j=i(d,a)}^{\infty} v_j^d.$$
(We are allowing $i(d,a)$ and $k(d,a)$ to be equal to ∞.) These definitions imply that $\Psi_d(a)$ is a reasonable approximation of $\Phi_*(a)$ up to $[1, k(d,a))$ for $a \in \mathcal{J}_{\mathrm{cont}}$, more precisely that for all a and $d \in \mathcal{X}$ we have
(*) $\min(\Psi_d(a)\Delta\Psi_d(a \cap d)) \geq k(d,a).$
For $a \subseteq \mathbb{N}$ let
$$T_a = \{\langle s, k\rangle : s \subseteq [1,k) \text{ and } \varphi((\Psi_d(a \cap d) \cap [1,k))\Delta s) \leq 2^{-n}$$
$$\text{for some } d \in \mathcal{X} \text{ such that } k(d,a) \geq k\},$$
$$\mathcal{Y}_a = \{b \subseteq \mathbb{N} : \langle b \cap [1,k), k\rangle \in T_a \text{ for all } k \in \mathbb{N}\}.$$

We would like to prove that $\mathcal{Y}_a$ is a set of good approximations for $\Theta(a)$, i.e., that any function Θ such that $\Theta(a) \in \mathcal{Y}_a$ for all $a \in \hat{\mathcal{H}} \cap \mathcal{J}_{\mathrm{cont}}$ is a 2^{-n+2}-approximation to Φ_* on that set. Since $\mathcal{X}$ is dense in $\mathcal{H}$ in a strong sense and $\mathcal{H}$ is L_1^n-homogeneous, we have $\Psi_{x(a)}(a) \in \mathcal{Y}_a$ for every $a \in \hat{\mathcal{H}} \cap \mathcal{J}_{\mathrm{cont}}$. By Claim 1 and $\mathcal{J}_2 \supseteq \mathcal{J}_{\mathrm{cont}}$ for every $a \in \hat{\mathcal{H}} \cap \mathcal{J}_{\mathrm{cont}}$ there is an m such that
$$\varphi((\Phi_*(a)\Delta\Psi_{x(a)}(a)) \cap [m, \infty)) \leq 2^{-n+1}.$$

Fix $a \in \hat{\mathcal{H}} \cap \mathcal{J}_{\text{cont}}$ and consider a $b \in \mathcal{Y}_a$. For $k \geq m$ find $d \in \mathcal{X}$ which witnesses $\langle b \cap [1, k), k \rangle \in T_a$, thus in particular $k(d, a) \geq k$. Hence the set $(b \Delta \Phi_*(a)) \cap [m, k)$ is included in the set (recall (*))

$$((b \Delta \Psi_d(a \cap d)) \cup (\Psi_d(a \cap d) \Delta \Psi_d(a))$$
$$\cup (\Psi_d(a) \Delta \Psi_{x(a)}(a)) \cup (\Psi_{x(a)}(a) \Delta \Phi_*(a))) \cap [m, k),$$

and therefore, since $\{d, x(a)\} \in L_1^n$, by using (*) we have

$$\varphi((b \Delta \Phi_*(a)) \cap [m, k)) \leq 2^{-n+2}.$$

This is true for all large enough k and φ is lower semicontinuous, hence every mapping $\Theta \colon \hat{\mathcal{H}} \cap \mathcal{J}_{\text{cont}} \to \mathcal{P}(\mathbb{N})$ such that $\Theta(a) \in \mathcal{Y}_a$ is a 2^{-n+2}-approximation to Φ_*. But the set $\mathcal{Y}_a$ is closed and $a \mapsto \mathcal{Y}_a$ is a Borel mapping of $\mathcal{P}(\mathbb{N})$ into its exponential space, and therefore the mapping Θ defined by $\Theta(a) = \inf \mathcal{Y}_a$ if $\mathcal{Y}_a \neq \emptyset$ and $\Theta(a) = \emptyset$ otherwise is a Baire-measurable function. This ends the construction. $\square$

Claim 3 implies that for every n the ideal $\mathcal{J}_{\text{cont}}$ can be covered by countably many hereditary sets $\hat{\mathcal{H}}_i^n \cap \mathcal{J}_{\text{cont}}$ and that on each one of them there is a Borel 2^{-n+2}-approximation to Φ_*. Therefore by Theorem 3.11.3 we have $\mathbb{N} \in \mathcal{J}_2$. $\square$

By OCA and Proposition 3.13.3, it only remains to prove there are no uncountable L_0^n-homogeneous subsets of $\mathcal{J}_2$ for every n. We shall apply MA to prove this. The proof that the partial ordering we use has the countable chain condition requires the following consequence of OCA.

LEMMA 3.13.4. *OCA implies that $\mathcal{J}_2$ is a P-ideal.*

PROOF. Assume $\mathcal{J}_2$ is not a P-ideal, and let $A_1 \subseteq A_2 \subseteq A_3 \subseteq \ldots$ be a sequence in $\mathcal{J}_2$ such that no $A \in \mathcal{J}_2$ almost includes all A_m. For the convenience of notation, assume $\mathcal{J}_{\text{cont}}$ and $\mathcal{J}_2$ are ideals on $\mathbb{N} \times \mathbb{N}$ and that $A_m = [1, m) \times \mathbb{N}$. For $f \in \mathbb{N}^\mathbb{N}$ let

$$\Gamma_f = \{\langle n, m \rangle : m \leq f(n)\}.$$

Note that our assumption that there is no $A \in \mathcal{J}_2$ which almost includes all A_n translates as $\mathbb{N}^2 \setminus \Gamma_f \notin \mathcal{J}_2$ for all $f \in \mathbb{N}^\mathbb{N}$. If for every f there is a $g \geq^* f$ such that $\Gamma_g \setminus \Gamma_f$ is not in $\mathcal{J}_{\text{cont}}$, then an easy recursive construction shows that $\mathcal{J}_{\text{cont}}$ is not ccc over Fin. Therefore using Proposition 3.12.1 and going to a set of the form $\mathbb{N}^2 \setminus \Gamma_f$ for some f we can assume Γ_g is in $\mathcal{J}_{\text{cont}}$ for all $g \in \mathbb{N}^\mathbb{N}$.

Fix $\bar{n}$. We claim there are no uncountable $L_0^{\bar{n}}$-homogeneous subsets of Γ_f ($f \in \mathbb{N}^\mathbb{N}$). Assume the contrary, let $I \subseteq \mathbb{N}^\mathbb{N}$ be of size $\aleph_1$ and such that $\{\Gamma_f : f \in I\}$ is $L_0^{\bar{n}}$-homogeneous. Then by OCA there is a $g_0 \in \mathbb{N}^\mathbb{N}$ which almost dominates all $f \in I$ (see Theorem 2.2.6). By going to a subset of I and possibly making a finite change to g, we can assume that Γ_{g_0} includes all $\{\Gamma_f : f \in I\}$, and that, therefore, there is an uncountable $L_0^{\bar{n}}(\Gamma_{g_0})$-homogeneous set. This contradicts Proposition 3.13.3. Applying OCA we conclude that $\{\Gamma_f : f \in \mathbb{N}^\mathbb{N}\}$ is covered by countably many $L_1^{\bar{n}}$-homogeneous sets. Since $\mathbb{N}^\mathbb{N}$ is σ-directed under $\leq^*$, for one of these sets, say $\mathcal{X} = \{\Gamma_f : f \in J\}$, the index set J is cofinal in $\langle \mathbb{N}^\mathbb{N}, \leq^* \rangle$ (Lemma 2.2.2). Then by Lemma 2.4.3 we can find m_0 such that for every $g \in \mathbb{N}^\mathbb{N}$ there is an $f \in J$ such that $g(n) \leq f(n)$ for all $n \geq m_0$.

CLAIM 4. *For every a in $\mathcal{J}_2$ there is an $m = m(a) \geq m_0$ such that for all $d \in \mathcal{X}$ and all $s \subseteq (a \cap d) \setminus A_m$ we have $\varphi(\Psi_d(s) \Delta \Psi_a(s)) \leq 2^{-\bar{n}+3}$.*

PROOF. Assume this fails for some $a \in \mathcal{J}_2$. Recursively find $d_k \in \mathcal{X}$ and finite $s_k \subseteq (a \cap d_k) \setminus A_k$, so that
$$\varphi(\Psi_a(s_k) \Delta \Psi_{d_k}(s_k)) > 2^{-\bar{n}+3}$$
for all k. There is a function $g \in \mathbb{N}^\mathbb{N}$ such that Γ_g almost includes $\bigcup_n s_n$, so we can find $d \in \mathcal{X}$ which almost includes $\bigcup_k s_k$, therefore $\bigcup_k s_k \in \mathcal{J}_{\text{cont}}$. By Claim 2, for all but finitely many k we have $\varphi(\Psi_a(s_k) \Delta \Psi_d(s_k)) \leq 2^{-\bar{n}}$, and therefore by the $L_1^{\bar{n}}$-homogeneity of $\mathcal{H}$ we have $\varphi(\Psi_a(s_k) \Delta \Psi_{d_k}(s_k)) \leq 2^{-\bar{n}+1}$, a contradiction. □

Since J dominates $\mathbb{N}^\mathbb{N}$ on $[m_0, \infty)$, by Claim 4 we can easily deduce that for every $m \geq m_0$ the set $\mathcal{H}_m = \{a \in \mathcal{J}_2 : m(a) = m\}$ is $L_1^{\bar{n}-2}(\mathbb{N}^2 \setminus A_m)$-homogeneous. Note that for $a \in \mathcal{H}_m$ we can replace the function Ψ_a by the function
$$b \mapsto \Psi_a(b \setminus A_m) \cup \Psi_{A_m}(b \cap A_m), \qquad \text{for } b \subseteq a$$
and still have that the ideal $\mathcal{K}_a$ (see definition of $\mathcal{J}_2$) is ccc over Fin. With this adjustment each $\mathcal{H}_m$ becomes $L_1^{\bar{n}-2}$-homogeneous. Hence we can assume that $\mathcal{J}_{\text{cont}}$ is covered by countably many hereditary $L_1^{\bar{n}-2}$-homogeneous sets for every n. From this and Proposition 3.13.3 we conclude that $\mathbb{N}^2$ belongs to $\mathcal{J}_2$, contradicting the assumption that no element of $\mathcal{J}_2$ almost includes every set A_n. □

LEMMA 3.13.5. *Martin's Axiom implies that for every $n \in \mathbb{N}$ there are no uncountable L_0^n-homogeneous subsets of $\mathcal{J}_2$.*

PROOF. Assume the contrary, that $\mathcal{H} = \{a_\alpha : \alpha < \omega_1\}$ is an uncountable L_0^n-homogeneous subset of $\mathcal{J}_2$ for some n. Since $\mathcal{J}_2$ is a P-ideal, we can find b_ξ ($\xi < \omega_1$) such that for all $\eta < \xi < \omega_1$ we have

(a) $a_\eta \subseteq^* b_\xi$,
(b) $b_\eta \subseteq^* b_\xi$.

Now redefine each Ψ_{b_ξ} so that it coincides with Ψ_{a_ξ} on $\mathcal{P}(a_\xi)$ and that it still witnesses that $b_\xi \in \mathcal{J}_2$. This can be done in a straightforward manner, using the original Ψ_{b_ξ}. Then the set $\{b_\xi : \xi < \omega_1\}$ is a L_0^n-homogeneous (where L_0^n is computed using the new Ψ_{b_ξ}'s) subset of $\mathcal{J}_2$. Therefore, from now on we can assume that the elements of $\mathcal{H}$ form an $\subseteq^*$-increasing ω_1-chain.

Our forcing terminology is standard. For the undefined notions see ([82]). The partial ordering we shall use is a descendant of Kunen's partial ordering for making gaps in $\mathcal{P}(\mathbb{N})/$Fin ccc-indestructible ([81]; see also [125, p. 74]). Let $\mathcal{P}$ be the poset whose conditions are of the form $p = \langle s, k, F \rangle$, where

($\mathcal{P}1$) $k \in \mathbb{N}$, $s \subseteq [1, k)$, and F is a finite $L_0^{n+2}(s)$-homogeneous subset of $\mathcal{H}$.

We shall add superscript "p" to elements of p when needed, like e.g., in ($\mathcal{P}2$) below. Ordering on $\mathcal{P}$ is defined by letting $p \leq q$ if

($\mathcal{P}2$) $k^p \geq k^q$, $s^p \cap [1, k^q) = s^q$, and $F^p \supseteq F^q$.

If G is a sufficiently generic filter in $\mathcal{P}$ and $D = \bigcup_{p \in G} s^p$, $\mathcal{X} = \bigcup_{p \in G} F^p$, then $\mathcal{X}$ is an $L_0^{n+2}(D)$-homogeneous subset of $\mathcal{H}$. Recall that $\mathcal{J}_2$ includes the ideal $\mathcal{J}_{\text{cont}}$ which is ccc over Fin, and is, therefore, ccc over Fin itself. We will apply MA to an amalgamation $\mathcal{P}_{\omega_1}$ of uncountably many copies of $\mathcal{P}$ to violate this fact, but first we have to prove $\mathcal{P}$ has a chain condition strong enough to assure the ccc-ness of $\mathcal{P}_{\omega_1}$.

DEFINITION 3.13.6. A pair of uncountable subsets X, Y of a poset $\mathcal{P}$ such that every $p \in X$ is incompatible with every $q \in Y$ is called an *uncountable rectangle of incompatible conditions*. One similarly defines the notion of an *uncountable rectangle of compatible conditions*.

Note that the conclusion of the following claim gives an uncountable rectangle of compatible conditions.

CLAIM 5. *For every uncountable rectangle $p_{0\xi}, p_{1\xi}$ ($\xi < \omega_1$) of conditions in $\mathcal{P}$ with equal working parts and such that sets $F_{0\xi}, F_{1\xi}$ from their third components are all pairwise disjoint there exist uncountable $X_0, X_1 \subseteq \omega_1$, $k \in \mathbb{N}$ and $s \subseteq [1, k)$ such that*

$$\langle s, k, F_{0\xi} \cup F_{1\eta} \rangle$$

is a joint extension of $p_{0\xi}$ and $p_{1\eta}$ for all $\xi \in X_0$ and $\eta \in X_1$.

PROOF. For $\bar{s} \subseteq [1, \bar{k})$ let

$$\mathcal{P}_{\bar{s}, \bar{k}} = \text{the set of all } p \in \mathcal{P} \text{ of the form } \langle \bar{s}, \bar{k}, F \rangle.$$

We shall first prove that no subset $\{p_\xi\}$ ($\xi < \omega_1$) of $\mathcal{P}_{\bar{s}, \bar{k}}$ such that F^ξ's are pairwise disjoint includes an uncountable rectangle of incompatible conditions. Assume the contrary, and let $p_{0\xi}, p_{1\xi}$ ($\xi < \omega_1$) be a counterexample. We can assume all $F_{0\xi}$ are of the same size l and all $F_{1\xi}$ are of the same size m, say

(1) $F_{0\xi} = \{a_{0\xi}^1, \ldots, a_{0\xi}^l\}$
(2) $F_{1\xi} = \{a_{1\xi}^1, \ldots, a_{1\xi}^m\}$.

For $\xi < \omega_1$ and $j \leq l$ let $n_i^{0\xi j}$, $v_i^{0\xi j}$ and $H_i^{0\xi j}$ ($i \in \mathbb{N}$) be the parameters of the function $\Psi_{0\xi}^j = \Psi_{a_{0\xi}^j}$, and for $j \leq m$ let $n_i^{1\xi j}$, $v_i^{1\xi j}$ and $H_i^{1\xi j}$ ($i \in \mathbb{N}$) be the parameters of the function $\Psi_{1\xi}^j = \Psi_{a_{1\xi}^j}$ (see Definition 1.5.1). Possibly by refining two sides of the rectangle of conditions by Claim 2 we can assume there is $k \geq \bar{k}$ such that for all ξ (recall that $a \subseteq^k b$ stands for $a \cap [k, \infty) \subseteq b$):

(3) $a_{0\xi}^1 \subseteq^k a_{0\xi}^2 \subseteq^k \cdots \subseteq^k a_{0\xi}^l \subseteq^k a_{1\xi}^1 \subseteq^k \cdots \subseteq^k a_{1\xi}^m$,

and for $\varepsilon = 0, 1$:

(4) $\varphi((\Psi_{\varepsilon\xi}^j(t) \Delta \Psi_{\varepsilon\xi}^{j'}(t)) \cap [k, \infty)) \leq 2^{-n-3}$ for all finite $t \subseteq a_{\varepsilon\xi}^j \cap a_{\varepsilon\xi}^{j'}$ and $j, j' \leq l$.

We can assume there is an $r \in \mathbb{N}$ and $\bar{n}_i^{0j}, \bar{v}_i^{0j}, \bar{H}_i^{0j}$ ($i \leq r, j \leq l$) such that for all ξ and all $i \leq r, j \leq l$ we have

(5) $n_i^{0\xi j} = \bar{n}_i^{0j}$, $v_i^{0\xi j} = \bar{v}_i^{0j}$, and $H_i^{0\xi j} = \bar{H}_i^{0j}$,
(6) $\bar{n}_r^{0j} > k$ for all $j \leq l$.

Fix $\xi < \eta < \omega_1$. Since $\mathcal{H}$ is L_0^n-homogeneous, there is a $t \subseteq a_{0\xi}^1 \cap a_{0\eta}^1$ such that

(7) $\varphi(\Psi_{0\xi}^1(t) \Delta \Psi_{0\eta}^1(t)) > 2^{-n}$.

Note that, by (5), we can assume that $\min t \geq k$. Assume

(8) $\varphi(\Psi_{1\xi}^1(t) \Delta \Psi_{0\xi}^1(t)) > 2^{-n-1}$.

We claim that $p_{0\xi}$ and $p_{1\eta}$ are compatible. First we prove that for all $j \leq m$ and $j' \leq l$ we have

(9) $\varphi(\Psi_{1\xi}^j(t) \Delta \Psi_{0\xi}^{j'}(t)) > 2^{-n-2}$.

Assuming (9) fails for some j and j', by
$$\Psi^1_{1\xi}(t)\Delta\Psi^1_{0\xi}(t) \subseteq (\Psi^1_{1\xi}(t)\Delta\Psi^j_{1\xi}(t)) \cup (\Psi^j_{1\xi}(t)\Delta\Psi^{j'}_{0\xi}(t)) \cup (\Psi^{j'}_{0\xi}(t)\Delta\Psi^1_{0\xi}(t)),$$
we have $\varphi(\Psi^1_{1\xi}(t)\Delta\Psi^1_{0\xi}(t)) \leq 2^{-n-3} + 2^{-n-2} + 2^{-n-3} < 2^{-n-1}$, contradicting (8). Therefore (8) implies $\langle \bar{s} \cup t, \max t+1, F_{1\xi} \cup F_{0\xi}\rangle$ is a joint extension of $p_{0\xi}$ and $p_{1\xi}$. Similarly, the assumption

(10) $\quad \varphi(\Psi^1_{1\xi}(t)\Delta\Psi^1_{0\eta}(t)) > 2^{-n-1}$.

implies, by the same argument as above, that $p_{1\xi}$ and $p_{0\eta}$ are compatible. If (8) and (10) both fail, then we have
$$\Psi^1_{0\xi}(t)\Delta\Psi^1_{0\eta}(t) \subseteq (\Psi^1_{0\xi}(t)\Delta\Psi^1_{1\xi}(t)) \cup (\Psi^1_{1\xi}(t)\Delta\Psi^1_{0\eta}(t))$$
and $\varphi(\Psi^1_{0\xi}(t)\Delta\Psi^1_{0\eta}(t)) \leq 2^{-n-1} + 2^{-n-1} = 2^{-n}$, contradicting (7). Therefore at least one of (8) or (10) happens, and this proves that $p_{0\xi}$, $p_{1\xi}$ ($\xi < \omega_1$) is not a rectangle of incompatible conditions.

To finish the proof, let $p_{0\xi}$, $p_{1\xi}$ ($\xi < \omega_1$) be conditions from $\mathcal{P}$ with equal working parts and such that sets $F_{0\xi}$, $F_{1\xi}$ are all pairwise disjoint. We can assume $F_{0\xi}$ ($F_{1\xi}$) are all of the same size l (m, respectively), and that the set of all $2l$-tuples
$$\langle a^1_{0\xi}, \Psi^1_{0\xi}, \ldots, a^l_{0\xi}, \Psi^l_{0\xi}\rangle$$
is an $\aleph_1$-dense in itself subset of $\mathcal{P}(\mathbb{N})^l \times (\mathbb{N}^\mathbb{N})^l$ (the functions $\Psi^j_{0\xi}$ are assumed to be asymptotically additive, and therefore naturally represented by elements of $\text{Fin}^\mathbb{N} \cong \mathbb{N}^\mathbb{N}$, the set which, as usually, we identify with the irrationals). Similarly we can assume the set of all $2m$-tuples
$$\langle a^1_{1\xi}, \Psi^1_{1\xi}, \ldots, a^m_{1\xi}, \Psi^m_{1\xi}\rangle$$
is an $\aleph_1$-dense in itself subset of $\mathcal{P}(\mathbb{N})^m \times (\mathbb{N}^\mathbb{N})^m$. Since we have already proved that $p_{0\xi}$ ($\xi < \omega_1$) and $p_{1\xi}$ ($\xi < \omega_1$) cannot be an uncountable rectangle of incompatible conditions, there are compatible $p_{0\xi}$ and $p_{1\eta}$. Let $\langle s, k, F\rangle$ be the joint extension of these two conditions. Since the compatibility relation of $\mathcal{P}$ is an open subset of $\mathcal{P} \times \mathcal{P}$ (where $\mathcal{P}$ is taken with the separable metric topology induced from $\text{Fin} \times \mathbb{N} \times \bigoplus_{i=1}^\infty (\mathcal{P}(\mathbb{N}) \times \mathbb{N}^\mathbb{N})^{2i}$ when $\langle s, k, \{a_1, \ldots, a_k\}\rangle$ is identified with $\langle s, k, a_1, \Psi_{a_1}, a_2, \Psi_{a_2}, \ldots, a_k, \Psi_{a_k}\rangle$), there are uncountable sets $X_0, X_1 \subseteq \omega_1$ such that $\langle s, k, F_{0\xi} \cup F_{1\eta}\rangle$ extends $p_{0\xi}$ and $p_{1\eta}$ for all $\xi \in X_0$ and $\eta \in X_1$. This completes the proof. $\square$

Let $\mathcal{P}_{\omega_1}$ be defined as follows: A typical condition is $p = \langle I, s_\xi, k_\xi, F_\xi : \xi \in I\rangle$, where

($\mathcal{P}3$) I is a finite subset of ω_1,
($\mathcal{P}4$) $p(\xi) = \langle s_\xi, k_\xi, F_\xi\rangle$ is in $\mathcal{P}$ for every $\xi \in I$.

The ordering is defined by letting $p \leq q$ if

($\mathcal{P}5$) $I^p \supseteq I^q$,
($\mathcal{P}6$) $p(\xi) \leq_\mathcal{P} q(\xi)$ for all $\xi \in I^q$, and
($\mathcal{P}7$) sets $s^p_\xi \cap [k^q_\xi, \infty)$ ($\xi \in I^q$) are pairwise disjoint.

If G is a sufficiently generic filter in $\mathcal{P}$ and $D_\xi = \bigcup_{p \in G} s^p_\xi$, then set $\mathcal{X}_\xi = \bigcup_{p \in G} F^p_\xi$ is $L^{n+2}_0(D_\xi)$-homogeneous and condition ($\mathcal{P}7$) assures that sets D_ξ are almost disjoint.

CLAIM 6. *The poset $\mathcal{P}_{\omega_1}$ is ccc.*

3.13. THE PROOF OF THE OCA LIFTING THEOREM

PROOF. Let p_α ($\alpha < \omega_1$) be an uncountable subset of $\mathcal{P}_{\omega_1}$. We can assume sets $I^\alpha = I^{p_\alpha}$ from their first coordinates form a Δ-system with root $\bar{I}$. It follows from the definitions that conditions p and q in $\mathcal{P}_{\omega_1}$ are compatible if and only if the conditions

$$\langle I, s_\xi^p, k_\xi^p, F_\xi^p : \xi \in I \rangle \quad \text{and} \quad \langle I, s_\xi^q, k_\xi^q, F_\xi^q : \xi \in I \rangle, \quad \text{where } I = I^p \cap I^q,$$

are compatible. Therefore we can assume that

$$I^\alpha = \bar{I} = \{\xi_1, \ldots, \xi_l\}$$

for all p_α. Uniformizing further, we can assume that for every fixed $i \in \{1, \ldots, l\}$ the working parts of all conditions $p_\alpha(\xi_i)$ ($\alpha < \omega_1$) are equal, say

$$s_{\xi_i}^\alpha = \bar{s}_i \text{ and } k_{\xi_i}^\alpha = \bar{k}_i, \text{ for all } \alpha \text{ and } i \in \{1, \ldots, l\}.$$

We can, moreover, assume that sets $F_{\xi_i}^\alpha = F_{\xi_i}^{p_\alpha}$ from the third coordinates of $p^\alpha(\xi_i)$'s form a Δ-system with root F_{ξ_i}. Now observe that by the definition of $\mathcal{P}$ the conditions $p, q \in \mathcal{P}$ are compatible if and only if the conditions

$$\langle s^p, k^p, F^p \setminus F \rangle \quad \text{and} \quad \langle s^q, k^q, F^q \setminus F \rangle, \quad \text{where } F = F^p \cap F^q,$$

are compatible. Therefore we can assume that the sets $F_{\xi_i}^\alpha$ ($\alpha < \omega_1$) are pairwise disjoint for every fixed $i \in \{1, \ldots, l\}$. We can apply Claim 5 to $p_\alpha(\xi_1)$ ($\alpha < \omega_1$) and get subsets X_1, Y_1 of ω_1 such that $p_\alpha(\xi_1)$ ($\alpha \in X_1$) and $p_\beta(\xi_1)$ ($\beta \in Y_1$) form an uncountable rectangle of compatible conditions. Let $\bar{k}_1' > \bar{k}_1$ and $\bar{s}_1' \subseteq [1, \bar{k}_1')$ be such that

$$\langle \bar{s}_1', \bar{k}_1', F_{\xi_1}^\alpha \cup F_{\xi_1}^\beta \rangle$$

is a joint extension of $p_\alpha(\xi_1)$ and $p_\beta(\xi_1)$ for all $\alpha \in X_1$ and $\beta \in Y_1$. Now we can extend every $p_\alpha(\xi_2)$ ($\alpha \in X_1 \cup Y_1$) to

$$p_\alpha'(\xi_2) = \langle \bar{s}_2, \bar{k}_2', F^\alpha(\xi_2) \rangle$$

so that $\bar{k}_2' > \bar{k}_1$. Another application of Claim 5 gives sets $X_2 \subseteq X_1$ and $Y_2 \subseteq Y_1$ such that $p_\alpha'(\xi_2)$ ($\alpha \in X_1$) and $p_\beta'(\xi_2)$ ($\beta \in Y_2$) is an uncountable rectangle of compatible conditions. Note that, if

$$\langle \bar{s}_2', k_2'', F_{\xi_2}^\alpha \cup F_{\xi_2}^\beta \rangle$$

is a joint extension of $p_\alpha'(\xi_2)$ and $p_\beta'(\xi)$, then by our choice of $\bar{k}_2'$ the condition $(\mathcal{P}7)$ is satisfied between $\bar{s}_1'$ and $\bar{s}_2'$. By continuing this construction for $i = 3, \ldots, l$, we get an uncountable rectangle of compatible conditions in $\mathcal{P}_{\omega_1}$. □

Let G be a sufficiently generic filter of $\mathcal{P}_{\omega_1}$ and for $\xi < \omega_1$ define

$$D_\xi = \bigcup_{p \in G} s_\xi^p \quad \text{and} \quad \mathcal{X}_\xi = \bigcup_{p \in G} F_\xi^p.$$

Then D_ξ's are pairwise almost disjoint and $\mathcal{X}_\xi$ is $L_0^{n+2}(D_\xi)$-homogeneous for all ξ. By ccc-ness of $\mathcal{P}_{\omega_1}$ there must be a condition p in it which forces there are uncountably many ξ's with uncountable $\mathcal{X}_\xi$'s. Then we can choose a suitable family of $\aleph_1$ many dense open subsets of $\mathcal{P}_{\omega_1}$ so that if G is a filter of $\mathcal{P}_{\omega_1}$ containing p and intersecting all these dense open sets, then for uncountably many ξ the set $\mathcal{X}_\xi$ is uncountable. (Note that this implies the corresponding D_ξ are infinite.) By Proposition 3.13.3, for such ξ the set D_ξ is not in $\mathcal{J}_2$ and therefore not in $\mathcal{J}_{\text{cont}}$. But this is in the contradiction with Proposition 3.12.1, and this completes the proof of Lemma 3.13.5. □

By Lemma 3.13.5 and OCA, the ideal $\mathcal{J}_2$ can be decomposed into countably many L_1^n-homogeneous subsets for all n. By Proposition 3.13.3 this implies $\mathbb{N} \in \mathcal{J}_2$. This ends the proof of Theorem 3.3.5. □

3.14. Remarks and questions

Let us start with a diagram of the embeddability relation between analytic quotients under OCA and MA.

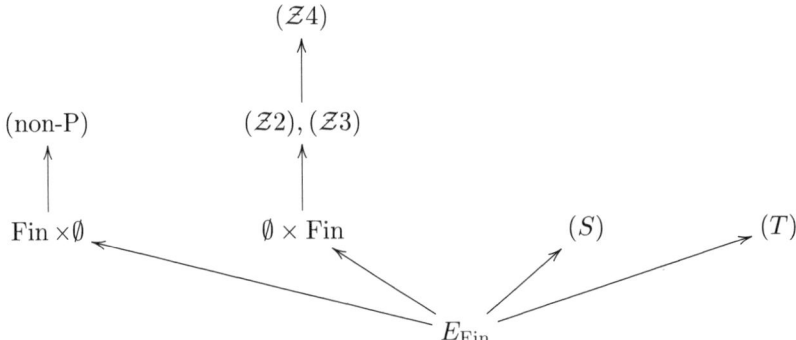

The ordering $\leq_{\mathrm{EM}}$ under OCA and MA.

This diagram is very similar to one given previously, which presents the $\leq_{\mathrm{BE}}$-ordering (see §1.14 for more details and definitions). The only visible differences between this diagram and one in §1.14 concern the classes of the summable ideals and the density ideals. Class (S) above consists of all summable ideals distinct from Fin; note that, by Claim 1 in the proof of Lemma 1.12.3, all these quotients are pairwise embeddable. This can be contrasted to the fact that there is an isomorphic copy of $\mathcal{P}(\mathbb{N})/\mathrm{Fin}$ inside the structure of summable ideals under the ordering $\leq_{\mathrm{BE}}$ of Baire-embeddability. A similar collapsing takes place in the domain of the density ideals, but the quotients over the ideals in ($\mathcal{Z}4$) are not embeddable into quotients over ideals in ($\mathcal{Z}3$). This can be proved by a proof similar to the proof of Theorem 1.13.3 (b). Note that the above diagram is even more similar to the one depicting the Borel-cardinality of analytic quotients (see [**75**]).

Motivated by the recent results of Kanovei and Reeken on the Radon–Nikodym property of analytic ideals which are not necessarily P-ideals (see §1.14), we pose the following.

PROBLEM 3.14.1. *Find more analytic ideals $\mathcal{I}$ with the property that under OCA and MA every homomorphism of $\mathcal{P}(\mathbb{N})/\mathrm{Fin}$ into $\mathcal{P}(\mathbb{N})/\mathcal{I}$ has a completely additive almost lifting.*

In particular, do the F_σ-ideals, or the ordinal ideals $\mathcal{I}_\alpha$ and $\mathcal{W}_\alpha$, (see §1.14) have this property?

It is also relevant to know that Just has proved ([**63**]) that a local version of the lifting theorem (Theorem 3.12.1) is consistent for all analytic ideals. It moreover follows from OCA and MA, by [**64**], for F_σ ideals (the open coloring in the proof of this fact is defined using Mazur's Theorem 1.2.5(a)).

There is an open question related to our work which can be regarded as a version of Todorcevic's conjecture for arbitrary connecting maps. Note that every additive lifting of a homomorphism is of the form $F_\mathcal{U}$ defined in Example 3.2.3, since if $F\colon \mathcal{P}(\mathbb{N}) \to \mathcal{P}(\mathbb{N})$ is a homomorphism then $\mathcal{U}_n = \{A : F(A) \ni n\}$ is an ultrafilter on $\mathbb{N}$. Since nonprincipal ultrafilters on $\mathbb{N}$ never have a property of Baire, a Baire-measurable additive lifting has to be completely additive and therefore Todorcevic's conjecture (see §1.4) can be rephrased as: Can every Baire-measurable homomorphism of quotients over analytic P-ideals be turned into an additive one? All examples of nontrivial homomorphisms of $\mathcal{P}(\mathbb{N})/\operatorname{Fin}$ constructed in §3.2 using the Axiom of Choice (or rather, the existence of a nonprincipal ultrafilter on $\mathcal{P}(\mathbb{N})$) have additive liftings.

QUESTION 3.14.2. *Do OCA and MA imply that every homomorphism*

$$\Phi\colon \mathcal{P}(\mathbb{N})/\operatorname{Fin} \to \mathcal{P}(\mathbb{N})/\operatorname{Fin}$$

has an additive lifting? Or, is this statement at least consistent with the usual axioms for Set theory?

A part of the motivation for asking this question comes from Chapter 4, (see Question 4.11.4). By Theorem 3.3.5 an answer to this question requires analysis of homomorphisms of $\mathcal{P}(\mathbb{N})/\operatorname{Fin}$ into itself whose kernel is ccc over Fin. The methods developed here do not seem to be relevant to this question.

Throughout this monograph we have been looking for the strongest possible rigidity results, and we have seen (§2.1) that OCA and MA form the right setting for such investigation. Let us say a word on some different lines of research. According to P. Erdős [28], the initial motivation for the question whether quotients over ideals $\mathcal{Z}_0$ and $\mathcal{Z}_{\log}$ are isomorphic (see §1.13) was the question of finding a large class of ideals on $\mathbb{N}$ whose quotients are pairwise nonisomorphic. This was later solved by Monk and Solovay ([98]) who have found such a class of $2^{2^{\aleph_0}}$ many ideals, which is the best possible. Of course, their ideals are not analytic.

QUESTION 3.14.3. *Are there infinitely (or even uncountably) many analytic P-ideals whose quotients are, provably in ZFC, pairwise nonisomorphic?*

In [65] the following was proved (see also the paragraph after Lemma 1.3.3).

THEOREM 3.14.4 (Just–Krawczyk).

(a) $\mathcal{P}(\mathbb{N})/\operatorname{Fin}$ *is never isomorphic to a quotient over an Erdős–Ulam ideal.*
(b) *All quotients over F_σ ideals are countably saturated and therefore isomorphic under CH.*
(c) *All quotients over Erdős–Ulam ideals are isomorphic under CH.* □

We know only a few analytic P-ideals whose quotients are provably pairwise nonisomorphic. Theorem 3.14.4 implies that quotients over the following ideals are pairwise nonisomorphic:

(i) Fin (whose quotient is under CH isomorphic to the quotient over any other F_σ ideal, by Theorem 3.14.4 (b)),
(ii) $\emptyset \times \operatorname{Fin}$ (whose quotient is isomorphic to $(\mathcal{P}(\mathbb{N})/\operatorname{Fin})^\mathbb{N}$,
(iii) $\mathcal{Z}_0$ (whose quotient is under CH isomorphic to a quotient over any other EU-ideal by Theorem 3.14.4 (c))),
(iv) $\mathcal{Z}_0 \oplus \operatorname{Fin}$ (whose quotient is isomorphic to $(\mathcal{P}(\mathbb{N})/\mathcal{Z}_0) \times (\mathcal{P}(\mathbb{N})/\operatorname{Fin})$),
(v) $\mathcal{Z}_0 \oplus \emptyset \times \operatorname{Fin}$ (whose quotient is isomorphic to $(\mathcal{P}(\mathbb{N})/\mathcal{Z}_0) \times (\mathcal{P}(\mathbb{N})/\operatorname{Fin})^\mathbb{N}$),
(vi) The ideal from Example 1.13.17.

The fact that the quotients over the above ideals are pairwise nonisomorphic follows from the facts that $\mathcal{Z}_0 \upharpoonright A$ is an Erdös–Ulam ideal for every positive set A (Corollary 1.13.5), that $\mathcal{P}(\mathbb{N})/\mathrm{Fin}$ is homogeneous, and Theorem 3.14.4 (a) above.

Recall that, by Corollary 1.13.8, CH implies that $(\mathcal{P}(\mathbb{N})/\mathcal{Z}_0)^{\mathbb{N}} \approx \mathcal{P}(\mathbb{N})/\mathcal{Z}_0$.

An another line of research which has attracted some attention is concerned with the possible structure of the automorphism group of $\mathcal{P}(\mathbb{N})/\mathrm{Fin}$, and in particular in finding models of set theory whose impact on this structure is intermediate between CH and OCA and MA in some prescribed way. For example, Velickovic proved in [**142**] that MA alone does not imply that all automorphisms of $\mathcal{P}(\mathbb{N})/\mathrm{Fin}$ are trivial. In [**108**] Shelah and Steprans proved that there can be nontrivial automorphisms of $\mathcal{P}(\mathbb{N})/\mathrm{Fin}$, while all automorphisms are locally trivial. For more recent results on the possible structure of the automorphism group of $\mathcal{P}(\mathbb{N})/\mathrm{Fin}$ see [**109, 119**]. Let us mention a question along similar lines.

QUESTION 3.14.5. *Is OCA a sufficient assumption for the lifting theorems of this chapter?*

Note that by the already mentioned result from [**142**], MA alone is not sufficient. It is known that OCA alone does imply some local versions of lifting theorems; see [**64**]. There is certain interest in finding OCA's limitations, but the author finds the program outlined in §2.1 much more attractive.

CHAPTER 4

Weak Extension Principle

4.1. Introduction

In this Chapter we use some results of Chapters 1, 2 and 3 to analyze Čech–Stone remainders of countable, locally compact spaces. By a classical result of Parovičenko ([**101**]), Continuum Hypothesis (CH) implies that all these spaces are homeomorphic and that their finite products are continuous images of ω^*. Van Douwen has proved that, if X is a countable locally compact space, then two finite powers of X^* are homeomorphic if and only if their exponents are equal ([**17**]). Later on, Just ([**61**]) has described a situation in which $(\omega^*)^{(d)}$ does not even map onto $(\omega^*)^{(d+1)}$ (this result was later reformulated into OCA-context, see [**64**]). More recently, A. Dow and K.P. Hart ([**23**]) have proved that under OCA the only Čech–Stone remainder of a σ-compact, locally compact, space which is a surjective image of ω^* is ω^* itself. The key part in their proof is the fact that $(\omega^2)^*$, the remainder of an ordinal ω^2, is not a surjective image of ω^*. (Note that in [**23**] the ordinal ω^2 is denoted by $\mathbb{D}$.) Before [**23**], van Mill ([**97**, Theorem 2.2.1]) proved that ω^* and $(\omega^2)^*$ are not homeomorphic assuming that all autohomeomorphisms of ω^* are trivial, the assumption which we now know to follow from OCA and MA ([**142**]).

We shall give a uniformized treatment of these results and extend them. Our main tool will be the following *weak Extension Principle* (α and γ will always denote countable ordinals and d, l are positive integers):

wEP(α,γ) For a continuous $F\colon (\alpha^*)^{(d)} \to (\gamma^*)^{(l)}$ there is a clopen partition $(\alpha^*)^{(d)} = U_0 \cup U_1$ such that $F''U_0$ is nowhere dense inside $(\gamma^*)^{(l)}$ and $F \restriction U_1$ extends to a continuous map $F_1\colon (\beta\alpha)^{(d)} \to (\beta\gamma)^{(l)}$ so that $F_1''\alpha^{(d)} \subseteq \gamma^{(l)}$.

By wEP we shall denote the statement "wEP(α,γ) for all countable ordinals α,γ." We shall prove that it follows from from OCA and MA.

The set of natural numbers is in this Chapter denoted by ω instead of $\mathbb{N}$, because we are working with countable ordinals. Note that a product of d many copies of X is denoted by $X^{(d)}$. This is done in order to avoid confusion and distinguish the ordinal ω^d from the finite power $\omega^{(d)}$.

A subset A of $\omega^{(\alpha)}$ is d-*compact* if it can be decomposed into sets $A_1,\ldots,A_d$ such that the i-th projection of A_i ($i=1,\ldots,d$) is compact. Let $\mathcal{R}^d_\alpha$ denote the algebra of subsets of $\alpha^{(d)}$ generated by clopen rectangles. We shall usually denote the algebra $\mathcal{R}^d_\omega$ by $\mathcal{R}^d$. Clearly $(\alpha^*)^{(d)}$ is the Stone space of the quotient of $\mathcal{R}^d_\omega$ over the ideal of d-compact sets. It is understood that, whenever we talk about the Čech–Stone remainder X^*, the space X is assumed to be non-compact. By $X \twoheadrightarrow Y$ we denote the fact that Y is a continuous image of X.

The organization of this chapter. In §4.2 we prove a combinatorial lemma about the maps $h\colon \omega^2 \to \omega$ which is later used to decompose maps $f\colon (\beta\omega)^{(2)} \to \beta\omega$.

(This lemma is generalized in [**42**].) In §4.4 we analyze continuous mappings between finite powers of $\beta\omega$, extending some work of van Douwen ([**17**]) and partially confirming two of his conjectures ([**17**, Conjectures 8.3 and 8.4]; see Conjectures 4.3.1 and 4.4.1 below). In particular we give a complete description of autohomeomorphisms of finite powers of $\beta\omega$. These results do not use any additional set-theoretic assumptions. Using wEP, we also give a description of autohomeomorphisms of a finite power of ω^*. In §4.5 wEP is used to characterize when α^* maps onto γ^* for countable ordinals α and γ, and to analyze the question when some finite power of α^* maps onto some finite power of γ^*. These results are extended to locally compact countable spaces in §4.7. In §4.8 we prepare for the proof of wEP from OCA and MA, given in §4.9. Some extensions of wEP are stated in §4.10, and other remarks and open questions can be found in §4.11.

The results of this Chapter can be put in proper mathematical context only in the light of §2.1.

4.2. Dependence of functions on their variables

In this section we prove a combinatorial result about the dependence of functions on their variables (see [**26**] for a related study of functions). This result is the essence of Theorem 4.3.2 from the following section. Let us introduce some definitions. For a finite power $X^{(d)}$ of a set X and $j \leq d$ let $\pi_j \colon X^{(d)} \to X$ be the projection mapping to the j-the coordinate. Mapping $F \colon X^{(d)} \to Y$ *depends on at most one coordinate*, (or *depends on at most one variable*) on $U \subseteq X^{(d)}$ if there is $i \leq d$ such that $F \restriction U = (G \circ \pi_i) \restriction U$ for some $G \colon X \to Y$.

THEOREM 4.2.1. *For every mapping $h \colon \omega^{(2)} \to \omega$ one of the following applies:*

(1) *There is a partition of $\omega^{(2)}$ into finitely many rectangles such that on each one of them h depends on at most one variable.*
(2) *There are infinite sets $X = \{m_i\}$ and $Y = \{n_i\}$ and $f \colon \omega \to \{0,1\}$ such that*
$$f \circ h(m_i, n_j) = \begin{cases} 0, & \text{if } i < j, \\ 1, & \text{if } i = j. \end{cases}$$

Before we start the proof, let us note that (1) and (2) exclude each other: If h satisfies (2) and $\omega^{(2)}$ is split into finitely many rectangles, then one of these rectangles contains $\langle m_i, n_i \rangle$ and $\langle m_j, n_j \rangle$ for some $i < j$. But $h(m_i, n_j)$ differs from both $h(m_i, n_i)$ and $h(m_j, n_j)$, and therefore h depends on both coordinates on this rectangle.

PROOF OF THEOREM 4.2.1. Fix for a moment a pair of nonprincipal ultrafilters $\mathcal{U}, \mathcal{V}$ on ω. If there are no sets $A \in \mathcal{U}$ and $B \in \mathcal{V}$ such that h depends on at most one coordinate on $A \times B$, we shall recursively pick sequences m_i, n_i satisfying (2). It will clearly suffice to assure

(3) $h(m_i, n_j) \neq h(m_k, n_k)$ whenever $i < j$ and $k \in \omega$.

Let us recall a standard notation. If $\mathcal{U}$ is an ultrafilter and $\tau(x)$ is a statement of the language of set theory, then
$$(\mathcal{U}x)\tau(x)$$
stands for "the set $\{x : \tau(x)\}$ is in $\mathcal{U}$."

4.2. DEPENDENCE OF FUNCTIONS ON THEIR VARIABLES

CASE 1. Suppose that
$$A_1 = \{m : (\mathcal{V}n)(\mathcal{V}n')h(m,n) \neq h(m,n')\} \in \mathcal{U}$$
and
$$B_1 = \{n : (\mathcal{U}m)(\mathcal{U}m')h(m,n) \neq h(m',n)\} \in \mathcal{V}.$$
Let $m_1 = \min A_1$, $B_1^1 = \{n \in B_1 : (\mathcal{V}n')h(m_1,n) \neq h(m_1,n')\} \in \mathcal{V}$, and $n_1 = \min B_1^1$. Assume $m_1 < m_2 < \cdots < m_{k-1}$ and $n_1 < n_2 < \cdots < n_{k-1}$ are chosen to satisfy (3), and moreover that $m_i \in A_1$ and $n_i \in B_1^1$ for all i. Let
$$B_k = \{n \in B_1 : h(m_i,n) \neq h(m_j,n_j) \text{ for } i,j < k\}.$$
Then $m_i \in A_1$ implies that B_k is in $\mathcal{V}$. [If B_k is not in $\mathcal{V}$, then for some fixed $i, j < k$ the set $C = \{n : h(m_i,n) = h(m_j,n_j)\}$ would be in $\mathcal{V}$ and we would therefore have $\mathcal{V}n\mathcal{V}n'(h(m_i,n) = h(m_i,n'))$, contradicting $m_i \in A_1$.] Let $n_k = \min(B_k \setminus \{1, \ldots, n_{k-1}\})$, so that we have $h(m_i,n_k) \neq h(m_j,n_j)$ for all $i, j < k$. Then $n_k \in B_1$ implies
$$A_k = \{m \in A_1 : h(m,n_k) \neq h(m_i,n_j) \text{ for all } i < j \leq k\} \in \mathcal{U}$$
so let $m_k = \min(A_k \setminus \{1, \ldots, m_{k-1}\})$. This describes the recursive construction of sequences $\{m_i\}$ and $\{n_i\}$ which satisfy (3).

CASE 2. Suppose that
$$A_1 = \{m : (\mathcal{V}n)(\mathcal{V}n')h(m,n) = h(m,n')\} \in \mathcal{U}$$
but there are no $A \in \mathcal{U}$ and $B \in \mathcal{V}$ such that $h \upharpoonright A \times B$ depends on at most one coordinate. Define $g \colon A_1 \to \omega$ by
$$g(m) = \lim_{n \to \mathcal{V}} h(m,n).$$
If $\lim_{m \to \mathcal{U}} g(m)$ is an integer, then let $A_2 \in \mathcal{U}$ be such that g is constant on A_2. Since for every $B \in \mathcal{V}$ the function h does not depend only on the first coordinate on the rectangle $A_2 \times B$, we can recursively find m_i, n_i so that
$$h(m_i,n_j) = \lim_{m \to \mathcal{U}} g(m) \quad \text{and} \quad h(m_i,n_i) \neq \lim_{m \to \mathcal{U}} g(m) \text{ for all } i < j,$$
therefore (3) is satisfied. If $\lim_{m \to \mathcal{U}} g(m)$ is not an integer, consider the function $g' \colon A_1 \to \omega$ defined by
$$g'(m) = h(m,n), \text{ if } n > m \text{ is minimal such that } h(m,n) \neq g(m).$$
If $\lim_{m \to \mathcal{U}} g'(m)$ is an integer, then we can recursively choose m_i, n_i such that
$$h(m_i,n_i) = \lim_{m \to \mathcal{U}} g'(m) \quad \text{and} \quad h(m_i,n_j) = g(m) \neq \lim_{m \to \mathcal{U}} g'(m) \text{ for all } i < j,$$
and this assures (3). Finally, if neither $\lim_{m \to \mathcal{U}} g(m)$ nor $\lim_{m \to \mathcal{U}} g'(m)$ is an integer then by construction similar to that of Case 1 we obtain sequences m_i and n_i as required.

CASE 3. Suppose that
$$A_1 = \{m : (\mathcal{V}n)(\mathcal{V}n')h(m,n) \neq h(m,n')\} \in \mathcal{U}$$
$$B_1 = \{n : (\mathcal{U}m)(\mathcal{U}m')h(m,n) = h(m',n)\} \in \mathcal{V}$$
and there are no $A \in \mathcal{U}$ and $B \in \mathcal{V}$ such that $h \upharpoonright A \times B$ depends on at most one coordinate. Assume $m_1 < m_2 < \cdots < m_{k-1}$ and $n_1 < n_2 < \cdots < n_{k-1}$ are chosen to satisfy (3) and moreover

(4) $m_i \in A_1$ and $n_i \in B_1$ for all i.

Let
$$B'_k = \{n \in B_1 : (\exists m \in (A_1 \setminus \{1, \ldots, m_{k-1}\}))$$
$$(h(m,n) \neq h(m_i, n)) \text{ for all } i < j < k\}.$$

If B'_k was not in $\mathcal{V}$, then we could easily find $A \in \mathcal{U}$ and $B \in \mathcal{V}$ such that h depends on only one coordinate on $A \times B$, so we may assume $B'_k \in \mathcal{V}$. Let
$$B_k = \{n \in B'_k : h(m_i, n) \neq h(m_j, n_j) \text{ for all } i, j < k\}.$$
Then $m_i \in A_1$ $(i = 1, \ldots, k-1)$ implies $B_k \in \mathcal{V}$. Let $n_k = \min(B_k \setminus \{1, \ldots, n_{k-1}\})$, and let m_k be a witness for $n_k \in B'_k$. Then $\{m_i, n_i\}_{i=1}^k$ satisfy both (3) and (4), so this describes the recursive construction.

We have proved that, if (2) fails, then for every pair $\mathcal{U}, \mathcal{V}$ of nonprincipal ultrafilters on ω we can find sets $A \in \mathcal{U}$ and $B \in \mathcal{V}$ such that h depends on only one coordinate on $A \times B$. Since such sets A, B exist when at least one of $\mathcal{U}, \mathcal{V}$ is a principal ultrafilter, compactness of $(\beta\omega)^2$ implies (1). This completes the proof. $\square$

Let us note a variation on Theorem 4.2.1. If $\mathcal{I}$ is an ideal on ω, then a subset A of $\omega^{(2)}$ is 2-$\mathcal{I}$-small if we can split A into two pieces A_1 and A_2 so that the i-th projection of A_i $(i = 1, 2)$ is in $\mathcal{I}$.

THEOREM 4.2.2. *Let $\mathcal{I}$ be a countably generated ideal on ω. Then for every mapping $h: \omega^{(2)} \to \omega$ one of the following applies:*
1. *There is a partition of $\omega^{(2)}$ into finitely many rectangles such that each one of them is either 2-$\mathcal{I}$-small or on it h depends on at most one variable.*
2. *There are infinite sets $X = \{m_i\}$ and $Y = \{n_i\}$ and $f: \omega \to \{0, 1\}$ such that*
$$f \circ h(m_i, n_j) = \begin{cases} 0, & \text{if } i < j, \\ 1, & \text{if } i = j, \end{cases}$$
and both sets $\{m_i : i \in \omega\}$ and $\{n_i : i \in \omega\}$ are in $\mathcal{I}^\perp$.

PROOF. Exactly like the proof of Theorem 4.2.1, only that the ultrafilters $\mathcal{U}$ and $\mathcal{V}$ have to be chosen to avoid $\mathcal{I}$. If (1) fails, then the required sequences m_i and n_i can be easily chosen to satisfy the requirements since the ideal $\mathcal{I}$ is countably generated. $\square$

Theorem 4.2.1 is generalized to maps $h: X^{(d)} \to X$, for an arbitrary X and a natural number d, in [**42**].

4.3. Prime mappings

In [**17**, Prime Mapping Lemma], van Douwen proved that for every continuous map $f: (\beta\omega)^{(d)} \to \beta\omega$ there is a clopen set U such that $f \upharpoonright U$ is elementary. He has actually proved this result for a larger class of so-called $\beta\omega$-*spaces*. In [**17**, Conjecture 8.4] (reproduced as [**52**, Question 42]), he has conjectured the following:

CONJECTURE 4.3.1 (van Douwen). *Let X be ω^* or $\beta\omega$. If f is any continuous binary operation on X, then there is a disjoint open cover $\mathcal{U}$ of X such that the restriction of φ to any set in $\mathcal{U}$ depends on at most one coordinate.*

The following theorem confirms van Douwen's conjecture in case when $X = \beta\omega$.

THEOREM 4.3.2. *Assume $F\colon (\beta\omega)^{(d)} \to (\beta\omega)^{(l)}$ is a continuous function. Then there is a partition $(\beta\omega)^{(d)} = U_1 \cup \cdots \cup U_k$ into clopen sets such that $\pi_j \circ F$ depends on at most one coordinate on each U_i ($i = 1, \ldots, k$) for all $j = 1, \ldots, l$.*

Before we prove Theorem 4.3.2, let us note that, together with wEP, it gives a partial positive result for the case $X = \omega^*$ of van Douwen's conjecture.

THEOREM 4.3.3. *Assume wEP and that $F\colon (\omega^*)^{(d)} \to (\omega^*)^{(l)}$ is a continuous function. Then there is a partition $(\omega^*)^{(d)} = U_1 \cup \cdots \cup U_k \cup V$ into clopen sets such that $\pi_j \circ F$ depends on at most one coordinate on each U_i ($i = 1, \ldots, k$) for all $j = 1, \ldots, l$ and the image of V is nowhere dense.*

PROOF. By wEP, we can decompose $(\omega^*)^{(d)}$ into two clopen pieces, U and V, such that the image of V is nowhere dense and $f \upharpoonright U$ continuously extends to a map from $(\beta\omega)^{(d)}$ into $\beta\omega$. Now apply Theorem 4.3.2 to this extension. $\square$

PROOF OF THEOREM 4.3.2. Define $h\colon \omega^{(2)} \to \beta\omega$ by $h = F \upharpoonright \omega^{(2)}$. If for some $A, B \subseteq \omega$ the map h depends on at most one variable on $A \times B$, then F depends on at most one variable on $\beta A \times \beta B$. Therefore by Theorem 4.2.1 it suffices to show that (2) does not apply.

Assume otherwise, and let m_i and n_i be the sequences such that $h(m_i, n_j) \neq h(m_k, n_k)$ for all $i < j$ and all k. Define a map $g\colon \omega^{(2)} \to \beta\omega$ by $g(i,j) = h(m_i, n_j)$. This map extends to a continuous map $G\colon (\beta\omega)^{(2)} \to \beta\omega$. We shall need a slight extension of [**17**, Lemma 14.1], but let us recall a bit of a notation first.

For $A \subseteq \omega$ define three subsets of $\omega^{(2)}$ as follows:

$$\Delta A = \{\langle n, n \rangle : n \in A\},$$
$$[A]^2 = \{\langle m, n \rangle : m < n \text{ are in } A\},$$
$$\langle A \rangle^2 = \{\langle m, n \rangle : m \leq n \text{ are in } A\}.$$

LEMMA 4.3.4. *If $f\colon (\beta\omega)^{(2)} \to \beta\omega$ is a continuous map, then the sets $f''[\omega]^2$ and $f''\Delta\omega$ are not disjoint.*

PROOF. The proof relies on the following fact, due to van Douwen.

CLAIM 1. *If s_n are disjoint finite subsets of ω such that $|s_n| \geq n$ for all n, then the closures of the sets $\bigcup [s_n]^2$ and $\Delta \bigcup s_n$ are not disjoint.*

PROOF. See [**17**, Fact 14.2]. $\square$

We shall consider three cases.

CASE 1. The closure of the set $X = f''\Delta\omega$ in $\beta\omega$ is countably infinite. Then this is a countable compact subset of $\beta\omega$, which is a contradiction.

CASE 2. The closure of the set $X = f''\Delta\omega$ in $\beta\omega$ is uncountable. Then we can find $\mathcal{U} \in \omega^*$ such that $f(\langle \mathcal{U}, \ldots, \mathcal{U} \rangle) \notin f''\Delta\omega$, and in particular

(**) $f(\langle \mathcal{U}, \ldots, \mathcal{U} \rangle) \notin f''\langle \omega \rangle^2$.

Let $U_1 \supset \overline{U_2} \supset U_2 \supset \overline{U_3} \supset \ldots$ be clopen neighborhoods of $\mathcal{U}$ such that $\bigcap_n U_n \cap f''\langle \omega \rangle^2 = \emptyset$. By the continuity there are sets $A_n \in \mathcal{U}$ such that $f''[A_n]^2 \subseteq U_n$ for every n. Let $s_n \subseteq A_n$ be pairwise disjoint and such that $|s_n| = n$ for all n. By Claim 1, the closures of the sets $X = \bigcup_n [s_n]^2$ and $Y = \Delta(\bigcup_n s_n)$ have a nonempty intersection, and therefore the closures of their f-images intersect as well. On the other hand, the set $T = f''X \cup f''Y$ is relatively discrete in $\beta\omega$ since $T \setminus U_n$ is finite for all n, and therefore all the limit points of T lie outside of $\bigcap_n U_n \supseteq f''[\omega]^2 \supseteq T$. But this is a contradiction.

CASE 3. The closure of the set $X = f''\Delta\omega$ in $\beta\omega$ is finite. By going to an infinite subset of ω, we can assume that for some fixed $\bar{z} \in \beta\omega$ we have $f(m,m) = \bar{z}$ for all m. Note that by Claim 1, closures of the sets $[A]^2$ and ΔA have a nonempty intersection for every infinite $A \subseteq \omega$. Therefore we can derive a contradiction by finding an infinite A such that the closures of $f''[A]^2$ and $f''\Delta A$ are disjoint.

Recall that for $A \subseteq \omega$ by $[A]^3$ we denote the set of all unordered triples. This can be identified with the set of all strictly increasing, ordered triples from A. By $\{l,m,n\}_<$ we denote a triple $\{l,m,n\}$ such that $l < m < n$.

Consider the partition $[\omega]^3 = K_0 \cup K_1 \cup K_2 \cup K_3$ defined by:

$$\{l,m,n\}_< \in \begin{cases} K_0, & \text{if } f(l,m) = f(l,n) \text{ and } f(l,n) = f(k,n), \\ K_1, & \text{if } f(l,m) = f(l,n) \text{ and } f(l,n) \neq f(k,n), \\ K_2, & \text{if } f(l,m) \neq f(l,n) \text{ and } f(l,n) = f(k,n), \\ K_3, & \text{if } f(l,m) \neq f(l,n) \text{ and } f(l,n) \neq f(k,n). \end{cases}$$

Recall that a set A is *homogeneous* for this partition if $[A]^3$ is included in K_i for some $i \leq 3$. Then we say that A is K_i-homogeneous. By Ramsey's theorem, let A be an infinite set homogeneous for this partition.

If A is K_0-homogeneous, then $f''[A]^2 = \{\bar{z}'\}$ and $f''\Delta A = \{\bar{z}\}$, and $\bar{z} \neq \bar{z}'$—a contradiction.

If A is K_1-homogeneous, then let $z_m = f(m,n)$, for $m < n$ in A (note that by the assumption z_m indeed does not depend on n). Note that $\bar{z}$ has an open neighborhood U such that the set $C = \{m : z_m \notin U\}$ is infinite. But this implies that the closures of $f''\Delta C$ and $f''[C]^2$ are disjoint, a contradiction.

If A is K_2-homogeneous, define $z_n = f(m,n)$ for $m < n$ in A. Like above, derive a contradiction by finding an infinite subsequence of $\{z_n\}$ which does not accumulate to $\bar{z}$.

Now assume A is K_3-homogeneous. Define $f_1 \colon \omega^{(2)} \to \beta\omega$ by

$$f_1(m,n) = f(m, n+1).$$

Then f_1 continuously extends to $f_2 \colon (\beta\omega)^{(2)} \to \beta\omega$. Since $f_2''\Delta\omega$ is infinite, either Case 1 or Case 2 applies to obtain a contradiction.

This completes the proof of Lemma 4.3.4. □

CLAIM 2. *Every point* $\langle \mathcal{U}_1, \mathcal{U}_2, \ldots, \mathcal{U}_{d+1}\rangle \in (\beta\omega)^{(d+1)}$ *has a clopen neighborhood U such that F depends on only one coordinate on U.*

PROOF. For every $n \in \omega$ consider the function $F_n \colon (\beta\omega)^{(d)} \to \beta\omega$ defined by

$$F_n(\mathcal{V}_1, \ldots, \mathcal{V}_d) = F(\mathcal{V}_1, \ldots, \mathcal{V}_d, \{n\}).$$

By the inductive assumption, there is an $i = i(n) \leq d$, sets $B_{nj} \in \mathcal{U}_j$ ($j = 2, \ldots, d$) and a function $g_n \colon \omega \to \beta\omega$ such that for all $\mathcal{V}_j \in B_{nj}^*$ ($j \leq d$) we have

$$F_n(\mathcal{V}_1, \ldots, \mathcal{V}_n) = \{A : g_n^{-1}(A) \in \mathcal{U}_{i(n)}\}.$$

If $i = \lim_{n \to \mathcal{U}_{d+1}} i(n)$, then in some clopen neighborhood of $\langle \mathcal{U}_1, \ldots, \mathcal{U}_{d+1}\rangle$ the function F depends only on two coordinates, i and $d+1$. Therefore the case $d = 2$ applies to reduce this further to a single coordinate on some clopen neighborhood of $\langle \mathcal{U}_1, \ldots, \mathcal{U}_{d+1}\rangle$. □

By the compactness, Claim 2 implies that $(\beta\omega)^{(d+1)}$ can be partitioned into finitely many clopen sets as required. This ends the proof of the case $l = 1$.

Now assume $l > 1$ and fix $F\colon (\beta\omega)^{(d)} \to (\beta\omega)^{(l)}$. Consider maps $F_i = \pi_i \circ F$, where $\pi_i\colon (\beta\omega)^{(l)} \to \beta\omega$ is the i-th projection ($i = 1,\ldots,l$). By the already proved case, for each i we can find a clopen partition $\beta\omega = U_1^i \cup U_2^i \cup \cdots \cup U_{k(i)}^i$ such that the restriction of F_i to U_j^i depends on at most one coordinate for $j \leq k(i)$. Then for $s \in \prod_{i=1}^{l}\{1,\ldots,k(i)\}$ the sets

$$U_s = \bigcap_{i=1}^{l} U_{s(i)}^i$$

form a clopen partition such that $\pi_j \circ F$ depends on at most one coordinate on U_s for all s and j. This completes the proof. □

4.4. Autohomeomorphisms of finite powers of $\beta\omega$ and ω^*

In this section we shall give a simple description of all autohomeomorphisms of some finite power $(\beta\omega)^{(d)}$. Every such autohomeomorphism F is determined by its restriction h to $\omega^{(d)}$, so it will suffice to describe those $h\colon \omega^{(d)} \to \omega^{(d)}$ whose unique extension to $(\beta\omega)^{(d)}$ is an autohomeomorphism. The simplest way to induce an autohomeomorphism is to take permutations $\varphi_1,\ldots,\varphi_d$ of ω, fix a permutation π of $\{1,\ldots,d\}$, and define $h\colon \omega^{(d)} \to \omega^{(d)}$ by

(1) $$h(n_1,\ldots,n_d) = (\varphi_1(n_{\pi(1)}),\ldots,\varphi_d(n_{\pi(d)})).$$

We say that a function h of this form is *elementary*. Not all autohomeomorphisms of $(\beta\omega)^{(d)}$ are induced in this way—for any positive integer m we can split $\omega^{(d)}$ into m many disjoint rectangles in two ways, say $\omega^{(d)} = R_1 \cup \cdots \cup R_m$ and $\omega^{(d)} = P_1 \cup \cdots \cup P_m$, and define h so that it maps R_i into P_i and that is elementary on R_i, for all $i \leq m$. Let us say that an autohomeomorphism of $(\beta\omega)^{(d)}$ is *piecewise elementary* if it is induced in this way.

In [**17**, Conjecture 8.3] (reproduced as [**52**, Question 41]), van Douwen has conjectured the following (see also Remark 11.4 of [**17**]):

CONJECTURE 4.4.1 (van Douwen). *All autohomeomorphisms of $(\beta\omega)^{(2)}$ and all autohomeomorphisms of $(\omega^*)^{(2)}$ are piecewise elementary.*

In terminology introduced above, this conjecture says that all autohomeomorphisms of $(\beta\omega)^{(d)}$ and all autohomeomorphisms of $(\omega^*)^{(d)}$ are piecewise elementary. The following theorem, confirms this conjecture in the case when $X = \beta\omega$.

THEOREM 4.4.2. *All autohomeomorphisms of $(\beta\omega)^{(d)}$ are piecewise elementary.*

PROOF. This follows immediately from Theorem 4.3.2. □

We shall now use wEP(ω,ω) to prove that all autohomeomorphisms of any finite power of ω^* are piecewise elementary. A mapping $\pi\colon A \to B$ between two infinite sets is an *almost bijection* if it is finite-to-one, and $h^{-1}(m)$ is a singleton for all but finitely many $m \in B$. Following the case of $\beta\omega$, we say that an autohomeomorphism F of $(\omega^*)^{(d)}$ is a *trivial autohomeomorphism of* $(\omega^*)^{(d)}$ if it is induced by an h which is defined as follows: We can split $\omega^{(d)}$ into m many disjoint rectangles in two ways, say $\omega^{(d)} = R_1 \cup \cdots \cup R_m$ and $\omega^{(d)} = P_1 \cup \cdots \cup P_m$ so that h maps R_i into P_i is an elementary map whose components φ_j^i are almost bijections.

THEOREM 4.4.3. *Assume wEP. Then all autohomeomorphisms of $(\omega^*)^{(d)}$ are trivial, and therefore piecewise elementary.*

PROOF. Let F be an autohomeomorphism of $(\omega^*)^{(d)}$, and let $(\omega^*)^{(d)} = U_0 \cup U_1 \cup \cdots \cup U_k \cup V$ be a clopen decomposition guaranteed by Theorem 4.3.3, such that F is elementary on each U_i and that $F''V$ is nowhere dense. Then V clearly has to be empty, and $(\omega^*)^{(d)} = U_0 \cup \cdots \cup U_k$ is the desired decomposition. □

Theorems 4.4.3 and 4.4.3 together confirm van Douwen's Conjecture 4.4.1, but only assuming that wEP(ω, ω) holds. The following result, appearing in [**41**], confirms both Conjecture 4.3.1 and Conjecture 4.4.1 (here d and l stand for arbitrary natural numbers).

THEOREM 4.4.4. *Every map $F: (\omega^*)^{(d)} \to (\omega^*)^{(l)}$ is piecewise elementary.* □

4.5. Čech–Stone remainders of countable ordinals

In the study of Čech–Stone remainders of countable ordinals we can restrict our attention to *indecomposable* ordinals, i.e., those α such that $\alpha = \alpha_0 + \alpha_1$ and $\alpha_1 < \alpha$ implies $\alpha_1 = 0$, without losing generality. This is because $\alpha = \alpha_0 + \alpha_1$ and $\alpha_1 \neq 0$ implies $\alpha^* \approx \alpha_1^*$, and every ordinal α can be written as $\alpha_0 + \alpha_1$, where α_1 is indecomposable. We have the following picture (see the proof of Theorem 4.5.1 below):

$$\ldots \twoheadrightarrow (\omega^\alpha)^* \twoheadrightarrow \ldots \twoheadrightarrow (\omega^\omega)^* \twoheadrightarrow \ldots \twoheadrightarrow (\omega^2)^* \twoheadrightarrow \omega^*.$$

(Recall that indecomposable ordinals are exactly those of the form ω^α.) In [**23**] it was proved that the rightmost arrow is irreversible under OCA and MA. We shall extend this result by using the weak Extension Principle. Recall that a continuous map between regular topological spaces is *perfect* if and only if preimages of compact sets are compact (see [**27**]).

THEOREM 4.5.1. *wEP implies that the following are equivalent for indecomposable countable ordinals α and γ:*

(1) $\gamma \leq \alpha$
(2) *There is a continuous perfect surjection $h: \alpha \to \gamma$.*
(3) $\alpha + 1 \twoheadrightarrow \gamma + 1$.
(4) $\alpha^* \twoheadrightarrow \gamma^*$

Before starting the proof, let us note that the only implication that requires wEP is (4) implies (1), (2) and (3), that under CH (4) is true for all α and γ (see [**101**]), that (2) implies (3) for any pair of topological spaces (by the proof below).

PROOF. (1) implies (2). Assume this implication fails, and let $\langle \alpha, \gamma \rangle$ be the lexicographically minimal pair for which this happens. We clearly have $\alpha > \gamma$. Let $\{\alpha_n\}$ be an increasing sequence of ordinals converging to α such that otp$([\alpha_n + 1, \alpha_{n+1}))$ is an indecomposable ordinal less than α for all n. Similarly let $\{\gamma_n\}$ be an increasing sequence of ordinals converging to γ such that otp$([\gamma_n + 1, \gamma_{n+1})) < \gamma$ for all n. By the assumption on α and γ, there is a continuous perfect surjection

$$h_n: [\alpha_n + 1, \alpha_{n+1}) \to [\gamma_n + 1, \gamma_{n+1})$$

for all n. Define $h\colon \alpha \to \gamma$ by
$$h(\xi) = \begin{cases} h_n(\xi), & \text{if } \xi \in \operatorname{dom} h_n, \\ \gamma_n, & \text{if } \xi = \alpha_n. \end{cases}$$

Then h satisfies (2). It is clearly continuous in all $\xi \neq \alpha_n$, and in α_n it is continuous because h_{n-1} maps sets unbounded in $[\alpha_{n-1}+1, \alpha_n)$ into sets unbounded in $[\gamma_{n-1}+1, \gamma_n)$. Also, h is perfect because it maps unbounded sets into unbounded sets and it is onto because $h''[\alpha_n, \alpha_{n+1}) = [\gamma_n, \gamma_{n+1})$.

(2) is obviously equivalent to (3).

(2) implies (4). Let $F_0\colon \beta\alpha \to \beta\gamma$ be the unique extension of h to $\beta\alpha$, and let $F = F_0 \upharpoonright \alpha^*$. Then F maps α^* into γ^* because h maps non-compact sets into non-compact sets. Then $F''\alpha^*$ covers γ^* because F_0 is onto and $F_0''\alpha = h''\alpha = \gamma$.

(4) implies (2). Fix $F\colon \alpha^* \to \gamma^*$. Let $\alpha^* = U_0 \cup U_1$ be a clopen decomposition as guaranteed by wEP. Since the set
$$\gamma^* \setminus F''U_1 \supseteq \gamma^* \cap F''U_0$$
is open and nowhere dense, it is empty. Therefore we can assume that F extends to $\beta\alpha$ (by possibly replacing α with α', the trace of U_1 on α, or with its end-segment if α' is not indecomposable). Set $\gamma \setminus h''\alpha$ is bounded in α because otherwise it includes an infinite closed discrete subset P, and P^* is disjoint from the image $F''\alpha$, contradicting the fact that F maps α^* onto γ^*. Therefore there is a $\gamma' < \gamma$ such that $h''\alpha \supseteq [\gamma', \gamma)$, and mapping $h\colon \alpha \to \gamma$ defined by
$$h(\xi) = \begin{cases} F(\xi), & \text{if } \xi > \gamma', \\ \gamma', & \text{if } \xi \leq \gamma', \end{cases}$$
verifies (2).

(3) implies (1). Recall that every indecomposable ordinal is of the form ω^ζ, and that $\omega^\xi + 1$ is a scattered space of rank $\xi + 1$ (see [**27**]). Finally, a continuous image of a compact scattered space is a scattered space whose rank is not bigger than that of the original space. □

A mapping $h\colon \alpha^{(d)} \to \gamma^{(l)}$ *induces* a continuous mapping $F\colon (\alpha^*)^{(d)} \to (\gamma^*)^{(l)}$ if there is a continuous extension F_0 of h to $(\beta\alpha)^{(d)}$ such that $F = F_0 \upharpoonright (\alpha^*)^{(d)}$. A subset A of $\alpha^{(d)}$ is d-compact if it can be written as $A = A_1 \cup A_2 \cup \cdots \cup A_d$ so that the i-th projection of A_i is a compact subset of α. A mapping $h\colon \alpha^{(d)} \to \gamma^{(l)}$ is (d,l)-*perfect* if preimages of l-compact sets are d-compact and images of d-compact sets are l-compact. It is $\mathcal{R}$-*measurable* if the preimages of sets in $\mathcal{R}_\gamma^l$ are in $\mathcal{R}_\alpha^d$.

LEMMA 4.5.2. *A map $h\colon \alpha^{(d)} \to \gamma^{(l)}$ induces a surjection $F\colon (\alpha^*)^{(d)} \to (\gamma^*)^{(l)}$ if and only if it satisfies the following three conditions:*

(A1) *h is $\mathcal{R}$-measurable,*

(A2) *h is (d,l)-perfect, and*

(A3) *the closure of the set $\gamma^{(l)} \setminus h''\alpha^{(d)}$ inside $\gamma^{(l)}$ is l-compact.*

PROOF. Condition (A1) is clearly equivalent to the fact that h continuously extends to a mapping $F_0\colon (\beta\alpha)^{(d)} \to (\beta\gamma)^{(l)}$. Condition (A2) (which essentially says that the preimages of compact sets are compact) is equivalent to the fact that $F_0''(\alpha^*)^{(d)}$ is included in $(\gamma^*)^{(l)}$. Finally, (A3) assures that $(\gamma^*)^{(l)}$ is covered by $F_0''(\alpha^*)^{(d)}$. □

COROLLARY 4.5.3. *wEP implies that the following are equivalent for countable indecomposable ordinals α and γ:*

(1) *there is a continuous surjection $F\colon (\alpha^*)^{(d)} \to (\gamma^*)^{(l)}$*
(2) *there is an $\mathcal{R}$-measurable, (d,l)-perfect surjection $h\colon \alpha^{(d)} \to \gamma^{(l)}$.*

PROOF. Let us prove only the nontrivial direction. If h satisfies (A1)–(A3), then let $\xi < \gamma$ be such that
$$\gamma^{(l)} \setminus h''\alpha^{(d)} \subseteq \{\langle \eta_1, \ldots, \eta_l\rangle : \min_{i\le l} \eta_i \le \xi\}$$
and define $h'\colon \alpha^{(d)} \to [\xi,\gamma)^{(l)}$ by
$$h'(\langle \eta_1, \ldots, \eta_d\rangle) = h(\langle \max(\eta_1,\xi), \ldots, \max(\eta_d,\xi)\rangle).$$
Since the interval $[\xi,\gamma)$ is homeomorphic to γ, h' satisfies the requirements. □

4.6. Some Parovičenko spaces under wEP

Let us now see what is the relationship between finite powers of Čech–Stone remainders of ω and ω^2. It is easy to see that we have the following relations:

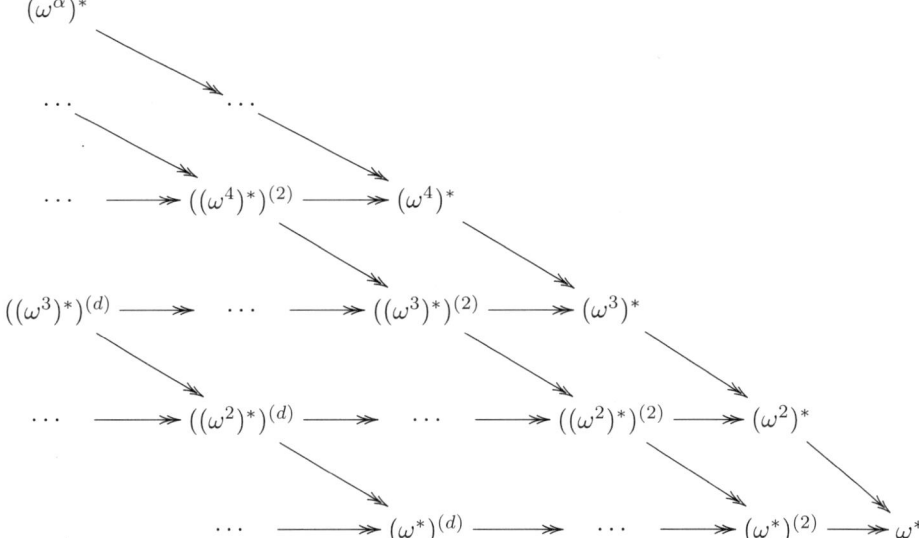

By a classical result of Parovičenko ([**101**]), the Continuum Hypothesis implies that all of these arrows are reversible; moreover for every fixed $d \in \omega$ spaces $((\omega^\alpha)^*)^{(d)}$ are homeomorphic for all countable α. We shall now use wEP to show that many of the above arrows are irreversible. Namely, we shall prove that for all $d \in \omega$ and all countable additively indecomposable $\alpha > \omega$:

(1) $(\omega^*)^{(d)} \not\twoheadrightarrow (\omega^*)^{(d+1)}$ (Theorem 4.6.1),
(2) $\alpha^* \not\twoheadrightarrow (\omega^*)^{(d)}$, (Lemma 4.6.2), and
(3) $(\omega^*)^{(d)} \not\twoheadrightarrow \alpha^*$ (Theorem 4.6.4).

These results are proved using Ramsey's theorem and Corollary 4.5.3. In September 1999 the author has proved, using different methods, that all arrows in the above diagram are irreversible (see [**41**]).

The following result was originally proved by Winfried Just ([**61**]; see also [**41**, Lemma 5.2] for a generalization).

THEOREM 4.6.1. *wEP(ω,ω) implies that the following are equivalent for all positive integers d and l:*
(1) $(\omega^*)^{(d)} \twoheadrightarrow (\omega^*)^{(l)}$,
(2) $d \geq l$.

PROOF. The direct implication is true because the projection
$$(\mathcal{U}_1, \ldots, \mathcal{U}_l, \mathcal{U}_{l+1}, \ldots \mathcal{U}_d) \to (\mathcal{U}_1, \ldots, \mathcal{U}_l)$$
is surjective. Note that by results of §4.4 variations of this map are essentially the only surjections.

(2) implies (1). By Lemma 4.5.2, it suffices to prove that there is no $\mathcal{R}$-measurable, $(d, d+1)$-perfect surjection $h\colon \omega^{(d)} \to \omega^{(d+1)}$. To prove this in case $d = 1$, note that if $h\colon \omega \to \omega^{(2)}$ is onto then the set $h^{-1}(\{1\} \times \omega)$ is infinite so h is not $(1,2)$-perfect. To prove that a continuous image of ω^* inside $(\omega^*)^{(d)}$ is nowhere dense, note that every clopen subset of $(\omega^*)^{(d)}$ is homeomorphic to $(\omega^*)^{(d)}$, and therefore a subset of $(\omega^*)^{(d)}$ with a nonempty interior which is a continuous image of ω^* gives rise to a surjection of ω^* to $(\omega^*)^{(d)}$, implying $d = 1$. To prove Theorem for arbitrary $d \geq 2$, assume that mapping $F\colon (\omega^*)^{(d)} \to (\omega^*)^{(d+1)}$ is continuous. By wEP we can assume that $F\colon (\beta\omega)^{(d)} \to (\beta\omega)^{(d+1)}$ and by Theorem 4.3.2 there is a partition $(\omega^*)^{(d)} = U_1 \cup U_2 \cup \cdots \cup U_l$ into finitely many clopen sets and
$$m\colon \{1, \ldots, d+1\} \times \{1, \ldots, l\} \to \{1, \ldots, d\}$$
such that $\pi_j \circ F$ depends only on $m(i,j)$-th coordinate on U_i, for all $i = 1, \ldots, l$ and $j = 1, \ldots, d+1$. Fix $i \leq l$. By the pigeonhole principle there are $j \neq j'$ such that $m(j,i) = m(j',i)$, and therefore the set $F''U_i$ is nowhere dense. Hence the image of $(\omega^*)^{(d)}$ under F is equal to the union of finitely many nowhere dense sets, and this finishes the proof. $\square$

LEMMA 4.6.2. *wEP implies that $(\omega^*)^{(d)}$ is not a continuous image of $(\omega^2)^*$.*

PROOF. It is sufficient to prove the case $d = 2$. Let us assume the contrary. By Lemma 4.5.2, there is an $\mathcal{R}$-measurable, $(1,2)$-perfect surjection $h\colon \omega^2 \to \omega^{(?)}$. Define a partition $[\omega]^3 = K_0 \cup K_1$ by letting $\{i,j,k\}_< \in K_0$ if and only if
$$h^{-1}(j,k) \subseteq \omega \cdot i.$$

Assume C is an infinite K_0-homogeneous subset of ω. Then if $i = \min C$ and $C' = C \setminus \{i\}$, we have $h^{-1}(C' \times C') \subseteq i \times \omega$, contradicting $(1,2)$-perfectness of h. Therefore by Ramsey's theorem there is an infinite K_1-homogeneous set C. Define a partition $[C]^2 = L_0 \cup L_1$ by letting $\{i,j\}_< \in L_0$ if and only if
$$h^{-1}(\{1, \ldots, i\} \times (C \setminus \{1, \ldots, i\})) \subseteq \omega \cdot j.$$

Since the preimages of 2-compact sets are compact, for every i there are only finitely many j such that $\{i,j\} \notin L_0$, and therefore there is an infinite L_0-homogeneous $C_1 \subseteq C$. This, together with the fact that C_1 is K_1-homogeneous, implies that if $m_1 < m_2 < m_3 < \ldots$ is an increasing enumeration of C_1, then for every i there is an $n_i \in [m_i, m_{i+1})$ such that $h^{-1}(m_i, k) \cap (\{n_i\} \times \omega)$ is nonempty for infinitely many $k \in C_1$. Now we can easily find $D \subseteq C_1$ such that there are infinitely many n for which both sets
$$(\omega \cdot (n-1), \omega \cdot n) \cap h^{-1}(D \times D) \quad \text{and} \quad (\omega \cdot (n-1), \omega \cdot n) \setminus h^{-1}(D \times D)$$

are infinite. Therefore $h^{-1}(D \times D)$ is not clopen, and this contradiction completes the proof. □

As pointed out in the introduction, in case when $d = 1$ a result stronger than the following was previously proved by A. Dow and K.P. Hart.

THEOREM 4.6.3. *wEP implies that the following are equivalent for a countable ordinal α and a positive integer d:*

(1) α^* *is a continuous image of* $(\omega^*)^{(d)}$
(2) $\alpha = \alpha_0 + \omega$ *for some ordinal α_0.*

In particular, if $\alpha > \omega$ is an indecomposable ordinal then $(\omega^*)^{(d)} \not\twoheadrightarrow \alpha^*$.

We shall prove a more precise version of this theorem, namely

THEOREM 4.6.4. *wEP implies that if $F \colon (\omega^*)^{(d)} \to \alpha^*$ is continuous, then the range of F is equal to the union of a set homeomorphic to ω^* and a nowhere dense set.*

PROOF OF THEOREM 4.6.3. Assume Theorem 4.6.4. For the nontrivial direction note that α^* is, by Lemma 4.5.2, homeomorphic to ω^* if and only if (2) applies. □

PROOF OF THEOREM 4.6.4. Let us prove the following lemma first.

LEMMA 4.6.5. *If $h \colon \omega^{(2)} \to \alpha$ is an $\mathcal{R}$-measurable, $(2,1)$-perfect map, then its range is equal to the union of a relatively compact set and a discrete set.*

PROOF. Assume this fails and let h be a counterexample. We can assume that h is onto. This is because the range of h is homeomorphic to some ordinal $\alpha_0 \leq \alpha$ and this ordinal is, by our assumptions, not of the form $\alpha_0' + \omega$. We can also assume that $\alpha_0 = \omega^2$, possibly by composing h with a continuous perfect surjection of α_0 to ω^2. For $\xi = \omega \cdot m + n < \omega^2$ fix $\langle x_{mn}, y_{mn} \rangle \in h^{-1}(\xi)$ so that the following condition is satisfied:

$$\min(x_{mn}, y_{mn}) = \max\{\min(x, y) : \langle x, y \rangle \in h^{-1}(\xi)\}.$$

Note that this maximum exists because set $h^{-1}(\xi)$ is 2-compact. Also note that the choice of $\langle x_{mn}, y_{mn} \rangle$ assures that (let $k = \max(x_{mn}, y_{mn})$):

(X) $h^{-1}(m \cdot \omega + n) \subseteq \{1, \ldots, k\} \times \omega \cup \omega \times \{1, \ldots, k\}$.

Define a partition $[\omega]^3 = K_0 \cup K_1 \cup K_1$ as follows:

$$\{m, n, k\}_< \in \begin{cases} K_0 & \text{if } x_{mn} = x_{mk}, \\ K_1 & \text{if } y_{mn} = y_{mk}, \\ K_2 & \text{otherwise.} \end{cases}$$

CLAIM 1. *There are no infinite K_2-homogeneous sets.*

PROOF. Assume otherwise, that A is an infinite K_2-homogeneous set and let $\bar{m} = \min(A)$. Then the set

$$h^{-1}inv\{\omega \cdot \bar{m} + n : \bar{m} < n \in A\}$$

is not relatively 2-compact, because it includes the set

$$\{\langle x_{\bar{m}n}, y_{\bar{m}n} \rangle : \bar{m} < n \in A\}$$

such that $\lim_n x_{\bar{m}n} = \lim_n y_{\bar{m}n} = \infty$, contradicting the assumption on h. □

4.6. SOME PAROVIČENKO SPACES UNDER WEP

Let us now assume A is an infinite K_0-homogeneous set. Define a partition $[A]^4 = L_0 \cup L_1 \cup L_2$ by

$$\{m,n,k,l\}_< \in \begin{cases} L_0, & \text{if } \langle x_{mn}, y_{kl} \rangle \in h^{-1}(\omega \cdot m + n), \\ L_1, & \text{if } \langle x_{mn}, y_{kl} \rangle \in h^{-1}(\omega \cdot k + l), \\ L_2, & \text{otherwise.} \end{cases}$$

CLAIM 2. *There are no infinite L_2-homogeneous sets.*

PROOF. Assume the contrary, that $B \subseteq A$ is such, let $B = \{m_i\}$ be its increasing enumeration and let

$$\xi_i = \omega \cdot m_{2i} + m_{2i+1}, \qquad x_i = x_{\xi_i}, \quad \text{and} \quad y_i = y_{\xi_i}.$$

Define a partition $[\omega]^3 = M_0 \cup M_1 \cup M_2 \cup M_3$ by

$$\{i,j,k\}_< \in \begin{cases} M_0, & \text{if } \langle x_i, y_j \rangle \in h^{-1}(\xi_k), \\ M_1, & \text{if } \langle x_j, y_k \rangle \in h^{-1}(\xi_i), \\ M_2, & \text{if } \langle x_i, y_k \rangle \in h^{-1}(\xi_j), \\ M_3, & \text{otherwise.} \end{cases}$$

Since the sets $h^{-1}(\xi_i)$ are pairwise disjoint, there are no infinite M_j-homogeneous sets for $j = 0, 1, 2$, so there is an infinite M_3-homogeneous set C. By the definition, all ξ_i are isolated points in ω^2, and therefore by $\mathcal{R}$-measurability $h^{-1}(\{\xi_i : i \in C\})$ is equal to the union of finitely many rectangles, say $Y_l \times Z_l$ ($l = 1, \ldots, k$). Let $i < j$ and $l \leq k$ be such that

$$\langle x_i, y_i \rangle \in Y_l \times Z_l \quad \text{and} \quad \langle x_j, y_j \rangle \in Y_l \times Z_l.$$

This implies $\langle x_i, y_j \rangle \in Y_l \times Z_l \subseteq h^{-1}(\{\xi_i : i \in C\})$. But by the L_2-homogeneity and the M_3-homogeneity $\langle x_i, y_j \rangle$ is not in $h^{-1}(\xi_n)$ for any $n \in C$—a contradiction. □

Therefore there is an infinite $B \subseteq A$ which is either L_0- or L_1-homogeneous. Let us first treat the case when B is L_0-homogeneous.

CLAIM 3. *The projection of $h^{-1}(\{\omega \cdot m + n\})$ to y axis is infinite whenever $m < n$ are in B.*

PROOF. Since the set $h^{-1}(\{\omega \cdot k + l : n < k < l \in B\})$ is not relatively 2-compact, (X) implies that set $\{y_{kl} : n < k < l \in B\}$ is infinite. By the L_0-homogeneity of B, this set is included in the projection of $h^{-1}(\{\omega \cdot m + n\})$ to y-axis, and this completes the proof. □

Let $\bar{m} = \min(B)$. Since $h^{-1}(\{\omega \cdot \bar{m} + n : \bar{m} < n \in B\})$ is relatively 2-compact, Claim 3 implies that we can find $n_1 < n_2$ in B such that $x_{\bar{m}n_1} = x_{\bar{m}n_2} = x$. Pick $l > k > \max(n_1, n_2)$ in B; then by the K_0-homogeneity we have

$$h^{-1}(\omega \cdot \bar{m} + n_1) \cap h^{-1}(\omega \cdot \bar{m} + n_2) \ni \langle x, y_{kl} \rangle,$$

contradicting the fact that these two preimages are disjoint. Therefore B cannot be L_0-homogeneous. Now assume B is L_1-homogeneous and let $\bar{m} < \bar{n}$ be the first two elements of B. Then

$$h''(\{x_{\bar{m}\bar{n}}\} \times \omega) \supseteq q\{\omega \cdot k + l : \bar{n} < k < l \text{ are in } B\},$$

contradicting the $(2,1)$-perfectness of h. This implies that the assumption that A is K_0-homogeneous was false.

To complete the proof of Lemma, it remains to prove there are no infinite K_1-homogeneous sets. But this reduces to the case of K_0-homogeneous sets by considering the function $h'(m,n) = h(n,m)$. □

Lemma clearly implies the case $d = 2$ of Theorem. The proof of general case uses the following variation of Lemma 4.6.5.

LEMMA 4.6.6. *If $h\colon \omega^{(d)} \to \omega^2$ is a $\mathcal{R}$-measurable, $(d,1)$-perfect map, then its range is equal to the union of a relatively compact set and a discrete set.*

PROOF. (Sketch) Assume h is a counterexample. We can assume that h is onto. For $\xi = \omega \cdot m + n$ we fix $\langle x^1_{mn}, \ldots, x^d_{mn}\rangle \in h^{-1}(\xi)$ so that
$$\min_{i \leq d}(x^i_{mn}) = \max\left\{\min_{i \leq d}(x^i) : \langle x^1, \ldots, x^d\rangle \in h^{-1}(\xi)\right\}.$$
Partitions K, L, M are defined analogously to the case $d = 2$. The only difference is that now we have more choices for the distinguished vertex of the hypercube with diagonal vertices $\langle x^1_{mn}, \ldots, x^d_{mn}\rangle$ and $\langle x^1_{kl}, \ldots, x^d_{kl}\rangle$, but proofs are the same. □

An application of Lemma 4.6.6 ends the proof. □

4.7. Remainders of locally compact, countable spaces

In this section we shall reformulate some of our results by using the following well-known fact.

PROPOSITION 4.7.1. *If X is a countable, locally compact, non-compact space, then it is homeomorphic to some ordinal.*

PROOF. Let $X \cup \{\infty\}$ be the one-point compactification of X; by a classical result of Mazurkiewicz and Sierpinski (see e.g., [**102**]), $X \cup \{\infty\}$ is homeomorphic to some ordinal, and therefore X itself is homeomorphic to an ordinal. □

In Theorem 4.9.1 we will prove that OCA and MA imply wEP(α, γ) for all countable ordinals α and γ.

COROLLARY 4.7.2. *OCA and MA imply wEP(X,Y) for all countable locally compact spaces X and Y.* □

COROLLARY 4.7.3. *wEP implies that the following are equivalent:*
(1) $X^* \twoheadrightarrow Y^*$
(2) *there is a continuous perfect surjection $h\colon X \to Y$.*
for all locally compact countable spaces X and Y. □

COROLLARY 4.7.4. *wEP implies that the following are equivalent:*
(1) $X^* \approx Y^*$
(2) *there are compact $K \subseteq X$ and $L \subseteq Y$ such that $X \setminus K$ and $Y \setminus L$ are homeomorphic*
for all locally compact countable spaces X and Y. □

Let us now consider a problem of having a surjectively universal remainder of a countable locally compact space.

COROLLARY 4.7.5. *wEP implies that for every countable locally compact X there is a countable locally compact space Y such that $Y^* \twoheadrightarrow X^*$ but $X^* \not\twoheadrightarrow Y^*$.* □

Note that under CH, ω^* is a counterexample to the conclusion of Corollary 4.7.5. Moreover, under CH all spaces of the form X^* for X a countable, locally compact, non-compact space are homeomorphic ([**101**]). Corollary 4.7.5 says that under wEP there is no surjectively universal object in the class of remainders of locally compact countable spaces. However, one does not have to go very far to find a space which maps onto all these remainders. Let G be a closed G_δ subset of ω^* which is not open. By the compactness and zero-dimensionality of ω^* we can write G as an intersection of some strictly decreasing sequence of clopen sets, and therefore G is homeomorphic to the Stone space of the algebra $\mathcal{P}(\omega^{(2)})/\operatorname{Fin}\times\emptyset$.

PROPOSITION 4.7.6. (a) *The family of continuous images of G includes all Čech–Stone remainders of countable locally compact spaces.*
(b) *This family is closed under taking finite products of its elements.*

PROOF. (a) For an ordinal α let $\operatorname{cl}(\alpha)$ denote the algebra of clopen subsets of α. Note that α^* is the Stone space of $\operatorname{cl}(\alpha)/\langle\alpha\rangle$, where $\langle\alpha\rangle$ stands for the ideal of bounded subsets of α, i.e., the ideal generated by $\alpha = \{\delta : \delta < \alpha\}$. Assume $\alpha > \omega$ is a countable ordinal and let $\{\alpha_n\}$ be an increasing sequence of ordinals converging to α so that every interval $[\alpha_n, \alpha_{n+1})$ is infinite, and let $h_\alpha\colon \omega^{(2)} \to \alpha$ be a bijection which sends $\{n\}\times\omega$ into $[\alpha_n, \alpha_{n+1})$. Let $\pi_\alpha\colon G \to \alpha^*$ be defined by

$$\pi_\alpha(\mathcal{U}) = \{h''_\alpha U : U \in \mathcal{U}\} \cap \operatorname{cl}(\alpha).$$

Then π_α is a continuous surjection.

(b) It will suffice to prove that $G \twoheadrightarrow G^{(2)}$. Let $h\colon \omega^{(2)} \to (\omega^{(2)})^{(2)}$ be a bijection such that $A \in \operatorname{Fin}\times\emptyset$ if and only if we can write $h^{-1}(A) = A_1 \cup A_2$ so that the first projection of A_1 and the second projection of A_2 are both in $\operatorname{Fin}\times\emptyset$. Define a mapping $F\colon (\omega^{(4)})^* \to (\omega^{(2)})^*$ by

$$F(\mathcal{U}) = \{h''U : U \in \mathcal{U}\} \cap \mathcal{R}^2_{\omega^{(2)}}.$$

Then F maps G onto $G^{(2)}$. □

COROLLARY 4.7.7. *If X is a countable, locally compact, non-compact space, then all finite powers of X^* are continuous images of G.* □

An another way to look at consequences of wEP is the following.

PROPOSITION 4.7.8. *If $F\colon \alpha^* \to \gamma^*$ extends to $F_1\colon \beta\alpha \to \beta\gamma$ so that $F_1''\alpha \subseteq \gamma$, then there is a continuous $F_*\colon G \to G$ such that the diagram*

$$\begin{array}{ccc} G & \xrightarrow{F_*} & G \\ \downarrow{\pi_\alpha} & & \downarrow{\pi_\gamma} \\ \alpha^* & \xrightarrow{F} & \gamma^* \end{array}$$

commutes.

PROOF. Let $h = F \upharpoonright \alpha$, and let $h_G\colon \omega^{(2)} \to \alpha$ be such that the diagram

$$\begin{array}{ccc} \omega^{(2)} & \xrightarrow{h_*} & \omega^{(2)} \\ \downarrow{h_\alpha} & & \downarrow{h_\gamma} \\ \alpha & \xrightarrow{h} & \gamma \end{array}$$

commutes, namely let $h_* = h_\gamma^{-1} \circ h \circ h_\alpha$. Then h_* sends the ideal Fin $\times \emptyset$ into itself, and therefore mapping $F_* \colon (\omega^{(2)})^* \to (\omega^{(2)})^*$ defined by

$$F_*(\mathcal{U}) = \{h_*'' U : U \in \mathcal{U}\}$$

sends G into G. Since the above diagram commutes, F_* is as required. $\square$

4.8. Almost liftings and duality

This section contains some prerequisites for the proof of wEP from OCA and MA, given in the next section. Let α be a countable indecomposable ordinal, and let $\langle \alpha \rangle^{(d)}$ denote the *ideal of d-bounded subsets of* $\alpha^{(d)}$, namely those $A \subseteq \alpha^{(d)}$ which can be partitioned as $A = A_1 \cup A_2 \cup \cdots \cup A_d$ so that the i-th projection of A_i is a bounded subset of α for all $i \leq d$. For a topological space X let $\operatorname{cl}(X)$ denote the algebra of clopen subsets of X, and let $\mathcal{K}(X)$ denote its ideal generated by compact open sets. One can also define the rectangle algebra of $\operatorname{cl}(X)^{(d)}$ and its ideal of d-compact sets, but we shall not need these objects.

Let us recall a basic fact about the topological duality. Let $\operatorname{Stone}(\mathcal{B})$ denote the *Stone space* of a Boolean algebra $\mathcal{B}$.

PROPOSITION 4.8.1.   (1) $(\alpha^*)^{(d)}$ *is homeomorphic to* $\operatorname{Stone}(\mathcal{R}_\alpha^d/\langle \alpha \rangle^{(d)})$.
(2) *If X is a zero-dimensional, locally compact space, then X^* is homeomorphic to* $\operatorname{Stone}(\operatorname{cl}(X)/\mathcal{K}(X))$.
Assume that $X = \operatorname{Stone}(\mathcal{B}_X)$ *and* $Y = \operatorname{Stone}(\mathcal{B}_Y)$.
(3) *Every continuous map* $f \colon X \to Y$ *uniquely determines a homomorphism* $\Phi_f \colon \mathcal{B}_Y \to \mathcal{B}_X$ *by*

$$\Phi_f(b) = f^{-1}(b).$$

Moreover, f is a surjection if and only if Φ_f is a monomorphism; f is an injection if and only if Φ_f is an epimorphism.
(4) *Every homomorphism* $\Phi \colon \mathcal{B}_X \to \mathcal{B}_Y$ *defines a continuous map* $f_\Phi \colon Y \to X$ *by*

$$f(y) = \bigcap_{a \in \mathcal{B}_Y, y \in a} \Phi_f^{-1}(a).$$

Moreover, f is a surjection if and only if Φ_f is a monomorphism; f is an injection if and only if Φ_f is an epimorphism.

PROOF. See [**14**]. $\square$

If $\mathcal{I}, \mathcal{J}$ are ideals of a Boolean algebra $\mathcal{B}$, then we say that $\mathcal{I}$ is *ccc over* $\mathcal{J}$ if there is no uncountable family $\{A_\xi\}$ of $\mathcal{I}$-positive sets in $\mathcal{B}$ such that $A_\xi \cap A_\eta$ is in $\mathcal{J}$ for all $\xi \neq \eta$. If $\Phi \colon \operatorname{cl}(X)/\mathcal{K}(X) \to \operatorname{cl}(Y)/\mathcal{K}(Y)$ is a homomorphism then a mapping $\Theta \colon \operatorname{cl}(X) \to \operatorname{cl}(Y)$ is an *almost lifting* of Φ if the family

$$\{A \in \operatorname{cl}(X) : \Phi_*(A) =^{\mathcal{K}(Y)} \Theta(A)\}$$

includes an ideal which is ccc over $\mathcal{K}(X)$.

LEMMA 4.8.2. *If* $\Phi \colon \mathcal{P}(\omega)/\operatorname{Fin} \to \mathcal{R}_\alpha^d/\langle \alpha \rangle^{(d)}$ *has a completely additive lifting* Φ_h, *then* $h^{-1}(A)$ *is equal to a set in* $\mathcal{R}_\alpha^d$ *for every* $A \subseteq \omega$. *In particular,* $\operatorname{dom}(h)$ *is equal to a finite union of rectangles in* $\alpha^{(d)}$.

PROOF. We shall prove that $h^{-1}(\omega)$ is equal to a set in $\mathcal{R}_\alpha^d$; the proof of the general case is identical.

Let us first prove the case when $d = 2$. If $E \subseteq \alpha^{(2)}$ and $h_0 \colon E \to \omega$ is a map such that Φ_{h_0} is a completely additive lifting of Φ, define $g \colon \alpha^{(2)} \to \{0, 1\}$ by

$$g(\xi, \eta) = \begin{cases} 1, & \text{if } \langle \xi, \eta \rangle \in E, \\ 0, & \text{if } \langle \xi, \eta \rangle \notin E. \end{cases}$$

Apply Theorem 4.2.2 to g and $\mathcal{I} = \langle \alpha \rangle$. If we can split $\alpha^{(2)}$ into finitely many rectangles $R_1, \ldots, R_k, R_{k+1}, \ldots, R_l$ so that $R_i \in \langle \alpha \rangle^{(2)}$ for $i = 1, \ldots, k$ and g depends on R_j on at most one coordinate for $j = k+1, \ldots, l$, then the restriction of h to $R_{k+1} \cup \cdots \cup R_l$ is as required.

Otherwise, there are sequences $X = \{\xi_i\}$ and $Y = \{\eta_i\}$ of ordinals converging to α and such that either

(a) $\langle \xi_i, \eta_i \rangle \in E$ for all i but $\langle \xi_i, \eta_j \rangle \notin E$ for all $i < j$, or
(b) $\langle \xi_i, \eta_i \rangle \notin E$ for all i and $\langle \xi_i, \eta_j \rangle \in E$ for all $i < j$.

If (a) applies, then $A = \{h(\langle \xi_i, \eta_i \rangle) : i \in \omega\}$ has the property that $h^{-1}(B)$ is not equal to a set in $\mathcal{R}_\alpha^2$ modulo a set in $\langle \alpha \rangle^{(2)}$, for every infinite $B \subseteq A$. But since Φ_{h_0} is an almost lifting of Φ, for one of these B we have $\Phi_*(B) \Delta \Phi_{h_0}(B) \in \langle \alpha \rangle^{(d)}$, a contradiction.

If (b) applies, then we find finite subsets s_i, t_i of ω satisfying the following conditions:

(1) $|s_i| = i$,
(2) $u_i = \{\xi_j : j \in s_i\}$, $v_i = \{\eta_j : j \in t_i\}$,
(3) the sets $Y_i = h''(u_i \times v_i)$ $(i \in \omega)$ are pairwise disjoint.

This can be done by a straightforward recursive construction, using the fact that $h^{-1}(\{n\})$ is in $\langle \alpha \rangle^{(d)}$ for every n and that $\xi_i \to \alpha$ and $\eta_i \to \alpha$ as $i \to \omega$.

Since Φ_{h_0} is an almost lifting of Φ, by Theorem 3.10.1 there is a set $B \subseteq \omega$ which is equal to an infinite union of u_i and such that $\Phi_*(B) \Delta \Phi_{h_0}(B) \in \langle \alpha \rangle^{(d)}$.

Let $X \subseteq \alpha^{(d)}$ be a set in $\mathcal{R}_\alpha^d$. Then X is equal to a finite union of rectangles, say $X = \bigcup_{i=1}^m R_i$. It will suffice to prove that $X \Delta \Phi_{h_0}(B)$ is not in $\langle \alpha \rangle^{(d)}$. But this follows from (b), the choice of u_i and v_i, and the following lemma.

LEMMA 4.8.3. *For every m there is an $m' = f(m)$ such that the set $D_{m'} = \{\langle i, j \rangle : 1 \leq i < j \leq m'\}$ cannot be covered by m many rectangles disjoint from the diagonal.*

PROOF. Let us first prove an infinitary version:

If the set $D_\infty = \{\langle i, j \rangle \in \omega^2 : 1 \leq i < j\}$ is covered by finitely many rectangles, then at least one of these rectangles intersects the diagonal.

Suppose the contrary, and let k be the minimal such that D_∞ can be covered by k many rectangles $A_i \times B_i$ $(i \leq k)$ each of which is disjoint from the diagonal. Since $\omega = \bigcup_{i=1}^k A_i$, at least one A_i has to be infinite. But this implies that $D_\infty \cap (A_i \times A_i)$ is covered by $k - 1$ rectangles each of which is disjoint from the diagonal, contradicting the minimallity of k.

Lemma 4.8.3 now follows by a standard compactness argument. □

This concludes the proof in the case when $d = 2$. The general case is proved by a simple induction. □

LEMMA 4.8.4. *If a homomorphism* $\Phi\colon \mathcal{P}(\omega)/\operatorname{Fin} \to \mathcal{R}_\alpha^d/\langle\alpha\rangle^{(d)}$ *has a completely additive almost lifting defined by* $h\colon \alpha^{(d)} \to \omega \cup \{\emptyset\}$, *then the set*
$$\mathcal{D} = \{s \in \alpha^{(d)} : h(s) \neq \emptyset\}$$
is equal to a set in $\mathcal{R}_\alpha^d$, *modulo a set in* $\langle\alpha\rangle^{(d)}$.

Also, $h^{-1}(A)$ *is equal to a set in* $\mathcal{R}_\alpha^d$ *(modulo* $\langle\alpha\rangle^{(d)}$*) for every* $A \subseteq \omega$.

PROOF. By applying Lemma 4.8.2, we can assume that h depends on only one coordinate, namely it will suffice to treat the case when $d = 1$. We need to prove that $\mathcal{D}$ is equal to a clopen subset, modulo a bounded subset of α.

Assume that $\mathcal{D}$ is not equal to an open set modulo $\langle\alpha\rangle^{(d)}$. Then we can find an increasing sequence of limit ordinals α_n converging to α such that
$$\alpha_n \in \mathcal{D} \text{ but } \sup(\alpha_n \setminus \mathcal{D}) = \alpha_n \text{ for all } n.$$
Consider the set
$$\{A \subseteq \omega : h(\alpha_n) \in A \text{ for at most finitely many } n\}.$$
This is a Borel ideal, and since
$$\{A \subseteq \omega : \Phi_*(A) =^{\langle\alpha\rangle} \Phi_h(A)\}$$
includes an ideal which is ccc over Fin, we can find $A \subseteq \omega$ such that the set $\Phi_*(A) \Delta \Phi_h(A)$ is in $\langle\alpha\rangle$, yet $h(\alpha_n) \in A$ for infinitely many n. But this implies that $\Phi_*(A)$ is not equal to a clopen subset of α modulo $\langle\alpha\rangle$, a contradiction.

Therefore it remains to prove that $\mathcal{D}$ is equal to a closed set modulo $\langle\alpha\rangle$; assume it is not. Then there is an increasing sequence of limit ordinals α_n such that
$$\alpha_n \notin \mathcal{D} \text{ yet } \sup(\mathcal{D} \cap \alpha_n) = \alpha_n \text{ for all } n.$$
Then
$$\{A \subseteq \omega : \sup(h^{-1}(A) \cap \alpha_n) = \alpha_n \text{ for at most finitely many } n\}$$
is a Borel ideal. Like before, we can find $A \subseteq \omega$ such that $\Phi_*(A) =^{\langle\alpha\rangle} \Phi_h(A)$ but $\Phi_h(A)$ is not equal to a closed subset of α modulo $\langle\alpha\rangle$; a contradiction.

The moreover part follows from the already proved part of the lemma, by applying it to the restriction of Φ to $\mathcal{P}(A)/\operatorname{Fin}$. □

LEMMA 4.8.5. *Assume that X is a locally compact zero-dimensional space such that every homomorphism* $\Phi\colon \operatorname{cl}(X)/\mathcal{K}(X) \to \mathcal{P}(\omega)/\mathcal{I}$ *has a completely additive almost lifting whenever $\mathcal{I}$ is a countably generated ideal on ω. Then* $\operatorname{wEP}(X, \alpha)$ *is true for all countable ordinals* α.

PROOF. Let us first treat the case $l = 1$ of wEP. Assume $F\colon \alpha^* \to (X^*)^{(d)}$ is continuous and let
$$\Phi\colon \operatorname{cl}(X)/\mathcal{K}(X) \to \mathcal{R}_\alpha^d/\langle\alpha\rangle^{(d)}$$
be the dual homomorphism. Also, the ideal $\langle\alpha\rangle^{(d)}$ is countably generated (by the sets $\{\xi : \xi < \eta\}$, for $\eta < \alpha$), and therefore the algebra $\operatorname{cl}^d(\alpha)/\langle\alpha\rangle^{(d)}$ is isomorphic to a subalgebra of $\mathcal{P}(\omega)/\mathcal{I}$, for a countably generated ideal $\mathcal{I}$. Let $h\colon \alpha^{(d)} \to X$ be such that Φ_h is a completely additive almost lifting of Φ, and let $A = h^{-1}(X) \subseteq \alpha^{(d)}$. By Lemma 4.8.4 we can assume that A is in $\mathcal{R}_\alpha^d$. Then $U_1 = A^*$ and $U_0 = (\alpha^*)^{(d)} \setminus A^*$ is the desired clopen decomposition.

Now assume $l > 1$ and $F\colon (X^*)^{(d)} \to (\gamma^*)^{(l)}$ and consider maps $F_i = \pi_i \circ F$, where $\pi_i\colon (\gamma^*)^{(l)} \to \gamma^*$ is the i-th projection ($i = 1, \ldots, l$). For each i we can find a partition $(X^*)^{(d)} = U_0^i \cup U_1^i$ into clopen sets such that $F_i \restriction U_0^i$ continuously extends to $(\beta X)^{(d)}$ and $F_i'' U_1^i$ is a nowhere dense subset of γ^*. This implies that for $U_0 = \bigcap_{i=1}^{(l)} U_0^i$ and $U_1 = \bigcup_{i=1}^{(l)} U_1^i$ the image $F''U_1$ is a nowhere dense subset of $(\gamma^*)^{(l)}$ and function $F \upharpoonright U_0$ continuously extends to $(\beta X)^{(d)}$. $\square$

4.9. OCA and MA imply wEP

In this section we shall deduce wEP from OCA and MA, using results of previous chapters. Recall that a topological space is *Polish* if it is separable and completely metrizable, and note that a space is Polish, locally compact and zero-dimensional if and only if it is second countable, zero-dimensional and locally compact ([**27**]). Such spaces include all countable ordinals, the space $\omega \times 2^\omega$, and countable topological sums of such spaces.

THEOREM 4.9.1. *OCA and MA imply wEP(α, X) for every zero-dimensional locally compact Polish space X and every countable ordinal α.*

PROOF. By Lemma 4.8.5, it will suffice to prove Theorem 4.9.2 below. $\square$

THEOREM 4.9.2. *Assume OCA and MA, and let X, d and α be as in Theorem 4.9.1.*

(1) *If $\mathcal{I}$ is a countably generated ideal on ω, then every homomorphism*
$$\Phi \colon \mathrm{cl}(X)/\mathcal{K}(X) \to \mathcal{P}(\omega)/\mathcal{I}$$
has a completely additive almost lifting.

(2) *Every homomorphism $\Phi \colon \mathrm{cl}(X)/\mathcal{K}(X) \to \mathcal{R}_\alpha^d/\langle\alpha\rangle^{(d)}$ has a completely additive almost lifting Φ_h for a function h whose domain is moreover an element of $\mathcal{R}_\alpha^d$.*

PROOF. We shall prove (1) and (2) simultaneously. Since $\mathcal{R}_\alpha^d$ is a subalgebra of $\mathcal{P}(\alpha^{(d)})$ and and the ideal $\langle\alpha\rangle^{(d)}$ is countably generated, we can fix a natural injection
$$\Psi \colon \mathcal{R}_\alpha^d/\langle\alpha\rangle^{(d)} \to \mathcal{P}(\omega^{(2)})/\mathrm{Fin} \times \emptyset$$
which has a completely additive lifting, induced by a bijection $h \colon \omega^{(2)} \to \alpha^{(d)}$. This simple observation shows that (2) is a refinement of (1).

Fix a countable basis $\mathcal{B}$ of X which consists of compact sets and moreover has the following property:

($\mathcal{B}$1) For every set $U \in \mathcal{B}$ the set of all $V \in \mathcal{B}$ such that $V \supseteq U$ is finite and linearly ordered by the inclusion.

We also fix a partition
$$X = \bigcup_{i=0}^\infty W_i$$
of X into pairwise disjoint compact open sets from $\mathcal{B}$. We may assume that $\mathcal{B}$ has the following additional property:

($\mathcal{B}$2) For every set $U \in \mathcal{B}$ there is (the unique) n such that $U \subseteq W_n$.

Let $\mathcal{A}$ denote the family of all partitions of X into pairwise disjoint compact open sets from $\mathcal{B}$. Let $\mathbf{A}, \mathbf{B}$ be partitions in $\mathcal{A}$. We say that $\mathbf{A}$ *refines* $\mathbf{B}$, in symbols $\mathbf{A} \prec \mathbf{B}$, if for every $U \in \mathbf{A}$ there is (the unique) $V \in \mathbf{B}$ such that $U \subseteq V$. A partition $\mathbf{A}$ *almost refines* $\mathbf{B}$, in symbols $\mathbf{A} \prec^* \mathbf{B}$, if for all but finitely many $U \in \mathbf{A}$ there is (the unique) $V \in \mathbf{B}$ such that $U \subseteq V$. For every $\mathbf{A} \in \mathcal{A}$ consider the mapping $g_\mathbf{A} \colon X \to \mathbf{A}$ uniquely defined by
$$x \in g_\mathbf{A}(x) \qquad \text{for every } x \in X.$$

136 4. WEAK EXTENSION PRINCIPLE

This mapping defines a homomorphism $\Psi_{\mathbf{A}} \colon \mathcal{P}(\mathbf{A})/\operatorname{Fin} \to \operatorname{cl}(X)/\mathcal{K}(X)$ by its completely additive lifting, $\Phi_{g_{\mathbf{A}}}$. (Note that Fin here stands for an ideal of finite subsets of $\mathbf{A}$.) For $\mathbf{A} \prec \mathbf{B}$ consider a mapping $g_{\mathbf{AB}} \colon \mathbf{A} \to \mathbf{B}$ uniquely defined by

$$U \subseteq g_{\mathbf{AB}}(U), \qquad \text{for } U \in \mathbf{A}.$$

This mapping defines a homomorphism $\Psi_{\mathbf{AB}} \colon \mathcal{P}(\mathbf{B})/\operatorname{Fin} \to \mathcal{P}(\mathbf{A})/\operatorname{Fin}$ by its completely additive lifting, $\Phi_{g_{\mathbf{AB}}}$. These definitions make the following diagram commute (for all $\mathbf{A} \prec \mathbf{B}$ in $\mathcal{A}$):

$$\begin{array}{ccc} \mathbf{B} & \xleftarrow{g_{\mathbf{AB}}} & \mathbf{A} \\ & \nwarrow{\scriptstyle g_{\mathbf{B}}} & \uparrow{\scriptstyle g_{\mathbf{A}}} \\ & & X \end{array}$$

Note that $\Phi_{\mathbf{A}} = \Phi \circ \Psi_{\mathbf{A}}$ is a homomorphism of $\mathcal{P}(\mathbf{A})/\operatorname{Fin}$ into $\mathcal{P}(\omega^{(2)})/\operatorname{Fin} \times \emptyset$. For $\mathsf{A} \in \emptyset \times \operatorname{Fin}$ let $\Phi_{\mathbf{A}}^{\mathsf{A}} \colon \mathcal{P}(\mathbf{A})/\operatorname{Fin} \to \mathcal{P}(\mathbf{A})/\operatorname{Fin}$ be defined as usual, by its lifting

$$(\Phi_{\mathbf{A}}^{\mathsf{A}})_*(C) = (\Phi_{\mathbf{A}})_*(C) \cap \mathsf{A},$$

and let $\operatorname{id}^{\mathsf{A}} \colon \mathcal{P}(\omega^{(2)})/\operatorname{Fin} \to \mathcal{P}(\mathbf{A})/\operatorname{Fin}$ be a homomorphism determined by its lifting, $C \mapsto C \cap \mathsf{A}$. These definitions make the following diagram commute (for $\mathbf{A} \prec \mathbf{B}$ in $\mathcal{A}$ and $\mathsf{A} \supseteq \mathsf{B}$ in $\emptyset \times \operatorname{Fin}$):

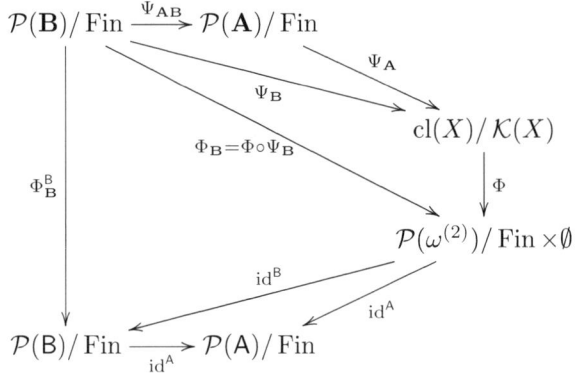

By Theorem 3.8.1, for all $\mathbf{A} \in \mathcal{A}$ and $\mathsf{A} \in \emptyset \times \operatorname{Fin}$, the homomorphism $\Phi_{\mathbf{A}}^{\mathsf{A}}$ has a completely additive almost lifting, given by a finite-to-one mapping $h_{\mathbf{A}}^{\mathsf{A}} \colon E_{\mathbf{A}} \to \mathbf{A}$. We shall apply the OCA uniformization theorem (more precisely, its Corollary 2.3.2) to $\{h_{\mathbf{A}}^{\mathsf{A}}\}$ (or more precisely, to a coherent family $\{f_{\mathbf{A}}^{\mathsf{A}}\}$ defined below) to obtain the desired completely additive lifting.

For $U \in \mathcal{B}$ let C_U be its upwards cone,

$$C_U = \{V \in \mathcal{B} : V \supseteq U\}.$$

Note that every C_U is finite, by ($\mathcal{B}1$). Consider the ideal $\mathcal{I}$ on $\mathcal{B}$ generated by the sets

$$\hat{\mathbf{A}} = \bigcup_{U \in \mathbf{A}} C_U$$

for $\mathbf{A} \in \mathcal{A}$.

CLAIM 1. *The ideal $\mathcal{I}$ is isomorphic to $\emptyset \times \operatorname{Fin}$.*

PROOF. Since $\emptyset \times \mathrm{Fin}$ is equal to the orthogonal of a countably generated ideal $\mathrm{Fin} \times \emptyset$, it will suffice to show that $\mathcal{I} = \mathcal{J}^\perp$, if $\mathcal{J}$ is the ideal generated by the sets

$$Z_k = \left\{ U \in \mathcal{B} : U \subseteq \bigcup_{i=1}^{k} W_i \right\}.$$

We first show that $\mathcal{I} \subseteq \mathcal{J}^\perp$. Fix $\mathbf{A} \in \mathcal{A}$. Then for every n by compactness there are at most finitely many $U \in \mathbf{A}$ such that $U \subseteq \bigcup_{i=1}^{n} W_i$. Since each C_U is finite, $\hat{\mathbf{A}}$ is in $\mathcal{J}^\perp$.

Now we show that $\mathcal{J}^\perp \subseteq \mathcal{I}$. Fix a set C in $\mathcal{J}^\perp$, and fix $n \in \omega$. Let $U_1^n, \ldots, U_{k(n)}^n$ enumerate all sets in C included in W_n (recall that by ($B2$) every set in C is equal to some U_j^n). We can find a natural number $l(n)$ and a disjoint partition $W_n = \bigcup_{j=1}^{l(n)} V_j^n$ such that for every U_i^n there is (the unique) $V_j^n \subseteq U_i^n$. Then $\mathbf{A} = \{V_j^n\}$ is an element of $\mathcal{A}$ such that $\hat{\mathbf{A}} \supseteq C$, and this completes the proof. □

For $\mathbf{A} \in \mathcal{A}$ and $\mathsf{A} \in \emptyset \times \mathrm{Fin}$ define a mapping $f_\mathbf{A}^\mathsf{A} \colon \hat{\mathbf{A}} \times \mathsf{A} \to \{0,1\}$ (we shall think of the ideal $\emptyset \times \mathrm{Fin}$ as an ideal on ω instead of ω^2, to simplify the notation)

$$f_\mathbf{A}^\mathsf{A}(U, i) = \begin{cases} 1, & \text{if } h_\mathbf{A}^\mathsf{A}(i) \subseteq U, \\ 0, & \text{if } h_\mathbf{A}^\mathsf{A}(i) \neq U. \end{cases}$$

We claim that the functions $f_\mathbf{A}^\mathsf{A}$ ($\mathbf{A} \in \mathcal{A}$, $\mathsf{A} \in \emptyset \times \mathrm{Fin}$) form a coherent family. By Lemma 2.3.3, it will suffice to check the coherence condition for $\mathbf{A} \prec \mathbf{B}$ (note that this is equivalent to $\hat{\mathbf{A}} \supseteq \hat{\mathbf{B}}$) and $\mathsf{A} \supseteq \mathsf{B}$.

Note that

$$(g_{\mathbf{AB}} \circ h_\mathbf{A}^\mathsf{A}) \restriction \mathsf{B} =^* h_\mathbf{B}^\mathsf{B},$$

since $\Phi_{(g_{\mathbf{AB}} \circ h_\mathbf{A}^\mathsf{A}) \restriction \mathsf{B}}$ and $\Phi_{h_\mathbf{B}^\mathsf{B}}$ are completely additive almost liftings of the same homomorphism, $\Phi_\mathbf{B}^\mathsf{B}$. For all i such that $g_{\mathbf{AB}}(h_\mathbf{A}^\mathsf{A}(i)) = h_\mathbf{A}^\mathsf{A}(i)$ and all $U \in \mathbf{B}$ we have

$$f_\mathbf{A}^\mathsf{A}(U, i) = f_\mathbf{B}^\mathsf{B}(U, i).$$

On the other hand, note that for every $j \in \mathsf{B}$ there are at most finitely many $U \in \hat{\mathbf{B}}$ such that $f_\mathbf{B}^\mathsf{B}(U, j) \neq 0$ (these are the elements of the finite set $C_{h_\mathbf{B}^\mathsf{B}(j)}$). Similarly, for every j and all but finitely many $U \in \hat{\mathbf{B}}$ we have $f_\mathbf{A}^\mathsf{A}(U, j) = 0$. But we can have $f_\mathbf{B}^\mathsf{B}(U, j) \neq f_\mathbf{A}^\mathsf{A}(U, j)$ only when one of them is equal to 0 and $j \leq n$, and we have just proved that there are at most finitely many such pairs (U, j).

Therefore $\{f_\mathbf{A}^\mathsf{A}\}$ is a coherent family of functions indexed by $(\emptyset \times \mathrm{Fin}) \times (\emptyset \times \mathrm{Fin})$ and by Corollary 2.3.2 there is $f \colon (\omega^{(2)} \times \mathcal{B}) \to \{0,1\}$ such that for all $\mathbf{A}, \mathsf{A}$, for all but finitely many $i \in \mathsf{A}$ and $U \in \hat{\mathbf{A}}$ we have

$$f_\mathbf{A}^\mathsf{A}(U, i) = f(U, i).$$

Consider a partition $[\mathcal{B} \times \omega]^{<\omega} = K_0 \cup K_1$ defined by letting a nonempty $s \in K_0$ if and only if

(K1) $\bigcap \{U : (U, i) \in s \text{ for some } i\}$ is nonempty, and
(K2) $\{i : (U, i) \in s \text{ for some } U\}$ has exactly one element.

Then $(f_\mathbf{A}^\mathsf{A})^{-1}(\{i\})$ is either empty or a maximal K_0-homogeneous subset of $\mathrm{dom}(f_\mathbf{A}^\mathsf{A})$ for every i, and therefore by Lemma 2.4.1 we can assume that the trivializing function f is such that $f^{-1}(\{j\})$ is a maximal K_0-homogeneous subset of $\mathrm{dom}(f) = (\mathcal{B} \setminus Z_k) \times (\omega^{(2)} \setminus [k, \infty))$ for $j = 0, 1$.

For every i consider the set
$$S_i = \{U \in \mathcal{B} : f(U, i) = 1\}.$$
By the above, each nonempty S_i is a maximal centered subset of $\mathcal{B} \setminus Z_k$ for all i, and by the compactness, the set $\bigcap_i S_i$ is a singleton for all such i. Let us define a partial map h from a subset of ω into X by letting
$$h(i) \in \bigcap S_i$$
whenever S_i nonempty; note that in this case $\bigcap S_i$ is a singleton.

CLAIM 2. *The mapping $\Phi_h \colon \mathrm{cl}(X) \to \mathcal{P}(\omega)$ is an almost lifting of Φ.*

PROOF. Let us first note that for all $\mathbf{A}, \mathsf{A}$ and all $C \subseteq \mathbf{A}$, we have

(*) $$F(g_\mathbf{A}(C)) =^* \Phi_{h_\mathbf{A}^\mathsf{A}}(C).$$

Assume the assertion of the claim is false, and let $\mathcal{U}_\xi$ ($\xi < \omega_1$) be an uncountable family in $\mathrm{cl}(X)$ such that
$$\mathcal{U}_\xi \cap \mathcal{U}_\eta \in \mathcal{K}(X)$$
but
$$F(\mathcal{U}_\xi) \Delta \Phi_*(\mathcal{U}_\xi) \notin \mathrm{Fin} \times \emptyset$$
for all $\xi \neq \eta$. For every ξ pick $\mathbf{A}_\xi \in \mathcal{A}$ such that $\mathcal{U}_\xi$ is equal to the union of some subfamily of $\mathbf{A}_\xi$. Since by Claim 1 the ideal $\mathcal{I}$ generated by $\hat{\mathbf{A}}$ is isomorphic to $\emptyset \times \mathrm{Fin}$, there is $\mathbf{A} \in \mathcal{A}$ such that $\mathbf{A} \prec \mathbf{A}_\xi$ for uncountably many ξ (since OCA implies $\mathfrak{b} > \aleph_1$, by Theorem 2.2.6). For every one of these ξ find $\mathsf{A}_\xi \in \emptyset \times \mathrm{Fin}$ such that the set
$$(F(\mathcal{U}_\xi) \Delta \Phi_*(\mathcal{U}_\xi)) \cap \mathsf{A}_\xi$$
is infinite. Again, we can find a single A which includes uncountably many of these A_ξ. Since $\Phi_{h_\mathbf{A}^\mathsf{A}}$ is an almost lifting of $\Phi_\mathbf{A}^\mathsf{A}$, for one of these ξ we have
$$\Phi_{h_\mathbf{A}^\mathsf{A}}(\mathcal{U}_\xi) =^* (\Phi_\mathbf{A}^\mathsf{A})_*(\mathcal{U}_\xi).$$
But this contradicts (*) and therefore concludes the proof. □

This concludes the proof of (1).

To prove (2) we need to show that the set $\{s \in \alpha^{(d)} : h(s) \neq \emptyset\}$ is in $\mathcal{R}_\alpha^d$. In a special case when X is an infinite discrete space, this follows from Lemma 4.8.4.

To treat the general case, pick an arbitrary $\mathbf{A} \in \mathcal{A}$. Since $\{h_\mathbf{A}^\mathsf{A} : \mathsf{A} \in \emptyset \times \mathrm{Fin}\}$ is a coherent family of functions indexed by $\emptyset \times \mathrm{Fin}$, by Theorem 2.2.7 we can find a partial function $h_\mathbf{A}$ from a subset of ω into $\mathbf{A}$ which uniformizes this family. (Remember that we are thinking of $\emptyset \times \mathrm{Fin}$ as an ideal supported by ω to simplify the notation; the supporting set is actually equal to $\alpha^{(d)}$.) Then $\Phi_{h_\mathbf{A}}$ is a completely additive almost lifting of $\Phi_\mathbf{A}$ (by a proof identical to that of Claim 2). Therefore by Lemma 4.8.4 we can assume that the domain of $h_\mathbf{A}$ is in $\mathcal{R}_\alpha^d$. But since $\Phi_{g_\mathbf{A} h(i)}$ and $\Phi_{h_\mathbf{A}(i)}$ are almost liftings of the same homomorphism, $\Phi_\mathbf{A}$, we can assume that the domains of $g_\mathbf{A} h(i)$ and $h_\mathbf{A}$ are equal. Since $g_\mathbf{A}$ is a total function, we can assume that the domain of $h(i)$ is equal to the domain of $h_\mathbf{A}$, and the latter set is in $\mathcal{R}_\alpha^d$. This completes the proof. □

4.10. Versions of wEP

There is no a priori reason why wEP should be restricted only to a class of countable locally compact spaces. In this section we show that at least a little bit more can be said at this point. Let us first note the following generalization of Theorem 4.9.1.

THEOREM 4.10.1. *OCA and MA imply wEP(X,Y) for all zero-dimensional locally compact Polish spaces X and Y.* □

The proof of this result will appear elsewhere. Let us only note that it requires analyzing the liftings of homomorphisms of $\mathcal{P}(\omega)/\operatorname{Fin}$ into $\operatorname{cl}(X)/\mathcal{K}(X)$ for a locally compact Polish space X, and this is done by mimicking proofs of Chapter 3.

We do not know for which spaces X, Y the principle wEP(X, Y) follows from OCA and MA, but we note that the assumption of local compactness is not necessary. Consider $S(\omega)$, a *sequential fan with countably many spines*, defined as follows. The underlying set of $S(\omega)$ is $\omega^{(2)} \cup \{\infty\}$, all points in $\omega^{(2)}$ are isolated and the neighborhood basis of ∞ is the family of all sets whose complements are in $\emptyset \times \operatorname{Fin}$ (see [**27**]).

PROPOSITION 4.10.2. *OCA and MA imply wEP$(\omega, S(\omega))$.*

PROOF. (Sketch) This proof goes along the same lines as the proof of Theorem 4.9.1, and we shall therefore only give a sketch. First note that the ideal of compact subsets of $S(\omega)$ coincides with the ideal of finite subsets of $\omega^{(2)}$ which we shall denote by Fin (note that $\{\infty\} \notin \operatorname{Fin}$). We have to prove that every homomorphism $\Phi \colon \operatorname{cl}(S(\omega))/\operatorname{Fin} \to \mathcal{P}(\omega)/\operatorname{Fin}$ has a completely additive almost lifting. This is proved by considering $\operatorname{cl}(S(\omega))$ as a limit of algebras isomorphic to $\mathcal{P}(\omega)/\operatorname{Fin}$, just like in proof of Theorem 4.9.1. For $A \in \emptyset \times \operatorname{Fin}$ let $\Psi_A \colon \mathcal{P}(A \cup \{\infty\})/\operatorname{Fin} \to \mathcal{P}(\omega)/\operatorname{Fin}$ be given by its lifting,

$$\Psi_{A*}(B) = \begin{cases} B \cup (\omega^{(2)} \setminus A) \cup \{\infty\}, & \text{if } \infty \in B, \\ B, & \text{if } \infty \notin B \end{cases}$$

Then $\Phi \circ \Psi_A$ is a homomorphism of $\mathcal{P}(A \cup \{\infty\})/\operatorname{Fin}$ into $\mathcal{P}(\omega)/\operatorname{Fin}$, therefore it has an almost lifting Φ_{h_A} for some $h_A \colon \omega \to A \cup \{\infty\}$. Since OCA implies that the ideal $\emptyset \times \operatorname{Fin}$ is $\aleph_1$-directed under $\subseteq^*$, we can uniformize h_A ($A \in \emptyset \times \operatorname{Fin}$) to get $h \colon \omega \to \omega^{(2)} \cup \{\infty\}$ such that Φ_h is an almost lifting of Φ. □

It should be clear that the above proof works when $\emptyset \times \operatorname{Fin}$ is replaced with any analytic P-ideal, or any ideal which is $\aleph_1$-directed under the inclusion modulo finite.

Following the terminology of §4.4, we say that a function $F \colon \alpha^* \to \gamma^*$ is *induced by* some continuous $h \colon \alpha \to \gamma$ if $F = F_1 \upharpoonright \alpha^*$, where $F_1 \colon \beta\alpha \to \beta\gamma$ is the unique extension of h to $\beta\alpha$.

PROPOSITION 4.10.3. *wEP implies that if α is a countable ordinal then every autohomeomorphism of α^* is induced by a perfect continuous map $h \colon \alpha \to \alpha$.* □

Therefore our results reduce the question of describing autohomeomorphism group of α^* (and possibly its finite powers) to a question about functions on the integers.

We can consider a version of weak Extension Principle for closed subspaces X of ω^* and other Čech–Stone remainders. Let *weak Extension Principle for X* be wEP(X) If $F\colon X \to \omega^*$ is a continuous map, then there is a partition $X = U_0 \cup U_1$ into clopen sets such that $F''U_0$ is nowhere dense and $F \upharpoonright U_1$ can be continuously extended to $F'\colon \beta\omega \to \beta\omega$.

A similarly-looking extension principle was proved by van Douwen and van Mill ([19]), who extended a classical result of W. Rudin ([104]). They proved that CH implies that every homeomorphism between nowhere dense P-subsets of ω^* can be extended to an autohomeomorphism of ω^*. Rudin has proved this in the case when these two sets are singletons. This is incompatible with OCA and MA, because MA implies that there are $2^{\mathfrak{c}}$ many P-points in ω^*, while there are only continuum many autohomeomorphisms of ω^*. But wEP (and therefore OCA and MA) does imply this principle for many "simply definable" closed nowhere dense P-subsets of ω^*, i.e., those $X \subseteq \omega^*$ such that (by $\overline{A}$ we denote the closure of the set A)

$$\{A \subseteq \omega : \overline{A} \cap X = \emptyset\}$$

is a non-pathological analytic P-ideal. This follows from and Theorem 3.8.1. Similarly, by Theorem 3.3.6, MA and OCA imply wEP(G). However, the existence of a Baire homomorphism with no completely additive lifting (Theorem 1.9.5) implies the following.

PROPOSITION 4.10.4. *There is a continuous surjection F of a closed P-subset of ω^* onto ω^* which cannot be continuously extended to ω^*.* □

Let us remark that our OCA lifting results of Chapter 3 have been recently extended by A. Dow and K.P. Hart ([24]), who used them to prove that under OCA the Lebesgue measure algebra does not embed into $\mathcal{P}(\omega)/\operatorname{Fin}$, answering a question of Fremlin ([49]). They have observed the following useful extension of Theorem 3.8.1.

THEOREM 4.10.5. *Assume OCA and MA. If $\Phi\colon \mathcal{P}(\omega) \to \mathcal{P}(\omega)/\operatorname{Fin}$ is a homomorphism, then Φ has a continuous almost lifting.* □

It is worth remarking that the Lebesgue measure algebra (moreover any measure algebra of size at most continuum) embeds into $\mathcal{P}(\omega)/\mathcal{Z}_0$, by a result of Fremlin (see [48]). Since such an embedding cannot have a completely additive lifting and $\mathcal{Z}_0$ is a non-pathological ideal, this result imposes some limitations to possible extensions to our results about liftings.

4.11. Remarks and questions

The following theorem is a natural generalization of Corollary 4.5.3; its proof can be found in [41].

THEOREM 4.11.1. *wEP implies that the following are equivalent*
(1) $(\alpha^*)^{(d)} \approx (\gamma^*)^{(l)}$,
(2) $\alpha = \gamma$ and $d = l$,
for all countable indecomposable ordinals α and γ and all positive integers d, l. □

By a result of van Douwen ([17, Theorem 7.2]) $(\alpha^*)^{(d)} \approx (\gamma^*)^{(l)}$ implies $d = l$. By Parovičenko's result, this is as much as can be said without using wEP or some other additional axiom. The following consequence of Theorem 4.11.1 also appears in [41].

THEOREM 4.11.2. *wEP implies that the following are equivalent*

(1) $(\alpha^*)^{(d)} \twoheadrightarrow (\gamma^*)^{(l)}$,
(2) $\alpha \geq \gamma$ and $d \geq l$,

for all countable indecomposable ordinals α and γ and all positive integers d, l. □

Note that wEP is indeed an optimal assumption in this context. By Corollary 4.5.3, wEP implies that (1) is equivalent to its consequence saying that there exists a mapping $h \colon \omega \to \omega$ with certain first-order properties, and the truth of such a statement does not depend on which Set-theoretic axioms we are assuming by Shoenfield's Absoluteness Theorem ([**110**]). Therefore if it is consistent that (1) and (2) of Theorem 4.11.2 are equivalent, then they are already equivalent under wEP.

Just's original proof that ω^* does not map onto $(\omega^*)^{(2)}$ ([**61**]) did not rely on an analysis of liftings of homomorphisms between analytic quotients like our proof from wEP. Instead, it used an axiom (introduced in [**63**] under then name AT) which states that every map between analytic quotients which merely preserves intersections is "trivial" on a large set. There is still an interest in analyzing liftings of such maps. By employing an idea due to M. Bell ([**6**]), we have used a version of this axiom which also follows from OCA to prove that the exponential space $\text{Exp}(\omega^*)$ is not a continuous image of ω^* (see [**38**, §8]).

In §4 of [**142**], Velickovic proved that the Proper Forcing Axiom implies all automorphisms of $\mathcal{P}(\kappa)/\text{Fin}$ are trivial. His methods suggest that the following (as well as its version for finite powers) should follow from Corollary 4.7.2.

CONJECTURE 4.11.3. *PFA implies that the following are equivalent*
(1) $X^* \twoheadrightarrow Y^*$,
(2) *there is a perfect* $h \colon X \to Y$ *such that the closure of* $Y \setminus h''X$ *is compact,*
for all locally compact, locally countable spaces X *and* Y.

It would be of an interest to find as large as possible class of spaces for which we can expect (1) and (2) to be equivalent. By Theorem 4.10.1 all zero-dimensional, locally compact, complete, separable metric spaces belong to this class.

The case $d = l = 1$ of the following question (or rather, its dual version) was already asked in Question 3.14.2. By the main result of [**41**], one of these questions has a positive answer if and only if the other one does.

QUESTION 4.11.4. *Do OCA and MA imply the following strong Extension Principle for finite powers of* ω^*:

(sEP) *If* $F \colon (\omega^*)^{(d)} \to (\omega^*)^{(l)}$ *is a continuous map then* F *can be continuously extended to* $F' \colon (\beta\omega)^{(d)} \to (\beta\omega)^{(l)}$?

An (expected) positive answer to this question would greatly improve our understanding of the continuous images of ω^*. In particular, it may be related to the following question asked by M. Bell which is still open:

QUESTION 4.11.5 (Bell). *Is it consistent that for every Boolean algebra* $\mathcal{B}$, *if* $\mathcal{P}(\omega)$ *embeds into* $\mathcal{B}/\mathcal{I}$ *for some countably generated ideal* $\mathcal{I}$, *then* $\mathcal{P}(\omega)$ *embeds into* $\mathcal{B}$ *itself?*

For more information and some partial results regarding this question (in particular, a negative CH-result) the reader can consult [**22**]. A related question was asked by A. Dow ([**24**]):

QUESTION 4.11.6 (Dow). *Is it consistent that a complete Boolean algebra $\mathcal{B}$ embeds into $\mathcal{P}(\omega)/\operatorname{Fin}$ if and only if it is σ-centered?*

Equivalently, is it consistent that every extremally disconnected continuous image of ω^* is separable?

Needless to say, under CH the answer is negative. It is clear that a version of strong extension principle (i.e., a strong lifting theorem) would give a positive answer. The following observation is due to K.P. Hart and the proof is included with his kind permission.

PROPOSITION 4.11.7. *If there is a nonseparable extremally disconnected continuous image X of ω^*, then Strong Extension Principle fails.*

PROOF. Let $f\colon \omega^* \to X$ be a continuous surjection. By a result of Simon ([**111**]), every compact extremally disconnected space can be embedded into ω^* as a weak P-set. (Recall that $P \subseteq \omega^*$ is a *weak P-set* if the closure of D is disjoint from P whenever $D \subseteq \omega^* \setminus P$ is countable.) We can assume that X itself is a weak P-subset of ω^*.

We claim that f cannot be continuously extended to $\beta\omega$. Assume the contrary, that $F\colon \beta\omega \to \beta\omega$ is such an extension, and let $D = F''\omega$. Then the closure $\overline{D}$ of D includes X. But since X is nonseparable, we have $\overline{D \cap X} \neq X$. Also, $\overline{D \setminus X}$ is disjoint from X, and therefore $X \setminus \overline{D}$ is nonempty; a contradiction. □

It is believed that PFA implies the positive answer to Question 4.11.6. It should be noted that under PFA every complete Boolean algebra which embeds into $\mathcal{P}(\omega)/\operatorname{Fin}$ has to have ccc (by [**125**, Theorem 8.8], which is a PFA-reformulation of a result of [**81**]). We should also note that the assumption that $\mathcal{B}$ is complete is necessary, since Bell ([**7**]) has shown that $\mathcal{P}(\omega)/\operatorname{Fin}$ has a rich supply of ccc subalgebras with versatile cellularity properties.

Results in this Chapter are formulated in terms of the existence of completely additive almost liftings for arbitrary homomorphisms assuming OCA and MA. However, our proofs give analogous results about the existence of completely additive liftings for homomorphisms with Baire-measurable liftings without any set-theoretic assumptions. Therefore we have the following picture for the Baire-embeddability ordering:

$$\mathcal{P}(\omega)/\operatorname{Fin} \to \operatorname{cl}(\omega^2)/\langle\omega^2\rangle \to \cdots \to \operatorname{cl}(\omega^\alpha)/\langle\omega^\alpha\rangle \to \cdots \to \mathcal{P}(\omega^{(2)})/\operatorname{Fin}\times\emptyset$$

where all the arrows are irreversible. This can be contrasted to a dichotomy result of Kechris and Louveau ([**76**]) which in particular implies that every quotient of a Borel subalgebra of $\mathcal{P}(\omega)$ over a Borel ideal which is of Borel-cardinality strictly less than $\mathcal{P}(\omega^{(2)})/\operatorname{Fin}\times\emptyset$ has to be of Borel-cardinality at most $\mathcal{P}(\omega)/\operatorname{Fin}$.

CHAPTER 5

Gaps and limits in analytic quotients

5.1. Introduction

One of the beautiful early discoveries in the study of $\mathcal{P}(\mathbb{N})/\operatorname{Fin}$ was Hausdorff's construction of a *gap* inside this quotient ([**54**], [**53**], see Theorem 5.2.1; all definitions are given at the end of this section). This theorem shows that the degree of saturatedness of $\mathcal{P}(\mathbb{N})/\operatorname{Fin}$ is not so large and that this algebra is a saturated structure only in the trivial case when the Continuum Hypothesis holds. A remarkable feature of Hausdorff's original gap is the fact that its construction did not require any additional set-theoretic axioms, a phenomenon which is rarely encountered at this level of complexity in investigating $\mathcal{P}(\mathbb{N})/\operatorname{Fin}$. After the invention of Forcing, Kunen ([**81**]) proved a rather surprising fact about gaps in $\mathcal{P}(\mathbb{N})/\operatorname{Fin}$: The ordinary axioms of Set Theory are not sufficient to give us any types of gaps other than those discovered by Hausdorff and Rothberger ([**54**], [**53**], [**103**]). Later, Todorcevic (see [**125**, Theorem 8.6]) deduced the same conclusion from OCA.

The question of the existence of gaps in arbitrary analytic quotients was addressed only relatively recently by Just ([**105**]). Mazur ([**95**]) constructed $\langle\omega_1,\omega_1\rangle$-gaps in every quotient over an F_σ ideal, and, using a weak form of CH, in every quotient over a meager (in particular, analytic) ideal. Then Todorcevic ([**130**], [**128**]) proved the following result that gave a new outlook to this kind of problems:

THEOREM 5.1.1. *Every monomorphism of $\mathcal{P}(\mathbb{N})/\operatorname{Fin}$ into a quotient $\mathcal{P}(\mathbb{N})/\mathcal{I}$ over an analytic ideal with a Baire-measurable lifting preserves all Hausdorff gaps. If $\mathcal{I}$ is moreover a P-ideal, then Φ preserves all gaps.*

Since by a theorem of Mathias $\mathcal{P}(\mathbb{N})/\operatorname{Fin}$ is Baire-embeddable into every analytic quotient (see §1.2), this implies the existence of Hausdorff gaps in every analytic quotient. Todorcevic ([**130**]) showed even more, that embeddings with completely additive liftings between analytic quotients (and Mathias' theorem does give an embedding of this form—see §1.2) preserve all gaps, and therefore the gap-spectrum of every analytic quotient always includes the gap-spectrum of $\mathcal{P}(\mathbb{N})/\operatorname{Fin}$. It is natural to ask whether analogues of Kunen's or Todorcevic's theorem are true for other analytic quotients (see [**130**, Problem 2], [**128**, Problems 3 and 4]). We shall see later (§5.7, §5.8, §5.10) that they both fail in certain analytic quotients.

Many authors consider only linear gaps, i.e., those gaps whose both sides are linearly ordered by the inclusion modulo the ideal. But some important gap phenomena that cannot be detected by linear gaps occur in the realm of nonlinear gaps (compare Proposition 5.3.2 to Lemmas 5.8.6 and 5.8.7, also see §5.7, §5.10 below). Any attempt on analysis and classification of gaps in analytic quotients requires a reasonable notion of a type of a gap. In the case when both sides $\mathcal{A}, \mathcal{B}$ of a gap are linear, the *type* of this gap is the pair of regular cardinals (see the end of this section). In an attempt to understand nonlinear gaps, in §5.6 we will suggest a definition of type of such a gap. This notion can be formulated using the language of *Tukey reductions* between partially ordered sets (see [**138**], [**124**], [**46**]), which we also use to describe the gap-preservation phenomena.

The organization of this chapter. In §5.2 we shall review some facts about gaps in $\mathcal{P}(\mathbb{N})/\text{Fin}$ relevant to our results in later sections. In §5.3 we prove that the (linear) gaps spectra of Fin and $\emptyset \times \text{Fin}$ coincide and that gaps are often preserved by Baire embeddings of $\mathcal{P}(\mathbb{N}^2)/\emptyset \times \text{Fin}$, answering [**128**, Problem 4]. In §5.4 we prove analogous results for $\text{Fin} \times \emptyset$, now using a consequence of OCA; this partially answers [**128**, Problem 3]. In §5.5 we prove that Hurewicz-like separation phenomenon occurring in $\mathcal{P}(\mathbb{N})/\text{Fin}$ discovered by Todorcevic in [**126**] also occurs in $\mathcal{P}(\mathbb{N}^2)/\text{Fin} \times \emptyset$, but it fails in a quotient over any analytic P-ideal other than Fin (§5.7). (This is where the nonlinear gaps enter the scene; their use is necessary by results of §5.3.) In §5.6 we introduce the notions of Tukey reduction and cofinal similarity and suggest how to define the type of a nonlinear gap. In §5.8 we consider the quotients over the ideals $\text{Fin} \times \text{Fin}$ and $\emptyset \times \text{Fin}$ as countable (reduced) powers of $\mathcal{P}(\mathbb{N})/\text{Fin}$ to obtain information about their gap spectra. The connection between the preservation of gaps and the (non)existence of analytic gaps, as well as the preservation of gaps under arbitrary embeddings are discussed in §5.9. In §5.10 we define an analytic Hausdorf gap in a quotient over an analytic P-ideal (this was rather surprising—cf. [**130**, page 96]). In §5.11 some easy facts about limits are mentioned, and in §5.12 a "multigap" from [**29**] is embedded into every analytic quotient. This Chapter closes with some remarks and questions given in §5.13.

Terminology and notation. Let us now define the basic objects of study in this Chapter. Families $\mathcal{A}, \mathcal{B}$ of sets of integers are $\mathcal{I}$-*orthogonal* ($\mathcal{A} \perp_\mathcal{I} \mathcal{B}$) if $A \cap B \in \mathcal{I}$ for all $A \in \mathcal{A}$ and $B \in \mathcal{B}$. (If $\mathcal{I} = \text{Fin}$, then we say that $\mathcal{A}$ and $\mathcal{B}$ are *orthogonal* and write $\mathcal{A} \perp \mathcal{B}$.) A pair of families $\mathcal{A}, \mathcal{B}$ which are orthogonal modulo some $\mathcal{I}$ is also called a *pregap* in $\mathcal{P}(\mathbb{N})/\mathcal{I}$. We say that $\mathcal{A}$ and $\mathcal{B}$ are *separated* if there is $C \subseteq \mathbb{N}$ such that $A \subseteq^\mathcal{I} C$ and $C \cap B \in \mathcal{I}$ for all $A \in \mathcal{A}$ and $B \in \mathcal{B}$. Such a set C *separates* $\mathcal{A}$ from $\mathcal{B}$. If a pair of $\mathcal{I}$-orthogonal families $\mathcal{A}, \mathcal{B}$ is not separated, then we say that $\mathcal{A}, \mathcal{B}$ is a *gap in* $\mathcal{P}(\mathbb{N})/\mathcal{I}$. If there is a sequence C_n ($n \in \mathbb{N}$) of subsets on $\mathbb{N}$ such that for every pair $A \in \mathcal{A}$, $B \in \mathcal{B}$ there is n such that $A \subseteq^\mathcal{I} C_n$ and $C_n \cap B \in \mathcal{I}$, then we say that the pregap $\mathcal{A}, \mathcal{B}$ is *countably separated*.

In order to define the type of a gap, we need some auxiliary definitions about the linearly ordered sets. Recall that $\text{cf}(\mathcal{P})$ denotes the *cofinality* of a partially ordered set $\mathcal{P}$, namely the smallest size of a *cofinal* subset X of $\mathcal{P}$, i.e., such that every element of $\mathcal{P}$ is below some element of X. For $\mathcal{A} \subseteq \mathcal{P}(\mathbb{N})$ by $\mathcal{A}/\mathcal{I}$ we denote the poset $\langle \mathcal{A}/\mathcal{I}, \subseteq^\mathcal{I} \rangle$ (where $\subseteq^\mathcal{I}$ stands for the inclusion modulo $\mathcal{I}$). If $\mathcal{A}, \mathcal{B}$ is a gap and both $\mathcal{A}/\mathcal{I}$ and $\mathcal{B}/\mathcal{I}$ are linear orderings of cofinality κ and λ, respectively, then we say that a gap $\langle \mathcal{A}, \mathcal{B} \rangle$ is *linear* and that its *type* is the pair of cardinals $\langle \kappa, \lambda \rangle$, or that it is a $\langle \kappa, \lambda \rangle$-gap. A *linear gap spectrum* of a quotient $\mathcal{P}(\mathbb{N})/\mathcal{I}$ is the family of all types $\langle \kappa, \lambda \rangle$ of its linear gaps. We analogously define a *Hausdorff gap spectrum* and a *gap spectrum* of a quotient.

5.2. Gaps in the quotient over Fin

The study of saturatedness properties of $\mathcal{P}(\mathbb{N})/\text{Fin}$ with the ordering $\subseteq^*$ of inclusion modulo finite started as early as in the second half of the last century, when du Bois-Reymond proved there are no $\langle \omega, 1 \rangle$-limits in $\mathcal{P}(\mathbb{N})/\text{Fin}$, namely whenever A_n is a strictly increasing sequence in $\mathcal{P}(\mathbb{N})/\operatorname{Fin}$ and $B \supseteq^* A_n$ or all n, then there is $C \subseteq^* B$ such that $B \setminus C$ is infinite and it almost includes all A_n. Later Hadamard ([**51**]) proved there are no $\langle \omega, \omega \rangle$-gaps in $\mathcal{P}(\mathbb{N})/\operatorname{Fin}$. The next step was made by Hausdorff.

THEOREM 5.2.1 (Hausdorff, [**53**]). *There is an $\langle \omega_1, \omega_1 \rangle$-gap in $\mathcal{P}(\mathbb{N})/\operatorname{Fin}$.* □

We shall call any pair of orthogonal, σ-directed families in some quotient which cannot be separated a *Hausdorff gap*. Compare Theorem 5.2.1 with another result of Hausdorff ([**54**]), rediscovered by Rothberger ([**103**]) (recall that $\mathfrak{b}$ is the minimal length of an unbounded sequence in $\langle \mathbb{N}, \leq^* \rangle$).

THEOREM 5.2.2 (Hausdorff, Rothberger). *There is an unbounded κ-chain of in $\langle \mathbb{N}^\mathbb{N}, \leq^* \rangle$ if and only if there is an $\langle \omega, \kappa \rangle$-gap in $\mathcal{P}(\mathbb{N})/\operatorname{Fin}$. In particular, there is an $\langle \omega, \mathfrak{b} \rangle$-gap in $\mathcal{P}(\mathbb{N})/\operatorname{Fin}$.* □

Any gap of type $\langle \omega, \kappa \rangle$ will be called a *Rothberger gap*, while any gap of type $\langle \kappa, \lambda \rangle$ where both κ and λ are uncountable and regular will be called a *Hausdorff gap*. (For the definitions of nonlinear Rothberger and Hausdorff gaps, see §5.6.) The cardinal $\mathfrak{b}$ equals ω_1 under CH, but it can assume different values in different models of Set theory (e.g., MA implies $\mathfrak{b} = \mathfrak{c}$, while OCA implies $\mathfrak{b} = \omega_2$; see Theorem 2.2.6). In particular, the existence of $\langle \omega, \omega_1 \rangle$-gaps is, unlike the existence of $\langle \omega_1, \omega_1 \rangle$-gaps, not decided by the usual axioms of Set theory. Are there any types of gaps in $\mathcal{P}(\mathbb{N})/\operatorname{Fin}$ other than Hausdorff's and Rothberger's? Under CH the answer is trivially negative, while there are situations in which gap-spectrum of $\mathcal{P}(\mathbb{N})/\operatorname{Fin}$ is rather rich (see e.g., Proposition 5.2.6 below). In ([**81**]), Kunen proved that it is impossible to construct any types of gaps other than Hausdorff's and Rothberger's without using some additional assumptions. Then Todorcevic ([**125**, §8]) formulated Theorem 5.2.4 below which not only gave an explanation to Kunen's theorem, but it raised considerably the confidence about the right formulation of OCA which motivated much of the further research on this axiom.

Todorcevic's result is best explained if we leave the domain of linear gaps. The following definition is extracted from [**89**] in [**126**]. Two orthogonal families $\mathcal{A}, \mathcal{B}$ form a *Luzin* gap if there is an uncountable $X \subseteq 2^\mathbb{N}$ and two surjections $f_0 \colon X \to \mathcal{A}$ and $f_1 \colon X \to \mathcal{B}$ such that for all $x, y \in X$ we have

$$f_0(x) \cap f_1(y) \neq \emptyset \qquad \text{if and only if} \qquad x \neq y.$$

LEMMA 5.2.3. *Every Luzin gap is a gap.*

PROOF. Assume $\mathcal{A}, \mathcal{B}$ is a Luzin gap, and X, f_0, f_1 are as above. Assume the contrary, that $C \subseteq \mathbb{N}$ separates $\mathcal{A}$ from $\mathcal{B}$. By a counting argument, we can find an uncountable $X' \subseteq X$, $n \in \mathbb{N}$ and $s, t \subseteq n$ such that for all $x \in X$ we have $f_0(x) \setminus C = s$ and $f_1(x) \cap C = t$. This implies that $f_0(x) \cap f_1(x) = s \cap t = f_0(x) \cap f_1(y)$ for all $x, y \in X'$, contradicting the definition of the Luzin gap. □

The following result shows that the Luzin gaps are in some sense canonical (see also [**126**, Theorem 2] for a definable version).

THEOREM 5.2.4 (Todorcevic). *OCA implies that for every two orthogonal families $\mathcal{A}$ and $\mathcal{B}$ of subsets of $\mathbb{N}$ exactly one of the following applies:*

(1) *$\mathcal{A}$ and $\mathcal{B}$ can be countably separated from each other,*
(2) *the restriction of $\mathcal{A}$ and $\mathcal{B}$ to an end-segment of the integers contains an uncountable Luzin subgap.*

In particular, every linear gap in $\mathcal{P}(\mathbb{N})/\operatorname{Fin}$ is either an $\langle \omega_1, \omega_1 \rangle$-gap or a Rothberger gap.

PROOF. Recall that $\hat{\mathcal{A}}$ stands for the hereditary closure of $\mathcal{A}$, $\{A : A \subseteq A'$ for some $A' \in \mathcal{A}\}$. Let
$$\mathcal{X} = \{\langle A, B \rangle \in \hat{\mathcal{A}} \times \hat{\mathcal{B}} : A \cap B = \emptyset\},$$
and define a partition $[\mathcal{X}]^2 = K_0 \cup K_1$ by letting $\{\langle A, B \rangle, \langle A', B' \rangle\}$ in K_0 if
$$A \cap B' \neq \emptyset \text{ or } A' \cap B \neq \emptyset.$$
This partition is open (see the beginning of §2.2).

We claim that $\mathcal{X}$ is σ-K_1-homogeneous if and only if $\mathcal{A}$ is countably separated from $\mathcal{B}$. For $C \subseteq \mathbb{N}$ define
$$\mathcal{H}_C = \{\langle A, B \rangle \in \mathcal{X} : A \subseteq C \text{ and } B \cap C = \emptyset\}.$$
The set $\mathcal{H}_C$ is clearly always K_1-homogeneous. Also, if C_n ($n \in \mathbb{N}$) is a countable separation of $\mathcal{A}$ from $\mathcal{B}$, then the K_1-homogeneous sets $\mathcal{H}_{C_n \Delta s}$ ($n \in \mathbb{N}$, $s \in \operatorname{Fin}$) cover $\mathcal{X}$. On the other hand, if $\mathcal{H}$ is K_1-homogeneous, the set
$$C_\mathcal{H} = \bigcup_{\langle A, B \rangle \in \mathcal{H}} A$$
has the property that $A \setminus C_\mathcal{H} \subseteq A \cup B$ and $B \cap C_\mathcal{H} \subseteq A \cup B$, whenever $\langle A, B \rangle \in \mathcal{H}$. Therefore if $\mathcal{X}$ is σ-K_1-homogeneous, $\mathcal{A}$ and $\mathcal{B}$ are countably separated, and this concludes the proof that the two properties are equivalent.

Now assume that $\mathcal{A}$ is not countably separated from $\mathcal{B}$. Since the partition K_0 is open, OCA implies that there is an uncountable K_0-homogeneous subset of $\mathcal{X}$. Let $\mathcal{X}'$ be the set of all $\langle A, B \rangle$ such that for som $\langle A', B' \rangle \in \mathcal{X}$ we have $A \subseteq A'$ and $B \subseteq B'$. Then for some $n \in \mathbb{N}$ and for uncountably many $\langle A, B \rangle \in \mathcal{X}'$ we have $A \cap B \subseteq n$. Identify $2^\mathbb{N} \times 2^\mathbb{N}$ with $2^\mathbb{N}$, let $\mathcal{X} = \mathcal{X}'$, and let f_0, f_1 be the projection maps of X to $\mathcal{A}$ and $\mathcal{B}$ respectively. Then the restriction of the image of $\mathcal{X}$ to $[n, \infty)$ is the required Luzin gap.

It remains to prove that a linear gap can be either an $\langle \omega_1, \omega_1 \rangle$-gap or a Rothberger gap. But every subgap of a $\langle \kappa, \lambda \rangle$-gap, where both κ and λ are uncountable, has to be of type $\langle \kappa, \lambda \rangle$. But a Luzin gap is of size $\aleph_1$, and therefore the conclusion follows. □

Note that by identifying a pair $\langle A, B \rangle$ from $\mathcal{X}$ with a partial function $f_{AB} \colon A \cup B \to \{0, 1\}$ such that $f^{-1}(0) = A$ one can see that the above proof is identical to the proof of Theorem 2.2.1. Moreover, it is not difficult to see that the conclusion of Theorem 5.2.4 is equivalent to a slight strengthening of the conclusion of Theorem 2.2.1, saying that every coherent family of partial functions indexed by some **ideal** is either σ-trivial or contains an uncountable subset consisting of pairwise incompatible functions.

The following result which appears in [**125**, Theorem 8.6] shows that the gap-existence results of Theorem 5.2.2 and Theorem 5.2.1 are sharp.

COROLLARY 5.2.5 (Todorcevic). *Assume OCA and that $\mathfrak{c} = \aleph_2$. Then the only types of linear gaps in $\mathcal{P}(\mathbb{N})/\operatorname{Fin}$ are the following: $\langle \omega, \omega_2 \rangle$, $\langle \omega_1, \omega_1 \rangle$, and $\langle \omega_2, \omega \rangle$.*

PROOF. By Theorem 2.2.6, $\mathfrak{b} = \omega_2$, and by Theorem 5.2.4 and $\mathfrak{c} = \aleph_2$, the only gaps of type $\langle \omega, \kappa \rangle$ or $\langle \kappa, \omega \rangle$ are those for $\kappa = \omega_2$. Let $\mathcal{A}, \mathcal{B}$ form a linear Hausdorff gap. By Theorem 5.2.4 it has a subgap of size $\aleph_1$, and therefore it has to be an $\langle\omega_1,\omega_1\rangle$-gap. The existence of such gaps is guaranteed by 5.2.1. □

Let us note that Hausdorff gap spectrum of $\mathcal{P}(\mathbb{N})/\,\text{Fin}$ can be almost anything. The following (probably folklore) result follows immediately from [**30**, §6].

PROPOSITION 5.2.6. *Assume CH and let $\mathcal{G}$ be a set of regular uncountable cardinals including ω_1. Then there is a cardinal-preserving forcing extension in which there is a $\langle\kappa,\kappa\rangle$-gap in $\mathcal{P}(\mathbb{N})/\,\text{Fin}$ if and only if $\kappa \in \mathcal{G}$.* □

What is the minimal complexity of a gap in $\mathcal{P}(\mathbb{N})/\,\text{Fin}$, considered as a set of reals? It is known (see [**126**]) that in Gödel's constructible universe there are coanalytic Hausdorff gaps, and this is the best possible since by Kunen–Martin theorem (see [**74**]) no analytic set can have order type ω_1 under $\subseteq^*$. This, however, does not exclude the existence of two analytic families, σ-directed under $\subseteq^*$, inside which an $\langle\omega_1,\omega_1\rangle$-gap occurs. For example, a Rothberger gap is hidden inside a pair of orthogonal analytic ideals Fin $\times\emptyset$ and $\emptyset \times$ Fin, and there exists a Luzin gap which is a perfect subset of reals (see e.g., [**126**, page 57]). In [**126**], Todorcevic extended the classical Hurewicz separation property (see [**74**], §5.5, also [**77**]) and proved that it implies the following.

THEOREM 5.2.7. *If $\mathcal{A}$ and $\mathcal{B}$ are orthogonal σ-directed families of sets of integers such that $\mathcal{A}$ is analytic, then they can be separated.* □

This implies not only the nonexistence of analytic Hausdorff gaps, but also Theorem 5.1.1 (see Proposition 5.5.3). For more information about this exciting subject and some of its applications the reader can consult [**105**], [**126**], [**130**] and we shall now turn our attention to gaps in analytic quotients other than $\mathcal{P}(\mathbb{N})/\,\text{Fin}$.

5.3. Gaps in the quotient over $\emptyset \times$ Fin

A question of determining the (linear) gap spectra of $\emptyset \times$ Fin and the preservation of gaps via Baire-embeddings of its quotient into other analytic quotients was raised in [**128**, Problem 4]. In Proposition 5.3.2 and Proposition 5.3.3 below we give a complete answer to this question by using Lemma 5.3.1 below and some results of [**130**] and [**128**]. For a family $\mathcal{A}$ of sets of integers its *restriction* to some $C \subseteq \mathbb{N}$ is $\mathcal{A} \upharpoonright C = \{A \cap C : A \in \mathcal{A}\}$. Recall that the restriction of $\emptyset \times$ Fin to any $[1,n] \times \mathbb{N}$ is isomorphic to Fin.

LEMMA 5.3.1. *If $\mathcal{A}, \mathcal{B}$ is a gap in $\mathcal{P}(\mathbb{N}^2)/\emptyset \times$ Fin, then there is an n such that $\mathcal{A}, \mathcal{B}$ is still a gap when restricted to $[1,n] \times \mathbb{N}$.*

PROOF. Assume that there is C_n which separates $\mathcal{A}_n$ from $\mathcal{B}_n$ for all n. Then $C = \bigcup_n C_n$ separates $\mathcal{A}$ from $\mathcal{B}$, contradicting the assumption that this was a gap and therefore completing the proof. □

PROPOSITION 5.3.2. *The linear gap spectra of $\emptyset \times$ Fin and Fin coincide.*

PROOF. By Todorcevic's Theorem 5.1.1, it will suffice to prove that if there is a gap of certain linear or σ-directed type in $\mathcal{P}(\mathbb{N}^2)/\emptyset \times$ Fin, then there is such a gap in $\mathcal{P}(\mathbb{N})/\,\text{Fin}$. The restriction of $\emptyset \times$ Fin to any $\{n\} \times \mathbb{N}$ is isomorphic to Fin, and we shall use this to reflect gaps to a set of this form.

CLAIM 1. *If $\mathcal{A}, \mathcal{B}$ is a gap in $\mathcal{P}(\mathbb{N}^2)/\emptyset \times$ Fin, then there is a natural number n such that $\mathcal{A} \upharpoonright \{n\}\times\mathbb{N}$ ($\mathcal{B} \upharpoonright \{n\}\times\mathbb{N}$, respectively) is unbounded in $\mathcal{A}$ ($\mathcal{B}$, respectively).*

PROOF. Assume otherwise. Then families $\mathcal{A}$ and $\mathcal{B}$ are separated when restricted to $\{n\} \times \mathbb{N}$ for every n, contradicting Lemma 5.3.1. □

To prove (a), assume $\mathcal{A}, \mathcal{B}$ is a linear gap. Let n_1 and n_2 be integers guaranteed by Claim 1 and Claim 2 respectively, and let $n = \max(n_1, n_2)$. Then $\mathcal{A}, \mathcal{B}$ when restricted to $\{n\} \times \mathbb{N}$ is a gap in $\mathcal{P}(\{n\} \times \mathbb{N})/\operatorname{Fin}$ of the same type as the original gap. This completes the proof of theorem. □

We should note that it is not necessarily true that spectra of nonlinear gaps in quotients over Fin and $\emptyset \times$ Fin coincide. In §5.7 we will construct an analytic gap in a quotient over $\emptyset \times$ Fin of type which, by Theorem 5.2.7, cannot occur in the quotient over Fin.

PROPOSITION 5.3.3. *If* $\Phi \colon \mathcal{P}(\mathbb{N}^2)/\emptyset \times \operatorname{Fin} \to \mathcal{P}(\mathbb{N})/\mathcal{I}$ *is a monomorphism with a Baire-measurable lifting, then* Φ *preserves all Hausdorff gaps. If* $\mathcal{I}$ *is moreover a P-ideal, then* Φ *preserves all gaps.*

PROOF. If $\mathcal{A}, \mathcal{B}$ is a gap in $\mathcal{P}(\mathbb{N}^2)/\emptyset \times \operatorname{Fin}$, then by Lemma 5.3.1 there is n such that the restriction $\mathcal{A}_n, \mathcal{B}_n$ of $\mathcal{A}, \mathcal{B}$ to $[1, n] \times \mathbb{N}$ is still a gap. Let $C = \Phi_*([1, n] \times \mathbb{N})$ (Φ_* is some lifting of Φ). Then $\Phi_1 \colon \mathcal{P}([1,n] \times \mathbb{N})/\operatorname{Fin} \to \mathcal{P}(C)/\mathcal{I}$ is a homomorphism with a Baire-measurable lifting, so by Todorcevic's Theorem 5.1.1, Φ_1 carries $\mathcal{A}_n, \mathcal{B}_n$ into a gap of $\mathcal{P}(C)/\mathcal{I}$, which is clearly also a gap in $\mathcal{P}(\mathbb{N})/\mathcal{I}$. □

5.4. Gaps in the quotient over Fin $\times \emptyset$

This section is devoted to study of gap-spectra and preservation of gaps for quotient over the ideal Fin $\times \emptyset$, analogous to that of previous section. This question was raised in [**128**, Problem 3]. The quotient $\mathcal{P}(\mathbb{N}^2)/\operatorname{Fin} \times \emptyset$ can be considered as an inverse limit

$$\mathcal{P}(\mathbb{N}^2)/\operatorname{Fin} \times \emptyset = \varprojlim \langle \mathcal{P}(\Gamma_f)/\operatorname{Fin}, \pi_g^f, \mathbb{N}^\mathbb{N}\rangle,$$

where $\Gamma_f = \{\langle m, n\rangle : n \leq f(m)\}$, and where $\pi_g^f \colon \Gamma_f/\operatorname{Fin} \to \Gamma_g/\operatorname{Fin}$ (for $f \geq g$) is the restriction mapping. It is clear that the quotient over $\emptyset \times$ Fin can be considered as an inverse limit of quotients $\mathcal{P}([1,n] \times \mathbb{N})/\operatorname{Fin}$, and this fact was implicitly used in previous section.

Let us say that *Hausdorff gaps in the quotient over* Fin $\times \emptyset$ *reflect* if for every Hausdorff gap in the quotient over Fin $\times \emptyset$ there is $f \in \mathbb{N}^\mathbb{N}$ such that the restriction of this gap to Γ_f remains a gap.

Using an analogous terminology, Lemma 5.3.1 states that all gaps in the quotient over $\emptyset \times$ Fin reflect. However, the reflection of gaps in the quotient over Fin $\times \emptyset$ does not follow from the usual axioms of Set theory since it contradicts CH (by Lemma 5.4.1 below and [**90**] or [**25**]). Although the restriction to Hausdorff gaps perhaps looks unnatural, it appears that the Rothberger gaps do not reflect (see Example 5.4.5).

LEMMA 5.4.1. *If Hausdorff gaps in the quotient over* Fin $\times \emptyset$ *reflect, then every coherent family of partial functions indexed by* $\mathbb{N}^\mathbb{N}$ *is trivial.*

PROOF. Let $g_f \colon \Gamma_f \to \{0, 1\}$ ($f \in \mathbb{N}^\mathbb{N}$) be a family of coherent functions, and

$$\mathcal{A} = \{g_f^{-1}(\{1\}) : f \in \mathbb{N}^\mathbb{N}\},$$
$$\mathcal{B} = \{g_f^{-1}(\{0\}) : f \in \mathbb{N}^\mathbb{N}\}.$$

Then $\mathcal{A}, \mathcal{B}$ is a pair of orthogonal modulo Fin $\times \emptyset$ families. Both $\mathcal{A}$ and $\mathcal{B}$ are σ-directed (under the inclusion modulo $\emptyset \times$ Fin) since the ideal $\emptyset \times$ Fin is. Moreover, restrictions of $\mathcal{A}$ and $\mathcal{B}$ to any Γ_f are separated by $C_f = g_f^{-1}(\{1\})$. By the assumed reflection of Hausdorff gaps in the quotient over Fin $\times \emptyset$, $\mathcal{A}$ and $\mathcal{B}$ are separated by some $C \subseteq \mathbb{N}^2$, and taking h to be the characteristic function of C witnesses that the coherent family is trivial. □

We should note that the converse to Lemma 5.4.1 is not true. This follows from the proof of the main result of [**68**].

By a proof analogous to that of Proposition 5.3.3, if Hausdorff gaps in the quotient over Fin $\times \emptyset$ reflect, then every monomorphism of $\mathcal{P}(\mathbb{N}^2)/$ Fin $\times \emptyset$ into a quotient $\mathcal{P}(\mathbb{N})/\mathcal{I}$ over an analytic ideal with a Baire-measurable lifting preserves all Hausdorff gaps. We shall later see (Corollary 5.5.7) that this conclusion does not require any additional assumptions. Also, if we assume OCA and MA instead, then the assumption of the existence of a Baire-measurable lifting is unnecessary, and every monomorphism of the quotient over Fin $\times \emptyset$ into another analytic quotient preserves all Hausdorff gaps (see Corollary 5.9.2).

The following is essentially a reformulation of Theorem 8.7 of [**125**], which says that under OCA every family of functions coherent modulo Fin is either trivialized by countably many functions or it has a subfamily of size $\aleph_1$ of pairwise inconsistent functions (see §§2.2–2.3 for more on similar results).

PROPOSITION 5.4.2. *OCA implies that Hausdorff gaps in the quotient over* Fin $\times \emptyset$ *reflect*.

PROOF. Assume OCA, and let $\mathcal{A}, \mathcal{B}$ be a Hausdorff gap in the quotient over Fin $\times \emptyset$. Recall that for $f \colon \mathbb{N} \to \mathbb{N}$ we have $\Gamma_f = \{\langle m, n \rangle : n \leq f(m)\}$, and define

$$\mathcal{A}' = \{\Gamma_f \cap A : A \subset \mathcal{A}\} \quad \text{and} \quad \mathcal{B}' = \{\Gamma_f \cap B : B \in \mathcal{B}\}.$$

Assume $\mathcal{A}', \mathcal{B}'$ has a subgap of size $\aleph_1$, and let $\mathcal{F} \subseteq \mathbb{N}^\mathbb{N}$ be the set of all f appearing in the definition of $\mathcal{A}'$ and $\mathcal{B}'$. Since OCA implies that $\mathfrak{b} > \omega_1$ (Theorem 2.2.6), we can find $f \colon \mathbb{N} \to \mathbb{N}$ which eventually dominates $\mathcal{F}$. Therefore the restriction of $\mathcal{A}', \mathcal{B}'$ to Γ_f is a gap.

Otherwise, by Theorem 5.2.4 the families $\mathcal{A}'$ and $\mathcal{B}'$ can be countably separated. But since $\mathcal{A}', \mathcal{B}'$ is a Hausdorff gap and $\mathbb{N}^\mathbb{N}$ is σ-directed under $\leq^*$, Lemma 2.2.2 implies that $\mathcal{A}'$ and $\mathcal{B}'$ can be separated, by a set $C \subseteq \mathbb{N}^2$, say. We claim that C separates $\mathcal{A}$ and $\mathcal{B}$. Assume the contrary, then either there is an $A \in \mathcal{A}$ such that $A \setminus C \notin$ Fin $\times \emptyset$, or there is a $B \in \mathcal{B}$ such that $C \cap B \notin$ Fin $\times \emptyset$. Then there is an $f \in \mathbb{N}^\mathbb{N}$ such that $\Gamma_f \cap (A \setminus C)$ (respectively, $\Gamma_f \cap (C \cap B)$) is infinite; but this contradicts the assumption on C. □

PROPOSITION 5.4.3. *OCA implies that the linear gap-spectra of quotients over* Fin $\times \emptyset$ *and* Fin *coincide*.

PROOF. Assume $\mathcal{A}, \mathcal{B}$ is a linear gap in $\mathcal{P}(\mathbb{N}^2)/$ Fin $\times \emptyset$. If this is a Hausdorff gap, let $f \in \mathbb{N}^\mathbb{N}$ be such that $\mathcal{A}, \mathcal{B}$ is still a gap when restricted to Γ_f. Such an f exists by Proposition 5.4.2. Then none of the restrictions of $\mathcal{A}$ and $\mathcal{B}$ to Γ_f has the minimal element. Since both $\mathcal{A}$ and $\mathcal{B}$ are linear, these restrictions give us a gap in $\mathcal{P}(\Gamma_f)/$ Fin of the same type as $\mathcal{A}, \mathcal{B}$.

Now assume $\mathcal{A}, \mathcal{B}$ is a Rothberger gap, i.e., that at least one of $\mathcal{A}$ and $\mathcal{B}$ (say, $\mathcal{A}$) is countable. Let $\mathcal{A} = \{A_m : m \in \mathbb{N}\}$ and $\mathcal{B} = \{B_\xi : \xi < \kappa\}$, where κ is a regular

cardinal. We define an increasing unbounded κ-sequence $\{f_\xi\}$ in $\mathbb{N}^\mathbb{N}$ by the formula
$$f_\xi(n) = \min\{i \,:\, B_\xi \cap A_n \subseteq [1,i] \times \mathbb{N}\}.$$
Then $B_\xi \setminus B_\eta \in \text{Fin} \times \emptyset$ implies $f_\xi \geq^* f_\eta$. If some g eventually bounds all f_ξ, then the set
$$C = \bigcup_{m=1}^\infty (A_m \setminus [1, g(m)] \times \mathbb{N})$$
separates $\mathcal{A}$ and $\mathcal{B}$, and therefore the sequence $\{f_\xi\}$ is unbounded. By a classical result ([**53**], [**103**]), this implies the existence of an $\langle \omega, \kappa \rangle$-gap in $\mathcal{P}(\mathbb{N})/\text{Fin}$. $\square$

As we have mentioned earlier, the fact that every monomorphism of the quotient over $\text{Fin} \times \emptyset$ into any analytic quotient preserves all Hausdorff gaps follows already from the standard axioms of set theory (Corollary 5.5.7). We do not know whether the assumption of the reflection of Hausdorff gaps can be completely removed from Proposition 5.4.3, but let us now show that it is not needed in a special case. The case when $\kappa = \lambda$ of the following lemma is an immediate consequence of [**25**, Theorem 4.1].

LEMMA 5.4.4. *Assume there are an unbounded κ-chain and an unbounded λ-chain in $\langle \mathbb{N}^\mathbb{N}, \leq^* \rangle$. Then there is a $\langle \kappa, \lambda \rangle$-gap in a quotient over $\text{Fin} \times \emptyset$ if and only if there is a gap of this type in a quotient over Fin.*

PROOF. Let f_ξ ($\xi < \kappa$) and g_η ($\eta < \lambda$) be unbounded chains in $\langle \mathbb{N}^\mathbb{N}, \leq^* \rangle$ and let A_ξ, B_η ($\xi < \kappa$, $\eta < \lambda$) be a gap in the quotient over $\text{Fin} \times \emptyset$. Then the sets $A'_\xi = A_\xi \cap \Gamma_{f_\xi}$ and $B'_\eta = B_\eta \cap \Gamma_{g_\eta}$ form a $\langle \kappa, \lambda \rangle$-gap in $\mathcal{P}(\mathbb{N}^2)/\text{Fin}$. (See the proof of Proposition 5.4.2.) $\square$

Let us now show that Rothberger gaps in the quotient over $\text{Fin} \times \emptyset$ need not reflect even under OCA. Recall that a strictly $\leq^*$-increasing sequence f_ξ in $\mathbb{N}^\mathbb{N}$ is a *scale* if it is cofinal in $\langle \mathbb{N}^\mathbb{N}, \leq^* \rangle$. Since a scale exists whenever $\mathfrak{b} = \mathfrak{c}$ ($\mathfrak{c}$ is the size of the continuum), and since OCA implies that $\mathfrak{b} > \omega_1$ (Theorem 2.2.6), in a model in which OCA is true and there are no scales the size of the continuum would have to be at least $\aleph_3$, and at present no such model is known (see Question 2.5.4).

EXAMPLE 5.4.5. Assuming there is a scale $\{f_\xi : \xi < \kappa\}$, there is an $\langle \omega, \kappa \rangle$-gap in $\mathcal{P}(\mathbb{N}^2)/\text{Fin} \times \emptyset$ such that its restriction to any Γ_f is not a gap. Our working copy of $\text{Fin} \times \emptyset$ will be an ideal on $\mathbb{N}^3$ generated by sets $[1,n] \times \mathbb{N}^2$. This ideal is clearly isomorphic to $\text{Fin} \times \emptyset$, so we shall call it $\text{Fin} \times \emptyset$. For $\xi < \kappa$ define $h_\xi \in \mathbb{N}^\mathbb{N}$ by
$$h_\xi(n) = \min\{j \,:\, f_\xi(j) \geq n\}.$$
Let $\mathcal{A}$ be generated by sets $A_n = \mathbb{N}^2 \times [1, n]$, and for $f \in \mathbb{N}^\mathbb{N}$ let
$$E_\xi = \{\langle n, f_\xi(n), h_\xi(n) \rangle : n \in \mathbb{N}\}.$$
Let $\mathcal{B}$ be the family of all
$$B_\alpha = \bigcup_{\xi < \alpha} E_\xi \cap \{\langle n, i, j \rangle \,:\, j \geq h_\xi(n)\}.$$
Then $B_\xi \cap A_m \subseteq [1,k] \times \mathbb{N}^2$ where k is such that $f_\xi(k) > m$, therefore $\mathcal{A}$ and $\mathcal{B}$ are orthogonal. Also, if $\xi < \eta$ then $B_\xi \setminus B_\eta \subseteq [1,k] \times \mathbb{N}^2$ where k is such that $f_\eta \geq^k f_\xi$, therefore B_ξ is (strictly) increasing. We claim that $\mathcal{A}, \mathcal{B}$ form a gap. Assume some $C \subseteq \mathbb{N}^3$ includes (modulo $\text{Fin} \times \emptyset$) all $A_j \in \mathcal{A}$ define $g \colon \mathbb{N} \to \mathbb{N}$ by
$$g(j) = \min\{k \,:\, A_j \setminus C \subseteq [1,k] \times \mathbb{N}^2\}.$$

But if $f_\xi \not\leq^* g$ then $B_\xi \cap C$ has an infinite intersection with the set
$$\{\langle n, f_\xi(n), h_\xi(n)\rangle : n \in \mathbb{N}\},$$
therefore C is not orthogonal to $\mathcal{B}$ and $\mathcal{A}, \mathcal{B}$ is indeed a gap.

It remains only to show that the restrictions of $\mathcal{A}$ and $\mathcal{B}$ to any Γ_f can be separated. In a "standard" interpretation of Fin $\times \emptyset$, the family Γ_f ($f \in \mathbb{N}^\mathbb{N}$) generates the ideal $\emptyset \times $ Fin $= ($Fin $\times \emptyset)^\perp$. Therefore analogues of Γ_f's in our case are sets orthogonal to Fin $\times \emptyset$, and every such set is included in some
$$\Lambda_\xi = \{\langle n, i, j, \rangle : i, j \leq f_\xi(n)\}$$
since $\{f_\xi\}$ is a scale. Then on Λ_ξ families $\mathcal{A}$ an $\mathcal{B}$ are separated by
$$C_\xi = \{\langle n, i, j\rangle \in \Lambda_\xi : j \geq h_\xi(n)\},$$
for $(A_j \cap \Lambda_\xi) \setminus C_\xi \subseteq [1, f_\xi(j)] \times \mathbb{N}^2$ and $(B_\eta \cap \Lambda_\xi) \cap C_\xi \subseteq [1, k] \times \mathbb{N}^2$ if $f_\eta \geq^k f_\xi$. This ends the proof.

5.5. The Todorcevic separation property, TSP

The ideal Fin can be considered as a tree under the ordering $\sqsubset$ of end-extension. Recall that a tree is *superperfect* if above every node it has an infinitely branching node.

DEFINITION 5.5.1. A set $\Sigma \subseteq $ Fin is a *superperfect $\mathcal{B}$-tree* if it is a superperfect subtree of $\langle$Fin, $\sqsubset\rangle$ such that for every infinitely branching node s the set
$$\bigcup\{t \subseteq [\max s + 1, \infty) : s \cup t \text{ is an infinitely branching node of } \Sigma\}$$
is included in some set from $\mathcal{B}$.

An ideal $\mathcal{I}$ has the *Todorcevic separation property*, TSP, if for every analytic hereditary $\mathcal{A}$ and an arbitrary set $\mathcal{B}$ which is $\mathcal{I}$-orthogonal to $\mathcal{A}$ one of the following two conditions applies:

(T1) $\mathcal{A}$ and $\mathcal{B}$ can be countably separated over $\mathcal{I}$, or

(T2) there is a superperfect $\mathcal{B}$-tree all of whose branches are in $\mathcal{A}$ and $\mathcal{I}$-positive.

Clearly, if $\mathcal{I}$ has the TSP and families $\mathcal{A}, \mathcal{B}$ are $\mathcal{I}$-orthogonal and such that $\mathcal{A}$ is analytic and B is σ-directed, then $\mathcal{A}$ and $\mathcal{B}$ can be countably separated. In particular, TSP of $\mathcal{I}$ implies there are no analytic Hausdorff gaps in $\mathcal{P}(\mathbb{N})/\mathcal{I}$.

THEOREM 5.5.2 (Todorcevic, [126]). *The ideal* Fin *has the TSP.* □

The interest in TSP comes from the following result.

PROPOSITION 5.5.3 (Todorcevic, [126]). *Assume $\mathcal{I}$ has the TSP and $\mathcal{J}$ is an analytic ideal. Then every Baire monomorphism* $\Phi \colon \mathcal{P}(\mathbb{N})/\mathcal{I} \to \mathcal{P}(\mathbb{N})/\mathcal{J}$ *preserves all Hausdorff gaps. Equivalently, there are no analytic Hausdorff gaps in* $\mathcal{P}(\mathbb{N})/\mathcal{I}$.

PROOF. Let Φ_* be a continuous lifting of Φ and let $\mathcal{A}, \mathcal{B}$ be a Hausdorff gap in $\mathcal{P}(\mathbb{N})/\mathcal{I}$. If families $\Phi''_*\mathcal{A}$ and $\Phi''_*\mathcal{B}$ are split by some C, let
$$\mathcal{A}' = \{A : \Phi_*(A) \setminus C \in \mathcal{J}\}.$$
This is an analytic family which includes $\mathcal{A}$ and is $\mathcal{I}$-orthogonal to $\mathcal{B}$. Since $\mathcal{A}$ and $\mathcal{B}$ cannot be countably separated (for Hausdorff pregaps "countably separated" implies "separated"), TSP implies there is a superperfect $\mathcal{B}$-tree Σ all of whose branches are in $\mathcal{A}'$. But $\mathcal{B}$ is σ-directed, so there is $B \in \mathcal{B}$ which almost includes

(modulo $\mathcal{I}$) all branchings of Σ. Since Σ is superperfect, such B includes many of its infinite branches, contradicting the fact that $\mathcal{A}'$ and $\mathcal{B}$ are orthogonal. $\square$

We shall use the language of games to reformulate TSP and prove that ideal Fin $\times \emptyset$ possesses this desirable property. In a game $G(\mathcal{A}, \mathcal{B})$ for $\mathcal{A}, \mathcal{B} \subseteq \mathcal{P}(\mathbb{N})$ two players, I and II, play according to the following rules:

I	$s_1 \in$ Fin, $B_1 \in \mathcal{B}$		$s_2 \in$ Fin, $B_2 \in \mathcal{B}$		...
II		$k_1 \in \mathbb{N}$		$k_2 \in \mathbb{N}$	

It is additionally required that $s_{n+1} \subseteq B_n \cap [k_n, \infty)$ for all $n \in \mathbb{N}$. If both players obey the rules, player I wins if and only if $\bigcup_n s_n$ is in $\mathcal{A}$.

LEMMA 5.5.4. *Assume $\mathcal{A}$ and $\mathcal{B}$ are families of sets of integers. There is a superperfect $\mathcal{B}$-tree all of whose branches lie in $\mathcal{A}$ if and only if I has a winning strategy in $G(\mathcal{A}, \mathcal{B})$.*

PROOF. Assume first Σ is a superperfect $\mathcal{B}$-tree all of whose infinite branches lie in $\mathcal{A}$, and for an infinitely branching node $u \in \Sigma$ fix $B_u \in \mathcal{B}$ including the set

$$\bigcup \{s \subseteq [\max u + 1, \infty) : u \cup s \text{ is an infinitely branching node of } \Sigma\}.$$

A winning strategy τ for I is defined as follows: In his first move, I plays some infinitely branching node s_1 of Σ and $B_1 = B_s$. In his i-th move, I finds $s_i \subseteq B_i \cap [k_i, \infty)$ such that $t_i = \bigcup_{j=1}^i s_j$ is an infinitely branching node of Σ and $B_i = B_{t_i}$. This is clearly a winning strategy.

Now assume I has a winning strategy τ for $G(\mathcal{A}, \mathcal{B})$. By playing against τ, II can easily construct a tree satisfying (T2). $\square$

Lemma 5.5.4 immediately implies the following.

COROLLARY 5.5.5. *A Borel ideal $\mathcal{I}$ has the TSP if and only if player I has a winning strategy in $G(\mathcal{A} \setminus \mathcal{I}, \mathcal{B})$ whenever $\mathcal{A}$ and $\mathcal{B}$ are $\mathcal{I}$-orthogonal, not countably separated families such that $\mathcal{A}$ is analytic.* $\square$

THEOREM 5.5.6. *The ideal* Fin $\times \emptyset$ *has the TSP.*

PROOF. For technical convenience, let us assume Fin $\times \emptyset$ is an ideal on $\mathbb{N}$ determined by $g \colon \mathbb{N} \to \mathbb{N}$ as Fin $\times \emptyset = \{A : g''A \text{ is bounded}\}$. Let $\mathcal{A}$ and $\mathcal{B}$ be two Fin $\times \emptyset$-orthogonal sets which cannot be countably separated and such that $\mathcal{A}$ is analytic. Let T be a subtree of Fin $\times$ Fin such that $\mathcal{A}$ is equal to its projection. By Corollary 5.5.5, it will suffice to describe a winning strategy τ for player I in $G(\mathcal{A}, \mathcal{B})$ for $\mathcal{A}, \mathcal{B}$ as above. First pick $(s_1, t_1) \in T$ so that $\mathcal{A}(s_1, t_1) = p[T(s_1, t_1)]$ is not countably separated from $\mathcal{B}$ over Fin $\times \emptyset$ and that set

$$X_1 = \{n : \mathcal{A}(s_1 \cup \{n\}, t'_n) \text{ is not countably separated from } \mathcal{B}$$
$$\text{for some } t'_n \text{ extending } t_n\}$$

is unbounded, and let $B_1 \in \mathcal{B}$ be such that $B \cap \bigcup \mathcal{A}_1$ is positive. In latter moves, I chooses (s_n, t_n) and B_n so that X_n is unbounded, $B_n \cap \bigcup \mathcal{A}(s_n, t_n)$ is positive, and $\max(g''(s_n \setminus s_{n-1})) \geq n$. It is clear that this is possible and $\bigcup_n s_n$ is positive. $\square$

Finally, Proposition 5.5.3 and Theorem 5.5.6 imply the following.

COROLLARY 5.5.7. *Every Baire monomorphism of $\mathcal{P}(\mathbb{N}^2)/$ Fin $\times \emptyset$ into some analytic quotient preserves all Hausdorff gaps.* $\square$

5.6. Tukey reductions of nonlinear gaps

In this section we suggest a natural definition of a *type* of a nonlinear gap. Let us first review some important definitions. In this section $\mathcal{P}, \mathcal{Q}$ will always denote directed partially ordered sets. A map $f\colon \mathcal{P} \to \mathcal{Q}$ is a *Tukey map* if it sends unbounded sets into unbounded sets. If such an f exists, then we say that $\mathcal{Q}$ is *cofinally finer* than $\mathcal{P}$ and write $\mathcal{P} \leq \mathcal{Q}$. We write $\mathcal{P}_1 \equiv \mathcal{P}_2$ if both $\mathcal{P} \leq \mathcal{Q}$ and $\mathcal{Q} \leq \mathcal{P}$. Two partially ordered sets $\mathcal{P}_1$ and $\mathcal{P}_2$ are said to be *cofinally similar* if there is a third partially ordered set $\mathcal{Q}$ such that both $\mathcal{P}_1$ and $\mathcal{P}_2$ are order-isomorphic to some cofinal subset of $\mathcal{Q}$. The following result is due to J.W. Tukey ([**138**]).

THEOREM 5.6.1. *Two directed sets $\mathcal{P}$ and $\mathcal{Q}$ are cofinally similar if and only if $\mathcal{P} \leq \mathcal{Q}$ and $\mathcal{Q} \leq \mathcal{P}$.* □

(For more details on these notions the reader can consult [**124**] and [**46**].) For example, note that
(1) $\langle \mathbb{N}^{\mathbb{N}}, \leq^* \rangle \equiv \langle \emptyset \times \mathrm{Fin}, \subseteq^* \rangle$,
(2) $\langle \mathbb{N}^{\mathbb{N}}, \leq \rangle \equiv \langle \emptyset \times \mathrm{Fin}, \subseteq \rangle$,
(3) $\langle \mathbb{N}, \leq \rangle \equiv \langle \mathrm{Fin}, \subseteq \rangle$.

A much deeper result proved recently by Todorcevic is (we consider ideals as partially ordered by the inclusion; in Definition 1.2.7 it was defined when two ideals are isomorphic):

THEOREM 5.6.2. *If $\mathcal{I}$ is an analytic P-ideal, then either $\mathcal{I}$ is isomorphic to Fin or $\mathbb{N}^{\mathbb{N}} \leq \mathcal{I} \leq \mathcal{I}_{1/n}$. This implies that $\mathbb{N}^{\mathbb{N}}/\mathrm{Fin} \leq \mathcal{I}/\mathrm{Fin} \leq \mathcal{I}_{1/n}/\mathrm{Fin}$.* □

The proofs can be found in [**126**] and [**2**]. Since $\mathcal{I}_{1/n}/\mathrm{Fin} \equiv \mathcal{N}$, where $\mathcal{N}$ stands for the ideal of Lebesgue measure zero sets ordered by the inclusion, this can be seen as a supplement to the well-known relation between the measure and category investigated by Tomek Bartoszynski and others (see [**46**], [**2**]). Recall that $\mathrm{add}(\mathcal{N})$ is the additivity of the Lebesgue measure.

LEMMA 5.6.3. *Every subset of an analytic P-ideal $\mathcal{J}$ of size less than $\mathrm{add}(\mathcal{N})$ is $\subseteq^*$-bounded in $\mathcal{J}$. In particular, $MA_{\aleph_1}$ implies that every subset of size $\aleph_1$ is bounded.*

PROOF. By the second part of Todorcevic's Theorem 5.6.2, $\langle \mathcal{J}, \subseteq^* \rangle$ is Tukey-reducible to $\langle \mathcal{I}_{1/n}, \subseteq^* \rangle$. Therefore if a subset of $\mathcal{J}$ of size $< \mathrm{add}(\mathcal{N})$ is $\subseteq^*$-unbounded, then a subset of $\mathcal{I}_{1/n}$ of size $< \mathrm{add}(\mathcal{N})$ is also $\subseteq^*$-unbounded. It is well-known that $\langle \mathcal{I}_{1/n}, \subseteq^* \rangle$ is Tukey-equivalent to $\mathcal{N}$ (see [**46**], [**2**]), therefore the first part of the statement follows.

Since $MA_{\aleph_1}$ implies that the additivity of measure is strictly bigger than $\aleph_1$, the lemma follows. □

If $\mathcal{A} \subseteq \mathcal{P}(\mathbb{N})$ and $\mathcal{I}$ is an ideal on $\mathbb{N}$, then by $\mathcal{A}/\mathcal{I}$ we shall denote the partially ordered set $\langle \mathcal{A}/\mathcal{I}, \subseteq^{\mathcal{I}} \rangle$.

DEFINITION 5.6.4. If $\mathcal{A}, \mathcal{B}$ is a gap in $\mathcal{P}(\mathbb{N})/\mathcal{I}$ it is of *type* $\langle \mathcal{P}, \mathcal{Q} \rangle$ (or, it is a $\langle \mathcal{P}, \mathcal{Q} \rangle$-*gap*) if
  (a) $\mathcal{P}, \mathcal{Q}$ are directed sets such that $\mathcal{A}/\mathcal{I} \equiv \mathcal{P}$, $\mathcal{B}/\mathcal{I} \equiv \mathcal{Q}$, and
  (b) if $X \subseteq \mathcal{A}$ is unbounded and $Y \subseteq \mathcal{B}$ is unbounded, then X cannot be separated from $\mathcal{B}$ and Y cannot be separated from $\mathcal{A}$.

The families $\mathcal{A}$ and $\mathcal{B}$ are the *sides* of this gap.

We note that requiring both sides of the gap to be directed under the inclusion modulo $\mathcal{I}$ is not a loss of generality, since we can always close $\mathcal{A}$ and $\mathcal{B}$ under the finite union of their elements and still have a pair of families orthogonal modulo $\mathcal{I}$.

Definition 5.6.4 clearly generalizes the *linear* gaps, i.e., $\langle \kappa, \lambda \rangle$-gaps when κ and λ are regular cardinals. We now update our terminology and say that a (nonlinear) gap is *Hausdorff* if both of its sides are σ-directed, and that it is *Rothberger* if one of its sides is cofinally similarly to $\mathbb{N}$. Let us write $\mathcal{B}^\perp/\mathcal{I}$ for an orthogonal of $\mathcal{B}$ modulo $\mathcal{I}$, i.e.,

$$\mathcal{B}^\perp/\mathcal{I} = \{A : A \cap B \in \mathcal{I} \text{ for all } B \in \mathcal{B}\}.$$

LEMMA 5.6.5. *A pair $\mathcal{A}, \mathcal{B}$ of $\mathcal{I}$-orthogonal families satisfies the condition (b) above if and only if the identity map from $\mathcal{A}$ into $\mathcal{B}^\perp/\mathcal{I}$ is a Tukey map.*

PROOF. Both conditions are equivalent to: A subset of $\mathcal{A}$ is unbounded in $\mathcal{A}/\mathcal{I}$ if and only if it is unbounded in $\mathcal{B}^\perp/\mathcal{I}$. □

The condition (b) is not necessarily satisfied by all gaps and therefore not every gap has a type. However for any gap $\mathcal{A}, \mathcal{B}$ we can find, by using Zorn's lemma, a gap $\mathcal{A}', \mathcal{B}'$ such that $\mathcal{A}' \supseteq \mathcal{A}$, $\mathcal{B}' \subseteq \mathcal{B}$ and (b) is satisfied. The purpose of (b) is to avoid having substantially different gaps having the same type. For example, if we did not require (b) then there would always be gaps of type $\langle \omega_1 \times \omega, \omega_1 \rangle$ in $\mathcal{P}(\mathbb{N})/\operatorname{Fin}$, but the existence of a gap of type $\langle \omega \times \omega_1, \omega_1 \rangle$ in the sense of Definition 5.6.4 is equivalent to the existence of an $\langle \omega, \omega_1 \rangle$-gap in $\mathcal{P}(\mathbb{N})/\operatorname{Fin}$. We can even strengthen (b) to require

(b') if $X \subseteq \mathcal{A}$ is unbounded and $Y \subseteq \mathcal{B}$ is unbounded, then X cannot be separated from Y.

We can say that $\mathcal{A}, \mathcal{B}$ has *strong type* $\mathcal{P}, \mathcal{Q}$ if (a) and (b') are satisfied. But the condition (b') seems to be too strong, in particular it is unclear whether every gap can be extended to a gap satisfying it. This is the reason why we use (b) instead. We note here that understanding of the relationship between (b) and (b') is tightly connected to the question of whether a gap constructed in §5.10 includes a linear gap. An important point in favor of (b) in comparison to (b') is that if a pair $\langle \mathcal{A}, \mathcal{B} \rangle$ satisfies (b'), then the families $\mathcal{A}$ and $\mathcal{B}$ cannot be hereditary (i.e., closed under taking subsets of their elements). On the other hand, if $\mathcal{A}, \mathcal{B}$ satisfy (b), then so do their hereditary closures $\hat{\mathcal{A}}, \hat{\mathcal{B}}$.

A characterization of preservation of gaps by a homomorphism related to the above is given in Proposition 5.6.6 below. The condition (3) has been first isolated in [**130**, Theorem 12], where it is used to show that the natural homomorphisms of $\mathcal{P}(\mathbb{N})/\operatorname{Fin}$ into the quotients over *Mazur ideals* (a large class introduced in [**95**]) preserve all gaps, i.e., where it is shown that it implies the condition (1).

PROPOSITION 5.6.6. *For a homomorphism $\Phi \colon \mathcal{P}(\mathbb{N})/\mathcal{I} \to \mathcal{P}(\mathbb{N})/\mathcal{J}$ the following are equivalent:*

(1) *Φ preserves gaps,*
(2) *for every $\mathcal{B} \subseteq \mathcal{P}(\mathbb{N})$ the restriction of Φ to $\mathcal{B}^\perp/\mathcal{I}$ is a Tukey map into $(\Phi''_* \mathcal{B})^\perp/\mathcal{J}$.*
(3) *There is a map $F \colon \mathcal{P}(\mathbb{N}) \to \mathcal{P}(\mathbb{N})$ such that*
  (i) *$\Phi_*(A) \cap B \in \mathcal{J}$ implies $A \cap F(B) \in \mathcal{I}$,*
  (ii) *$\Phi_*(A) \setminus B \in \mathcal{J}$ implies $A \setminus F(B) \in \mathcal{I}$.*

PROOF. Note that (2) says that every gap of the form $\mathcal{A}, \mathcal{B}$ for some $\mathcal{A}$ is preserved, therefore (1) and (2) are equivalent. To see that (2) and (3) are equivalent, note that F is a witness that Φ is a Tukey map. □

An evidence that Definition 5.6.4 is natural is given in:

COROLLARY 5.6.7. *If a monomorphism* $\Phi\colon \mathcal{P}(\mathbb{N})/\mathcal{I} \to \mathcal{P}(\mathbb{N})/\mathcal{J}$ *preserves gaps, then it also preserves their types.*

PROOF. We shall check the condition (2) from Proposition 5.6.6. We have to prove that the identity map
$$\mathrm{id}\colon (\Phi''(\mathcal{B}^\perp/\mathcal{I}))/\mathcal{J} \to (\Phi''\mathcal{B})^\perp/\mathcal{J}$$
is a Tukey map. Note that

$$\begin{array}{ccc} \mathcal{B}^\perp/\mathcal{I} & \xrightarrow{\Phi} & \Phi''(\mathcal{B}^\perp/\mathcal{J}) \\ & \Phi \searrow & \downarrow \mathrm{id} \\ & & (\Phi''\mathcal{B})^\perp/\mathcal{J} \end{array}$$

and that the map Φ is an order-isomorphism between $\mathcal{B}^\perp/\mathcal{I}$ and $\Phi''(\mathcal{B}^\perp/\mathcal{I})$. Therefore the map
$$\mathrm{id} \circ \Phi^{-1} \colon \Phi''(\mathcal{B}^\perp/\mathcal{I}) \to (\Phi''\mathcal{B})^\perp/\mathcal{J}$$
is well-defined and equal to a composition of two Tukey maps, and thus a Tukey map itself. □

Although the nonlinear gaps have been considered very early (see e.g., [**89**]) and they do have some use (see e.g., [**83**]), many authors (see [**105**] for a survey) consider only linear gaps. One of the reasons for doing so is that any attempt to describe spectra of nonlinear gaps would very likely require a classification of directed sets of size continuum, a task which we know to be impossible (see [**124**]). (Even the restriction of this problem to Borel directed sets seems to be quite difficult (see [**88**]).) Our reason for going into this generality is to point out the new gap phenomena in quotients beyond $\mathcal{P}(\mathbb{N})/\mathrm{Fin}$ which witnesses that results about gaps in $\mathcal{P}(\mathbb{N})/\mathrm{Fin}$ cannot be transferred to arbitrary quotients over analytic ideals (see §5.7).

5.7. TSP in quotients over analytic P-ideals

We now prove that Theorem 5.5.2 has a converse. Recall that an ideal $\mathcal{I}$ is isomorphic to Fin (see Definition 1.2.7) if and only if $\mathcal{I}$ is equal to Fin or of the form $\mathrm{Fin} \oplus A$ for some $A \subseteq \mathbb{N}$.

THEOREM 5.7.1. *If $\mathcal{I}$ is an analytic P-ideal, then $\mathcal{I}$ has the Todorcevic Separation Property if and only if $\mathcal{I}$ is isomorphic to* Fin.

This theorem will be proved by using the following proposition, due to Todorcevic.

PROPOSITION 5.7.2. *Assume that the ideal $\mathcal{I}$ has the TSP and $\langle \mathcal{A}, \mathcal{B} \rangle$ is a gap in $\mathcal{P}(\mathbb{N})/\mathcal{I}$ such that $\mathcal{B}$ is σ-directed and $\mathcal{A}$ is analytic. Then $\mathcal{A}$ can be countably separated from $\mathcal{B}$.*

PROOF. Since for every $\mathcal{B}$-tree there exists a $B \in \mathcal{B}$ which includes all of its branching nodes (and therefore many of the branches) modulo $\mathcal{I}$, TSP implies that such a gap is countably separated by some sequence $\{C_n\}$. □

Recall that $[\mathbb{R}]^{<\omega}$ is a poset of all finite sets of reals ordered by the inclusion. This is a Tukey-maximal directed set of size at most $\mathfrak{c}$.

COROLLARY 5.7.3. *If there is a gap $\langle \mathcal{A}, \mathcal{B} \rangle$ in $\mathcal{P}(\mathbb{N})/\mathcal{I}$ such that $\mathcal{A} \equiv [\mathbb{R}]^{<\omega}$, $\mathcal{A}$ is analytic, $\mathcal{B}$ is σ-directed and no unbounded subset of $\mathcal{A}/\mathcal{I}$ is separated from $\mathcal{B}$, then $\mathcal{I}$ does not have the TSP.*

PROOF. If $\mathcal{I}$ has the TSP, then $\mathcal{A}$ and $\mathcal{B}$ can be σ-separated by Proposition 5.7.2. Therefore an infinite subset of $\mathcal{A}$ can be separated from $\mathcal{B}$, contradicting the assumptions. □

PROOF OF THEOREM 5.7.1. We shall prove that in every quotient over an analytic P-ideal other than Fin there is an analytic gap $\langle \mathcal{A}, \mathcal{B} \rangle$ satisfying the assumptions of Corollary 5.7.3. This is done for F_σ P-ideals in Lemma 5.7.4 and for other P-ideals in Lemma 5.8.7 below. The conclusion will then follow by Proposition 5.7.2. □

LEMMA 5.7.4. *If $\mathcal{I}$ is an F_σ P-ideal other than Fin, than there is an analytic $\langle [\mathbb{R}]^{<\omega}, \mathcal{Q} \rangle$-gap in its quotient for some σ-directed poset $\mathcal{Q}$.*

PROOF. First we give an example of a summable quotient with this kind of a gap, and then use this example as a template in proving the general case. Let $f \colon \{0,1\}^{<\mathbb{N}} \to \mathbb{R}^+$ be defined by $f(s) = 1/|s|$, and define a submeasure (see Definition 1.2.1) φ by

$$\varphi(B) = \sup_{x \in \{0,1\}^{\mathbb{N}}} \mu_f(A_x \cap B), \quad \text{where} \quad A_x = \{x \restriction n : n \in \mathbb{N}\}.$$

Let $\mathcal{B}$ be the analytic P-ideal defined by

$$\mathcal{B} = \mathrm{Exh}(\varphi) = \Big\{ B \subseteq \mathbb{N} : \lim_n \varphi(B \setminus \{0,1\}^{\leq n}) = 0 \Big\},$$

(this ideal was introduced in Example 1.13.17 as an ideal rather different from both the summable and the density ideals). Now let

$$\mathcal{A} = \{A_x : x \in \{0,1\}^{\mathbb{N}}\}.$$

If $\langle \mathcal{A} \rangle$ is an ideal generated by $\mathcal{A}$, then the triple $\mathcal{I}_f, \langle \mathcal{A} \rangle, \mathcal{B}$ is, in some sense, analogous to the triple $\mathrm{Fin}, \mathrm{Fin} \times \emptyset, \emptyset \times \mathrm{Fin}$. It is clear from the definitions that $\mathcal{A}/\mathcal{I}_f \equiv [\mathbb{R}]^{<\omega}$, and since $\mathcal{B}$ is an analytic P-ideal, $\mathcal{B}/\mathcal{I}_f$ is σ-directed.

CLAIM 1. *No infinite subset of $\mathcal{A}$ can be separated from $\mathcal{B}$.*

PROOF. Assume $\{A_n\}$ is a subset of $\mathcal{A}$, and let C be such that $C \supseteq^{\mathcal{I}_f} A_n$ for all n. We can assume that A_n forms a converging sequence in $\{0,1\}^{\mathbb{N}}$ with $\lim_n A_n = A_\infty$ so that $\Delta(A_n, A_\infty) = \Delta(A_n, A_{n+1}) < \Delta(A_{n+1}, A_\infty)$ for all n. Let $u_n \subseteq A_n \cap C$ be disjoint from A_∞ (and therefore all A_m, $m \neq n$) and such that $1/n \leq \mu_f(u_n) \leq 2/n$—it exists since $\mu_f(A_n \cap C)$ is infinite. Then $B = \bigcup_n u_n$ is in $\mathcal{B}$, but $\mu_f(B \cap C) = \infty$ and therefore C does not separate $\{A_n\}$ from $\mathcal{B}$. □

Therefore the pair $\mathcal{A}, \mathcal{B}$ satisfies the assumptions of Corollary 5.7.3. This completes the proof in the case of $\mathcal{I}_f$.

Now we prove Lemma 5.7.4 for an arbitrary F_σ P-ideal $\mathcal{I}$ by mimicking the above construction. Let s_u ($u \in \{0,1\}^{<\mathbb{N}}$) be a family of finite sets such that the union of any of its infinite subfamilies is $\mathcal{I}$-positive (as guaranteed by [**93**]). Let $\mathcal{A}$ be the ideal generated by sets

$$A_x = \bigcup_{k=1}^{\infty} s_{x\restriction k}$$

for $x \in \{0,1\}^{\mathbb{N}}$ and let $\mathcal{B} = \mathrm{Exh}(\varphi)$, where φ is defined by (let $\varphi_\mathcal{I}$ be such that $\mathcal{I} = \mathrm{Exh}(\varphi_\mathcal{I})$)

$$\varphi(B) = \sup_{x \in \{0,1\}^{\mathbb{N}}} \varphi_\mathcal{I}(B \cap A_x).$$

This is clearly a lower semicontinuous submeasure, therefore $\mathcal{B}$ is an analytic P-ideal, in particular, $\mathcal{B}/\mathcal{I}$ is σ-directed. Like in the case of $\mathcal{I}_f$, we have $\mathcal{A}/\mathcal{I} \equiv [\mathbb{R}]^{<\omega}$.

CLAIM 2. *No infinite subset of $\mathcal{A}$ can be separated from $\mathcal{B}$.*

PROOF. Assume $A_n = A_{x_n} \in \mathcal{A}$ are pairwise distinct elements of $\mathcal{A}$ and let C be such that $A_n \setminus C \in \mathcal{I}$ for all n. Pick $u_n \in x_n \setminus \bigcup_{i \neq n} x_i$ so that

$$\varphi_\mathcal{I}(s_{u_n} \setminus C) < 2^{-n}$$

(this is possible since $\lim_n \varphi_\mathcal{I}(s_{u_n} \setminus C) = 0$). We claim there is $B \subseteq D = \bigcup_n s_{u_n} \setminus C$ such that $\lim_n \varphi_\mathcal{I}(B \cap s_{u_n}) = 0$ but $B \notin \mathcal{I}$. Otherwise we have

$$\mathcal{I} \restriction D = \{B \subseteq D : \lim_n \varphi_\mathcal{I}(B \cap s_{u_n}) = 0\} = \mathrm{Exh}(\sup_n \varphi_I(s_{u_n})),$$

and the ideal $\mathcal{I} \restriction D$ (and therefore $\mathcal{I}$) is not F_σ (by Proposition 1.13.14), contradicting our assumption. Let B be as above. Then $B \setminus C \in \mathcal{I}$, $B \in \mathcal{B}$, but $B \notin \mathcal{I}$, and therefore C does not separate $\{A_n\}$ from $\mathcal{B}$. This completes the proof. □

Therefore $\mathcal{A}$ and $\mathcal{B}$ satisfy the assumptions of Corollary 5.7.3, and this completes the proof of Lemma 5.7.4. □

We continue the proof of Theorem 5.7.1 in the following section.

5.8. Quotients as reduced products

To complete the proof of Theorem 5.7.1 it remains to construct a required gap in a quotient over every analytic P-ideal which is not F_σ. We shall do this first in the case when $\mathcal{I} = \emptyset \times \mathrm{Fin}$ and then "embed" this gap into every other quotient over an analytic P-ideal which is not F_σ. Before we continue the proof of Theorem 5.7.1, we shall construct gaps in the quotient over the ideal

$$\mathrm{Fin} \times \mathrm{Fin} = \{A \subseteq \mathbb{N}^2 : \{i : \{j : \langle i,j \rangle \in A\} \notin \mathrm{Fin}\} \in \mathrm{Fin}\}.$$

Recall that if $\langle \mathcal{P}_n, \leq_n \rangle$ ($n \in \mathbb{N}$) is a sequence of partially ordered sets, then their *reduced product*, $\prod_n \mathcal{P}_n / \mathrm{Fin}$ is the set of all $f \in \prod_n \mathcal{P}_n$ preordered by

$f \leq^* g$ if and only if $f(n) \leq_n g(n)$ for all but finitely many n.

The following Lemma follows immediately from the definitions.

LEMMA 5.8.1. *The quotient $\mathcal{P}(\mathbb{N}^2)/\mathrm{Fin} \times \mathrm{Fin}$ is isomorphic to the reduced product $(\prod_n \mathcal{P}(\mathbb{N}))/\mathrm{Fin}$.* □

LEMMA 5.8.2. *If for every n there is an (analytic) gap of type $\langle \mathcal{P}_n, \mathcal{Q}_n \rangle$ in the quotient over Fin, then there is an (analytic) gap of type $\langle \prod_n \mathcal{P}_n / \mathrm{Fin}, \prod_n \mathcal{Q}_n / \mathrm{Fin} \rangle$ in the quotient over $\mathrm{Fin} \times \mathrm{Fin}$.*

PROOF. Let $\langle \mathcal{A}_n, \mathcal{B}_n \rangle$ be a gap of type $\langle \mathcal{P}_n, \mathcal{Q}_n \rangle$ in $\mathcal{P}(\mathbb{N})/\mathrm{Fin}$ such that any unbounded subset of $\mathcal{A}_n$ cannot be separated from $\mathcal{B}_n$, and that any unbounded subset of $\mathcal{B}_n$ cannot be separated from $\mathcal{A}_n$. We can assume that $\mathcal{A}_n, \mathcal{B}_n$ is a gap in $\mathcal{P}(\{n\} \times \mathbb{N})/\mathrm{Fin}$. Define $\mathcal{A}, \mathcal{B}$ by (let $A(n) = A \cap \{n\} \times \mathbb{N}$):

$$\mathcal{A} = \{A \subseteq \mathbb{N}^2 : A(n) \in \mathcal{A}_n\}$$
$$\mathcal{B} = \{B \subseteq \mathbb{N}^2 : B(n) \in \mathcal{B}_n\}.$$

It is clear that $\mathcal{A}, \mathcal{B}$ are analytic if all $\mathcal{A}_n, \mathcal{B}_n$ are. We shall prove that $\mathcal{A}, \mathcal{B}$ is a gap of type $\langle \prod_n \mathcal{P}_n/\mathrm{Fin}, \prod_n \mathcal{Q}_n/\mathrm{Fin} \rangle$. Clearly $\mathcal{A}/(\mathrm{Fin} \times \mathrm{Fin}) \equiv \prod_n \mathcal{P}_n/\mathrm{Fin}$, $\mathcal{B}/(\mathrm{Fin} \times \mathrm{Fin}) \equiv \prod_n \mathcal{Q}_n/\mathrm{Fin}$, and $\mathcal{A}$ and $\mathcal{B}$ are $\mathrm{Fin} \times \mathrm{Fin}$-, and even $\emptyset \times \mathrm{Fin}$-, orthogonal. Assume $\mathcal{A}' \subseteq \mathcal{A}$ can be separated from $\mathcal{B}$ by some $C \subseteq \mathbb{N}^2$. Then the set

$$\{n : (B \setminus C)(n) \in \mathrm{Fin} \text{ for all } B \in \mathcal{B}\}$$

is cofinite (otherwise we can find a $B \in \mathcal{B}$ such that $B \setminus C \notin \mathrm{Fin} \times \mathrm{Fin}$). Therefore we can assume, possibly by adding a set in $\mathrm{Fin} \times \mathrm{Fin}$ to C, that $(B \setminus C)(n) \in \mathrm{Fin}$ for all n. Let

$$\mathcal{A}_n^0 = \{A \in \mathcal{A}_n : A \cap C \in \mathrm{Fin}\}.$$

Then $C(n)$ separates $\mathcal{A}_n^0$ from $\mathcal{B}_n$, and since on $\langle \mathcal{A}_n, \mathcal{B}_n \rangle$ satisfies (b) of Definition 5.6.4, the set $\mathcal{A}_n^0$ is bounded in $\mathcal{A}_n/\mathrm{Fin}$ by some A_n. Then $A_\infty = \bigcup_n A_n$ is in $\mathcal{A}$, and for every $A \in \mathcal{A}'$ we have $A(n) \in \mathcal{A}_n^0$ for all but finitely many n, therefore the set

$$\{n : (A(n)) \setminus A_\infty \notin \mathrm{Fin}\}$$

is finite and finally $A \setminus A_\infty \in \mathrm{Fin} \times \mathrm{Fin}$. This means that $\mathcal{A}'$ is bounded in $\mathcal{A}/(\mathrm{Fin} \times \mathrm{Fin})$ by A_∞. We similarly prove that $\mathcal{B}' \subseteq \mathcal{B}$ can be separated from $\mathcal{A}$ if and only if it is bounded in $\mathcal{B}/\mathrm{Fin} \times \mathrm{Fin}$, and therefore $\langle \mathcal{A}, \mathcal{B} \rangle$ is a gap of the desired type. $\square$

PROPOSITION 5.8.3. *There is an analytic gap of type $\langle \mathbb{N}^\mathbb{N}/\mathrm{Fin}, \mathbb{N}^\mathbb{N}/\mathrm{Fin} \rangle$ in the quotient $\mathcal{P}(\mathbb{N}^2)/(\mathrm{Fin} \times \mathrm{Fin})$. In particular, $\mathrm{Fin} \times \mathrm{Fin}$ does not have the TSP.*

PROOF. The ideals $\mathrm{Fin} \times \emptyset$ and $\emptyset \times \mathrm{Fin}$ form a gap of type $\langle \mathbb{N}, \mathbb{N}^\mathbb{N}/\mathrm{Fin} \rangle$ in the quotient over Fin. By Lemma 5.8.2, in the quotient over $\mathrm{Fin} \times \emptyset$ there is a gap of type $\langle \prod_n \mathbb{N}/\mathrm{Fin}, (\prod_n \mathbb{N}^\mathbb{N}/\mathrm{Fin})^\mathbb{N}/\mathrm{Fin} \rangle \equiv \langle \mathbb{N}^\mathbb{N}/\mathrm{Fin}, \mathbb{N}^\mathbb{N}/\mathrm{Fin} \rangle$. $\square$

PROPOSITION 5.8.4. *If there are $\langle \omega, \kappa \rangle$- and $\langle \omega, \lambda \rangle$-gaps in $\mathcal{P}(\mathbb{N})/\mathrm{Fin}$, then there is a $\langle \kappa, \lambda \rangle$-gap in $\mathcal{P}(\mathbb{N}^2)/\mathrm{Fin} \times \mathrm{Fin}$. In particular, there is a $\langle \mathfrak{b}, \mathfrak{b} \rangle$-gap in this quotient.*

PROOF. By Lemma 5.8.2, there is a gap of type

$$\langle \kappa^\mathbb{N}/\mathrm{Fin}, \mathbb{N}^\mathbb{N}/\mathrm{Fin} \rangle \equiv \langle \kappa, \mathbb{N}^\mathbb{N}/\mathrm{Fin} \rangle$$

in this quotient. By Rothberger's result (Theorem 5.2.2), there is an unbounded λ-chain in $\mathbb{N}^\mathbb{N}/\mathrm{Fin}$ and this gives us the desired $\langle \kappa, \lambda \rangle$-gap. $\square$

In §26 of [**32**] we have constructed a model of set theory in which there is an $\langle \omega_2, \omega_2 \rangle$-gap in the quotient over $\mathrm{Fin} \times \mathrm{Fin}$ yet there are no such gaps in $\mathcal{P}(\mathbb{N})/\mathrm{Fin}$ (an interested reader familiar with the methods of [**30**], in particular the model of §2 and Theorems 6.1 and 9.2, should be able to reproduce this result). The above proposition, and its consequence Theorem 5.8.5 below, show that the same conclusion even follows from OCA.

THEOREM 5.8.5. *Assume OCA. Then there is an $\langle \omega_2, \omega_2 \rangle$-gap in the quotient $\mathcal{P}(\mathbb{N}^2)/\operatorname{Fin} \times \operatorname{Fin}$.*

PROOF. By Proposition 5.8.4 there is a $\langle \mathfrak{b}, \mathfrak{b} \rangle$-gap in this quotient, and by Theorem 2.2.6 we have $\mathfrak{b} = \omega_2$. □

This result sheds some light on Problem 2 of [**130**], asking to determine the gap spectrum of $\mathcal{P}(\mathbb{N})/\mathcal{I}$ for every analytic ideal $\mathcal{I}$ on $\mathbb{N}$. It also shatters the hope expressed in the discussion following the problem that under OCA or PFA $\{\langle \omega_2, \omega \rangle, \langle \omega_1, \omega_1 \rangle, \langle \omega, \omega_2 \rangle\}$ might be the gap spectrum of every analytic quotient (see [**130**, page 95]). However, it is still possible that OCA (or PFA) is sufficient for determining the gap spectrum of any analytic quotient which could have been the primary idea behind the discussion in [**130**]. (See also Corollary 5.10.3.)

Note that a $\langle \kappa, \lambda \rangle$-gap given by Proposition 5.8.4 is a "subgap" of the gap constructed in Proposition 5.8.3. If this latter gap satisfied the condition (b') instead of (b) (see §5.6), then we would be able to extract a $\langle \kappa, \lambda \rangle$-gap from it directly. However, it is easy to see that (b') fails for the gap constructed in the proof of Proposition 5.8.3. (Choose families $\mathcal{A}' \subseteq \mathcal{A} \cap \mathcal{P}(A \times \mathbb{N})$ and $\mathcal{B}' \subseteq \mathcal{B} \times \mathcal{P}(B \times \mathbb{N})$ unbounded in $\mathcal{A}, \mathcal{B}$ respectively for some disjoint infinite sets $A, B \subseteq \mathbb{N}$.) A gap witnessing Proposition 5.8.3 and satisfying (b') is not difficult to construct, but at present we do not have an application of such a gap.

Let us now go back to the proof of Theorem 5.7.1 and the quotient $\mathcal{P}(\mathbb{N}^2)/\emptyset \times \operatorname{Fin}$. In the following lemma the product $\prod_n \mathcal{P}_n$ is taken with its coordinatewise ordering, $f \leq g$ if $f(n) \leq_n g(n)$ for all n.

LEMMA 5.8.6. *If there are (analytic) gaps of type $\langle \mathcal{P}_n, \mathcal{Q}_n \rangle$ in $\mathcal{P}(\mathbb{N})/\operatorname{Fin}$, then there is an (analytic) gap of type $\langle \prod_n \mathcal{P}_n, \prod_n \mathcal{Q}_n \rangle$ in $\mathcal{P}(\mathbb{N}^2)/\emptyset \times \operatorname{Fin}$.*

PROOF. Starting with gaps $\langle \mathcal{A}_n, \mathcal{B}_n \rangle$ of type $\langle \mathcal{P}_n, \mathcal{Q}_n \rangle$, define their product $\langle \mathcal{A}, \mathcal{B} \rangle$ like in the proof of Lemma 5.8.2. Then $\mathcal{A}$ and $\mathcal{B}$ are $\emptyset \times \operatorname{Fin}$-orthogonal, and their types are $\prod_n \mathcal{P}_n$ and $\prod_n \mathcal{Q}_n$, respectively. Let $\mathcal{A}' \subseteq \mathcal{A}$ be unbounded in $\mathcal{A}/\emptyset \times \operatorname{Fin}$; let $\mathcal{A}'(n) = \{A(n) : A \in \mathcal{A}'\}$.

CLAIM 1 *Either $\mathcal{A}'(n)$ is unbounded in $\mathcal{A}_n/\operatorname{Fin}$ for some n or $\mathcal{A}'$ is unbounded in $\mathcal{A}/\operatorname{Fin} \times \operatorname{Fin}$.*

PROOF. Assume $\mathcal{A}'(n)$ is bounded in $\mathcal{A}_n/\operatorname{Fin}$ by some A_n, for every n, and that $\mathcal{A}'$ is bounded in $\mathcal{A}/\operatorname{Fin} \times \operatorname{Fin}$ by some C. Then $C \cup \bigcup_n A_n$ bounds $\mathcal{A}'$ in $\mathcal{A}/\emptyset \times \operatorname{Fin}$. □

Therefore $\mathcal{A}'$ cannot be separated from $\mathcal{B}$ over $\emptyset \times \operatorname{Fin}$, by the assumptions on $\mathcal{A}_n, \mathcal{B}_n$ and Lemma 5.8.2. We similarly prove that $\mathcal{B}' \subseteq \mathcal{B}$ is separated from $\mathcal{A}$ if and only if it is bounded in $\mathcal{B}/\emptyset \times \operatorname{Fin}$, and therefore the gap $\mathcal{A}, \mathcal{B}$ is as required. □

LEMMA 5.8.7. *If $\mathcal{I}$ is an analytic P-ideal which is not F_σ, then there is an analytic gap of type $\langle \mathbb{N}^\mathbb{N}, \mathbb{N}^\mathbb{N}/\operatorname{Fin} \rangle$ in the quotient over $\mathcal{I}$.*

PROOF. Let us first assume that $\mathcal{I} = \emptyset \times \operatorname{Fin}$. By Lemma 5.8.6, since there is a $\langle \mathbb{N}, \mathbb{N}^\mathbb{N}/\operatorname{Fin} \rangle$-gap in $\mathcal{P}(\mathbb{N})/\operatorname{Fin}$, there is a gap of type $\langle \mathbb{N}^\mathbb{N}, (\mathbb{N}^\mathbb{N}/\operatorname{Fin})^\mathbb{N} \rangle$ in $\mathcal{P}(\mathbb{N}^2)/\emptyset \times \operatorname{Fin}$. Note that this gap is analytic. To prove that $\mathcal{P} = (\mathbb{N}^\mathbb{N}/\operatorname{Fin})^\mathbb{N} \equiv \mathbb{N}^\mathbb{N}/\operatorname{Fin}$, note that if $f, g \in (\mathbb{N}^\mathbb{N})^\mathbb{N}$ then

$f \leq_\mathcal{P} g$ if and only if $f(\cdot, n) \leq^* g(\cdot, n)$ for all n

if and only if $f(m, n) \leq g(m, n)$ for all but finitely many pairs m, n

and since $\mathbb{N}^{\mathbb{N}^2}/\operatorname{Fin} \cong \mathbb{N}^{\mathbb{N}}/\operatorname{Fin}$, this completes the proof of lemma.

Now let $\mathcal{I}$ be any analytic P-ideal which is not F_σ. By a result of Solecki ([**112**]) we have $\emptyset \times \operatorname{Fin} \leq_{\mathrm{RB}} \mathcal{I}$, and therefore $\emptyset \times \operatorname{Fin} \leq_{\mathrm{BE}} \mathcal{I}$. By Proposition 5.3.3, gaps are preserved by this embedding, and by Corollary 5.6.7, their types are also preserved. But we have already proved that an analytic gap of the desired type exists in $\emptyset \times \operatorname{Fin}$, and since the continuous lifting sends it into an analytic gap, this completes the proof. □

5.9. Preservation of gaps

The study of gaps in arbitrary analytic quotients was given a new impetus after Todorcevic ([**130**]) proved that gaps are frequently preserved by embeddings of $\mathcal{P}(\mathbb{N})/\operatorname{Fin}$ into other analytic quotients. Namely, it appears that the existence of a certain kind of analytic gap in $\mathcal{P}(\mathbb{N})/\mathcal{I}$ is equivalent to non-preservation result of gaps by a monomorphism with a completely additive lifting of $\mathcal{P}(\mathbb{N})/\mathcal{I}$ into a quotient over some analytic ideal (see proofs of Proposition 5.7.2 and Proposition 5.5.3). It appears that gaps can be preserved by arbitrary monomorphisms (as always, this fails under CH, since monomorphisms constructed using saturatedness of $\mathcal{P}(\mathbb{N})/\operatorname{Fin}$ clearly need not preserve gaps). We have an analogue of Theorem 3.4.1 in this context.

PROPOSITION 5.9.1. *Let $\mathcal{I}, \mathcal{J}$ be analytic P-ideals or $\operatorname{Fin} \times \emptyset$. Then the following are equivalent:*

(1) *Every Baire monomorphism $\Phi \colon \mathcal{P}(\mathbb{N})/\mathcal{I} \to \mathcal{P}(\mathbb{N})/\mathcal{J}$ preserves all (Hausdorff) gaps.*
(2) *OCA and MA imply that every monomorphism $\Phi \colon \mathcal{P}(\mathbb{N})/\mathcal{I} \to \mathcal{P}(\mathbb{N})/\mathcal{J}$ preserves all (Hausdorff) gaps.*

PROOF. (2) implies (1) is trivial. Assume (1) and let $\mathcal{A}, \mathcal{B}$ be a gap in $\mathcal{P}(\mathbb{N})/\mathcal{I}$ and let Φ be a monomorphism. By Theorem 3.3.5 (or 3.3.6), Φ is an amalgamation of a Baire homomorphism Φ_1 and a homomorphism Φ_2 with a a large kernel. Since Φ is a monomorphism, $\Phi \neq \Phi_2$, so the conclusion follows from the fact that Φ_1 preserves the gap $\mathcal{A}, \mathcal{B}$. □

By Proposition 5.9.1 and Theorem 5.1.1, we have the following.

COROLLARY 5.9.2. *Under OCA and MA every monomorphism Φ of $\mathcal{P}(\mathbb{N})/\operatorname{Fin}$ into a quotient $\mathcal{P}(\mathbb{N})/\mathcal{I}$ over an analytic ideal preserves all Hausdorff gaps. If $\mathcal{I}$ is moreover a P-ideal, then Φ preserves all gaps.* □

By Shoenfield's Absoluteness theorem and Proposition 5.9.1, we obtain the following (see §2.1 for a discussion of such statements).

PROPOSITION 5.9.3. *For every analytic P-ideal $\mathcal{I}$ the statement "all gaps are preserved by embeddings of $\mathcal{P}(\mathbb{N})/\mathcal{I}$ into other quotients over analytic P-ideals" is either false or it follows from OCA and MA.* □

The following (and an analogous theorem for arbitrary gaps and analytic P-ideals) was proved by Todorcevic in [**130**], but for the convenience of the reader we will sketch the proof.

PROPOSITION 5.9.4. *Let $\mathcal{I}$ be an analytic ideal. Then the following are equivalent:*

(1) *Every Baire monomorphism* $\Phi\colon \mathcal{P}(\mathbb{N})/\mathcal{I} \to \mathcal{P}(\mathbb{N})/\mathcal{J}$ *preserves all Hausdorff gaps for every analytic ideal* $\mathcal{J}$.
(2) *There are no analytic Hausdorff gaps in* $\mathcal{P}(\mathbb{N})/\mathcal{I}$.

PROOF. Assume that (2) fails, and let $\mathcal{A}, \mathcal{B}$ be an analytic Hausdorff gap in $\mathcal{P}(\mathbb{N})/\mathcal{I}$. Let $\mathcal{J}$ be the ideal $\mathbb{N} \times \{0,1\}$ generated by $(\mathcal{I} \cup \mathcal{A}) \times \{0\}$ and $(\mathcal{I} \cup \mathcal{B}) \times \{1\}$. This is an analytic ideal, and if $h\colon \mathbb{N} \times \{0,1\} \to \mathbb{N}$ is the projection $h(i,j)=i$, then the Φ_h-image of $\mathcal{A}, \mathcal{B}$ is separated by $\mathbb{N} \times \{1\}$, therefore (1) fails.

Assume that (1) fails, $\mathcal{A}, \mathcal{B}$ is a Hausdorff gap whose image under a monomorphism $\Phi\colon \mathcal{P}(\mathbb{N})/\mathcal{I} \to \mathcal{P}(\mathbb{N})/\mathcal{J}$ is split by some C, and that Φ_* is a continuous lifting of Φ. Then $\mathcal{A}' = \{A : \Phi_*(A) \cap C \in \mathcal{J}\}$ and $\mathcal{B}' = \{B \cap \Phi_*(B) \setminus C \in \mathcal{J}\}$ is an analytic Hausdorff gap in $\mathcal{P}(\mathbb{N})/\mathcal{I}$ (for details see [**130**]). □

5.10. An analytic Hausdorff gap

In [**130**, page 96] Todorcevic writes that in view of the results from his paper it is natural to expect that OCA (or PFA) imply that there are only $\langle \omega, \omega_2 \rangle$, $\langle \omega_2, \omega \rangle$ and $\langle \omega_1, \omega_1 \rangle$-gaps in $\mathcal{P}(\mathbb{N})/\mathcal{I}$ for any analytic ideal $\mathcal{I}$. (Recall that this is true for $\mathcal{I} = \text{Fin}$, by a result of Todorcevic ([**125**], see also Theorem 5.2.4)). We have already seen in Corollary 5.8.5 that this fails in case when $\mathcal{I}$ is $\text{Fin} \times \text{Fin}$. Moreover, the "definable" version also fails: There exists an analytic Hausdorff gap in $\mathcal{P}(\mathbb{N}^2)/\text{Fin} \times \text{Fin}$ (the gap constructed in Lemma 5.8.2 is analytic). On the other hand, in $\mathcal{P}(\mathbb{N})/\text{Fin}$ there are no analytic Hausdorff gaps ([**126**]). Velickovic's proof that all automorphisms of $\mathcal{P}(\mathbb{N})/\text{Fin}$ are trivial under OCA and MA is related to the analysis of gap-spectra of this structure under OCA. Therefore there was a hope that some elements of our analysis of homomorphisms between quotients over analytic P-ideals under OCA (Theorem 3.3.5) would help us determine gap spectra of $\mathcal{P}(\mathbb{N})/\mathcal{I}$ at least in the case of analytic P-ideals, and perhaps even disprove the existence of analytic Hausdorff gaps in such quotients. We shall now see that it is not so. Let us concentrate on analytic gaps for a moment. A submeasure φ is *absolutely continuous with respect to* φ if $\varphi(A) = 0$ implies $\psi(A) = 0$ for all $A \subseteq \mathbb{N}$. Recall that $\|A\|_\varphi = \lim_n \varphi(A \setminus n)$.

PROPOSITION 5.10.1. *The following are equivalent:*
(1) *There are no analytic Hausdorff gaps in quotients over analytic P-ideals.*
(2) *For all lower semicontinuous submeasures* φ, ψ *on* $\mathbb{N}$ *there is* $C \subseteq \mathbb{N}$ *such that* $\|\cdot\|_\varphi$ *is absolutely continuous with respect to* $\|\cdot\|_\psi$ *on* C *and* $\|\cdot\|_\psi$ *is absolutely continuous with respect to* $\|\cdot\|_\varphi$ *on* $\mathbb{N} \setminus C$.

PROOF. Assume (2), and let $\mathcal{A}, \mathcal{B}$ be σ-directed $\mathcal{I}$-orthogonal families, where $\mathcal{I}$ is an analytic P-ideal. We can assume that $\mathcal{A}, \mathcal{B}$ are ideals, and therefore by a result of Solecki (see Theorem 1.2.5) there are lower semicontinuous submeasures φ and ψ such that $\mathcal{A} = \text{Exh}(\varphi)$ and $\mathcal{B} = \text{Exh}(\psi)$. Let C be as in (2). We claim that C splits the pregap $\mathcal{A}, \mathcal{B}$: If $A \in \mathcal{A}$ and $A \setminus C \in \mathcal{A}$, then $\|A \setminus C\|_\varphi = 0$, therefore by the absolute continuity $\|A \setminus C\|_\psi = 0$ as well, and $A \in \mathcal{A} \cap \mathcal{B} \subseteq \mathcal{I}$. If on the other hand A is such that $A \cap C \in \mathcal{B}$, we similarly get that $A \cap C \in \mathcal{A} \cap \mathcal{B} \subseteq \mathcal{I}$. This completes the proof that (2) implies (1).

Now assume (1), and let φ, ψ be two lower semicontinuous submeasures on $\mathbb{N}$. Let ζ be the pointwise maximum of φ, ψ, let $\mathcal{I} = \text{Exh}(\zeta)$, $\mathcal{A} = \text{Exh}(\varphi)$ and $\mathcal{B} = \text{Exh}(\psi)$. Clearly $\mathcal{A}$ and $\mathcal{B}$ are $\mathcal{I}$-orthogonal, analytic and σ-directed. By (1) they can be separated, and there is C such that

for all $A \subseteq \mathbb{N} \setminus C$ if $A \in \mathcal{A}$ then $A \in \mathcal{I}$, and
for all $A \subseteq C$, if $A \in \mathcal{B}$ then $A \in \mathcal{I}$.

We claim that C is as required in (2). Note that
$$\|A\|_\zeta = \lim_{n\to\infty} \max(\varphi(A \setminus n), \psi(A \setminus n)) = \max(\|A\|_\varphi, \|A\|_\psi),$$
and therefore for $A \subseteq \mathbb{N} \setminus C$ the condition $\|A\|_\varphi = 0$ implies $\|A\|_\zeta = 0$ and $A \in \mathcal{I}$. Similarly if $A \subseteq C$ then $\|A\|_\psi = 0$ implies $\|A\|_\zeta = 0$, and this concludes the proof. □

Proposition 5.10.1 gives a very good hint on how to find Hausdorff gaps in quotients over analytic P-ideals, and the following result is not very surprising anymore.

THEOREM 5.10.2. *There exists an F_σ Hausdorff gap in a quotient over some F_σ P-ideal $\mathcal{I}$.*

PROOF. Let $\{u_n\}$ be a (disjoint) partition of $\mathbb{N}$ into consecutive intervals, so that $|u_n| = n^3$ for all n. On u_n we define three submeasures:
$$\varphi_n(s) = \frac{\min(|s|, n)}{n^2},$$
$$\nu_n(s) = \frac{|s|}{n^3},$$
$$\psi_n(s) = \max(\varphi_n(s), \nu_n(n)).$$
(Therefore ν_n is simply a uniform probability measure.) These sequences define lower semicontinuous submeasures on $\mathbb{N}$ by
$$\varphi(A) = \sum_{n=1}^\infty \varphi_n(A \cap u_n), \quad \nu(A) = \sum_{n=1}^\infty \nu_n(A \cap u_n), \quad \psi(A) = \sum_{n=1}^\infty \psi_n(A \cap u_n).$$
and the corresponding F_σ-P-ideals (recall that $\|A\|_\varphi = \lim_n \varphi(A \setminus [1, n])$ and $\mathrm{Exh}(\varphi) = \{A : \|A\|_\varphi = 0\}$):
$$\mathcal{A} = \mathrm{Exh}(\varphi), \qquad \mathcal{B} = \mathrm{Exh}(\nu), \qquad \mathcal{I} = \mathrm{Exh}(\psi).$$
These ideals are F_σ, for all three submeasures have the additional property
$$\|A\|_\varphi = 0 \quad \text{if and only if} \quad \varphi(A) < \infty.$$

CLAIM 1. *We have $\mathcal{I} = \mathcal{A} \cap \mathcal{B}$, in particular families $\mathcal{A}$ and $\mathcal{B}$ are orthogonal modulo $\mathcal{I}$ (in symbols $\mathcal{A} \perp_\mathcal{I} \mathcal{B}$).*

PROOF. This follows from
$$\max(\varphi(A), \nu(A)) \le \psi(A) = \sum_{n=1}^\infty \max(\varphi_n(A \cap u_n), \nu_n(A \cap u_n))$$
$$\le \varphi(A) + \nu(A). \qquad \square$$

Therefore, if $A \in \mathcal{A}$ and $B \in \mathcal{B}$, then $\psi((A \cap B)) \setminus [1, n)) \le \max(\varphi(A \setminus [1, n)), \nu(B \setminus [1, n)))$, and this expression approaches 0 as $n \to \infty$. □

CLAIM 2. *Families $\mathcal{A}$ and $\mathcal{B}$ form a gap inside the quotient $\mathcal{P}(\mathbb{N})/\mathcal{I}$.*

PROOF. It suffices to prove that for every $C \subseteq \mathbb{N}$ one of the following applies:
(1) C is too large, i.e., there is $B \subseteq C$ such that $B \in \mathcal{B}$ but $B \notin \mathcal{I}$, or
(2) C is too small, i.e., there is $A \subseteq \mathbb{N} \setminus C$ such that $A \in \mathcal{A}$ but $A \notin \mathcal{I}$.

So let us fix $C \subseteq \mathbb{N}$, and let $I = \{n : |C \cap u_n| \geq n\}$.

CASE 1. $\sum_{n \in I} 1/n = \infty$. By shrinking C, we may assume that $|C \cap u_n| = n$ for all $n \in I$. Then the set
$$B = \bigcup_{n \in I} (C \cap u_n).$$
is in $\mathcal{B}$ since $\nu(B) = \sum_{n \in I} \nu_n(C \cap u_n) \leq \sum_{n=1}^{\infty} 1/n^2 < \infty$. On the other hand we have $\varphi(B) = \sum_{n \in I} \varphi_n(C \cap u_n) = \sum_{n \in I} 1/n = \infty$, therefore B is not in $\mathcal{A}$ and (by Claim 1) not in $\mathcal{I}$.

CASE 2. $\sum_{n \in I} 1/n < \infty$. Then we can find an infinite $J \subseteq \mathbb{N} \setminus I$ such that
$$\sum_{n \in J} \frac{1}{n} < \infty.$$
Let $A = \bigcup_{n \in J} (u_n \setminus C)$. We have $\varphi(A) = \sum_{n \in J} \varphi_n(u_n \setminus C) \leq \sum_{n \in J} 1/n < \infty$, and on the other hand
$$\nu(A) = \sum_{n \in J} \nu_n(u_n \setminus C) \geq \sum_{n \in J} \frac{n^3 - n}{n^3} = \infty$$
therefore A is in $\mathcal{A} \setminus \mathcal{B}$, and (by Claim 1) not in $\mathcal{I}$. This completes the proof of Claim 2 ... □

... and Theorem 5.10.2. □

The following Corollary in particular shows that OCA does not imply that the gap-spectrum of $\mathcal{P}(\mathbb{N})/\mathcal{I}$ for the ideal $\mathcal{I}$ defined in Theorem 5.10.2 is equal to $\langle \omega, \omega_2 \rangle, \langle \omega_2, \omega \rangle, \langle \omega_1, \omega_1 \rangle$ (see the introduction to this section, also Theorem 5.8.5 and the discussion following it).

COROLLARY 5.10.3. *Assume MA, and let $\mathcal{I}$ be the F_σ P-ideal constructed in Theorem 5.10.2. Then there is a $\langle \mathfrak{c}, \mathfrak{c} \rangle$-gap in $\mathcal{P}(\mathbb{N})/\mathcal{I}$.*

PROOF. By Lemma 5.6.3, we can recursively construct unbounded $\mathfrak{c}$-chains inside the families $\mathcal{A}$ and $\mathcal{B}$ defined in Theorem 5.10.2, while taking care that the resulting pair forms a gap in the end. □

Note that Claim 2 is a statement about families $\mathcal{A}$ and $\mathcal{B}$, rather than a statement about the quotient $\mathcal{P}(\mathbb{N})/\mathcal{I}$ (see also Proposition 5.10.1). The above family $\mathcal{B}$ is a summable ideal, since ν is a measure, while $\mathcal{A}$ is equal to the first known example of a non-summable F_σ P-ideal (see §1.11). It is very easy to see that we cannot hope to get both families $\mathcal{A}$ and $\mathcal{B}$ as above to be summable ideals (if ν and μ are the corresponding measures, then the set $C = \{n : \mu(\{n\}) \geq \nu(\{n\})\}$ separates $\mathcal{A}$ from $\mathcal{B}$). We do not know much about the class $\mathcal{C}$ of all analytic P-ideals $\mathcal{I}$ such that in $\mathcal{P}(\mathbb{N})/\mathcal{I}$ there are analytic Hausdorff gaps; e.g., the ideals Fin and $\emptyset \times$ Fin are not in this class. Although the ideal $\mathcal{I} \in \mathcal{C}$ constructed above is non-pathological, it seems possible (plausible?) that no summable ideals and no Erdös–Ulam ideals belong to the class $\mathcal{C}$.

Theorem 5.10.2 was used in [**99**] (see also [**132**, Theorem 8.4]) to construct a topological space with interesting cellularity properties.

5.11. Limits in analytic quotients

In [112] Solecki proved the equivalence of conditions (1) and (2) below. We give a simple-minded extension of his result.

THEOREM 5.11.1. *The following are equivalent for an analytic P-ideal $\mathcal{I}$ on $\mathbb{N}$:*

(1) *$\mathcal{I}$ is not F_σ,*
(2) *$\emptyset \times \text{Fin} \leq_{\text{RB}} \mathcal{I}$,*
(3) *there is a disjoint partition $\mathbb{N} = \bigcup_{n=1}^{\infty} A_n$ into $\mathcal{I}$-positive sets such that*

$$\mathcal{I} = \{B : B \cap A_n \in \mathcal{I}\}$$

(4) *there is an $\langle \omega, 1 \rangle$-limit $\langle B_n, D \rangle$ in $\mathcal{P}(\mathbb{N})/\mathcal{I}$*

PROOF. The equivalence of (2) and (3) was established by Solecki in [114].

To prove that (2) implies (3), assume $h \colon \mathbb{N} \to \mathbb{N}^2$ is a Rudin–Blass reduction of $\mathcal{I}$ to $\emptyset \times \text{Fin}$, and let $A_n = h^{-1}(\{n\} \times \mathbb{N})$. To see that this sequence satisfies (3), let B be such that $B \cap A_n \in \mathcal{I}$ for all n. Since $\mathcal{I}$ is a P-ideal, there is $C \in \mathcal{I}$ which includes all $B \cap A_n$ modulo finite. Then there is a finite set $s_n \subseteq \{n\} \times \mathbb{N}$ such that

$$h^{-1}(s_n) \supseteq (B \cap A_n) \setminus C,$$

so the set $B \subseteq C \cup h^{-1}(\bigcup_n s_n)$ is in $\mathcal{I}$. To see that (3) implies (4), let $B_m = \bigcup_{n=1}^{m} A_n$ where $\{A_n\}$ is as in (3). Then every C such that $C \cap B_m$ is in $\mathcal{I}$ for all m is, by (3), in $\mathcal{I}$, so that $\langle B_m, \mathbb{N} \rangle$ is the required limit. To see that (4) implies (3), let $\langle B_m, D \rangle$ be an $\langle \omega, 1 \rangle$-limit. We can assume $D = \mathbb{N}$ by letting $B'_m = B_m \cup D$ for all m, and we can also assume that all $A_n = B_n \setminus B_{n-1}$ ($B_0 = \emptyset$) are $\mathcal{I}$-positive sets. To see that A_n satisfy (3), let C be such that $C \cap A_n \in \mathcal{I}$ for all n. Let D be such that $D \supseteq^* C \cap A_n$ for all n. The set

$$E = \bigcup_{n=1}^{\infty} (C \cap A_n) \setminus D$$

has a finite intersection with every A_n, so it is in $\mathcal{I}$ as well as $C \subseteq D \cap E$. For (3) implies (2), let $h_n \colon A_n \to \{n\} \times \mathbb{N}$ be a Rudin–Blass reduction of $\mathcal{I} \upharpoonright A_n$ to Fin guaranteed by Mathias' theorem. Then $h = \bigcup_n h_n \colon \mathbb{N}^2 \to \mathbb{N}$ is a Rudin–Blass reduction of $\mathcal{I}$ to $\emptyset \times \text{Fin}$.

This completes the proof. □

We should note that the implication from (1) to (4) in the case of EU-ideals (see §1.13) is implicit in [65] (see also the paragraph after Lemma 1.3.2).

5.12. A coherent family of functions

Inspired by a question in cohomology theory, D. Talayco ([123]) recently proved the existence of a new kind of gap in $\mathcal{P}(\mathbb{N})/\text{Fin}$:

THEOREM 5.12.1. *There are sets $A_{\xi n}$ ($\xi < \omega_1$, $n \in \mathbb{N}$) such that for every surjection $g \colon \mathbb{N} \to \{0, 1\}$ the families*

$$B_\xi = \bigcup_{g(n)=0} A_{\xi n}, \qquad C_\xi = \bigcup_{g(n)=1} A_{\xi n}$$

form a Hausdorff gap in $\mathcal{P}(\mathbb{N})/\text{Fin}$. □

Talayco ([**123**]) has also proved, using some additional assumptions, that there can be an uncountable coherent family of gaps in $\mathcal{P}(\mathbb{N})/\operatorname{Fin}$ (see Theorem 5.12.2 below). We have extended Theorem 5.12.1 in [**29**], and we would now like to point out a corollary of this result and Todorcevic's gap preservation theorem (Theorem 5.1.1). Let $\mathcal{I}$ be an ideal of sets of integers. A family $\mathcal{F}$ of partial functions on integers is *coherent modulo $\mathcal{I}$* if for all $f, g \in \mathcal{F}$ the set of all $n \in \operatorname{dom}(f) \cap \operatorname{dom}(g)$ such that $f(n) \neq g(n)$ is in $\mathcal{I}$. Obviously, the family of all subfunctions of a given $f \colon \mathbb{N} \to \mathbb{N}$ is always coherent, but the following theorem shows there can be a highly nontrivial coherent family.

THEOREM 5.12.2. *Let $\mathcal{I}$ be an analytic ideal. Then there is a family of functions $f_\alpha \colon A_\alpha \to \alpha$ ($A_\alpha \subseteq \mathbb{N}$, $\alpha < \omega_1$) coherent modulo $\mathcal{I}$ and such that for every $g \colon \omega_1 \to \{0, 1\}$ which is onto, the families*

$$B_\alpha = \{n \,:\, g(f_\alpha(n)) = 0\}, \qquad C_\alpha = \{n \,:\, g(f_\alpha(n)) = 1\}$$

form a Hausdorff gap in $\mathcal{P}(\mathbb{N})/\mathcal{I}$.

PROOF. In [**29**] we have proved there is such a family $\mathcal{F}$ of coherent functions modulo Fin. Let $\Phi \colon \mathcal{P}(\mathbb{N})/\operatorname{Fin} \to \mathcal{P}(\mathbb{N})/\mathcal{I}$ be a monomorphism with a Baire-measurable lifting. The image of $\mathcal{F}$ under Φ is coherent, and it is nontrivial by Theorem 5.1.1. □

The existence of a nontrivial coherent family of partial functions was proved in [**29**] in an indirect way, by using Keisler's completeness theorem for the ω-logic $L^\omega(Q)$ with the additional quantifier "there exist uncountably many" ([**78**]) and forcing. It would be interesting to see a direct construction of such a family supported by a given arbitrary $\subseteq^*$-increasing sequence A_ξ ($\xi < \omega_1$), similar to one given in [**5**, pages 96–98]. In this construction, due to Todorcevic, Hausdorff's gap is not constructed recursively as usual but it is given by an explicit description.

QUESTION 5.12.3. *Is there a direct construction of a nontrivial coherent family of partial functions as in Theorem 5.12.2? Can such a family be constructed inside an arbitrary $\subseteq^*$-increasing ω_1-sequence $\{A_\alpha\}$?*

5.13. Remarks and questions

The fact that we were able to analyze gap spectra and preservation of gaps in the case of quotients over ideals $\emptyset \times \operatorname{Fin}$ and $\operatorname{Fin} \times \emptyset$ is neither surprising nor new. These two ideals were, besides Fin, the first two ideals for which the Radon–Nikodym property was known to be true (see §1.6). On the other hand, extending these results to a larger class of ideals (see §1.9) required a much deeper analysis involving a notion of ε-approximate homomorphism (see §1.9 or [**33**]) connecting this area with finite combinatorics and analysis. It is therefore plausible that further analysis of gap-spectra and preservation of gaps in other analytic quotients will lead to similar discoveries and connections (as extension in another direction already did—see [**37**]). Let us therefore reproduce a problem first asked by Todorcevic in [**130**]:

PROBLEM 5.13.1. *Determine the gap-spectrum of $\mathcal{P}(\mathbb{N})/\mathcal{I}$ for every analytic ideal $\mathcal{I}$ on $\mathbb{N}$.*

In particular:

QUESTION 5.13.2. *Are there analytic ideals other than $\emptyset \times$ Fin whose linear gap-spectra are equal to that of* Fin*? In particular, does the linear gaps spectrum of* Fin $\times \emptyset$ *always coincide with that of* Fin*?*

By Proposition 5.4.3 the answer to the latter question is positive under OCA. A positive answer to the following question, essentially suggested by Todorcevic in [**130**, page 95], would give even more support that OCA is the 'right' axiom for the study of Hausdorff-completeness.

QUESTION 5.13.3. *Can the Hausdorff gap spectrum of every analytic quotient be determined under OCA?*

Another related question is whether some form of Martin's Axiom is necessary in Corollary 5.10.3. We conjecture that it is not, if Corollary 5.10.3 is stated in the right way. Recall that $\mathrm{add}(\mathcal{N})$ is the additivity of the Lebesgue measure.

CONJECTURE 5.13.4. *There is an $\langle \mathrm{add}(\mathcal{N}), \mathrm{add}(\mathcal{N}) \rangle$-gap in the quotient over an F_σ P-ideal defined in Theorem 5.10.2.*

A confirmation of this Conjecture would give a positive answer to Problem 2 of [**130**], where it was asked whether some standard cardinal invariants of the continuum other than ω, ω_1 and $\mathfrak{b}$ naturally occur in the (linear) gap spectra of some analytic quotient. Note that, using some form of MA linear gaps can be found inside every analytic gap we have constructed in this Chapter (see Lemma 5.8.2, Proposition 5.8.3, Lemma 5.8.7, and Theorem 5.10.2).

PROBLEM 5.13.5. *Describe the class of analytic ideals $\mathcal{I}$ such that there are no analytic Hausdorff gaps in its quotient.*

PROBLEM 5.13.6. *Describe the class of analytic ideals $\mathcal{I}$ such that OCA (or PFA) implies that there are only $\langle \omega_1, \omega_1 \rangle$, $\langle \omega, \omega_2 \rangle$, and $\langle \omega_2, \omega \rangle$-gaps in its quotient.*

QUESTION 5.13.7. *Do the classes of ideals defined in Problem 5.13.5 and Problem 5.13.6 coincide? Do $\mathcal{I}_{1/n}$ and $\mathcal{Z}_0$ belong to either of these two classes?*

If there is an analytic Hausdorff gap in a quotient over $\mathcal{I}$ then Proposition 5.7.2 implies that $\mathcal{I}$ fails the TSP, therefore Fin and $\emptyset \times$ Fin belong to this class. But these two conditions on ideals, having the TSP and not having analytic Hausdorff gaps in its quotient, are not equivalent. This is because the ideal $\emptyset \times$ Fin fails the TSP (by Theorem 5.7.1) yet by Proposition 5.3.3 and Proposition 5.9.4 there are no analytic Hausdorff gaps in its quotient.

QUESTION 5.13.8. *Is there an analytic ideal other than* Fin *and* Fin $\times \emptyset$ *which has the TSP?*

QUESTION 5.13.9. *Is there some other "natural" separation property for analytic ideals which the summable ideals, or $\mathcal{Z}_0$, satisfy?*

To motivate the above question, let us note that Todorcevic has shown in [**130**, page 59] that the TSP of Fin has both the classical Hurewicz separation property ([**56**]) and the one proved by Kechris–Louveau–Woodin ([**77**]) as its corollaries. Therefore the discovery of some new separation phenomena in analytic quotients over the ideals closely related to σ-ideals on reals like $\mathcal{Z}_0$ (see [**112**]) may also lead to discoveries of some new phenomena in the realm of classical Descriptive Set Theory.

Bibliography

[1] U. Abraham, M. Rubin, and S. Shelah. On the consistency of some partition theorems for continuous colorings, and the structure of $\aleph_1$-dense real order types. *Annals of Pure and Applied Logic*, 29:123–206, 1985.

[2] T. Bartoszynski. Invariants of measure and category. In M. Foreman and A. Kanamori, editors, *Handbook of Set Theory*. to appear.

[3] J.E. Baumgartner. All $\aleph_1$-dense sets of reals can be isomorphic. *Fundamenta Mathematicae*, 79:101–106, 1973.

[4] J.E. Baumgartner. Chains and antichains in $\mathcal{P}(\omega)$. *The Journal of Symbolic Logic*, 45:85–92, 1980.

[5] M. Bekkali. *Topics in Set Theory*, volume 1476 of *Lecture Notes in Mathematics*. Springer-Verlag, 1991.

[6] M. Bell. an email of October 1997.

[7] M. Bell. Two Boolean algebras with extreme cellular and compactness properties. *Canadian Journal of Mathematics*, XXXV:824–838, 1983.

[8] A. Blass. Near coherence of filters II: Applications to operator ideals, the Stone–čech remainder of a half-line, order ideals of sequences, and slenderness of groups. *Transactions of the American Mathematical Society*, 300:557–581, 1987.

[9] A. Blass and C. Laflamme. Consistency results about filters and the number of inequivalent growth types. *The Journal of Symbolic Logic*, 54:50–56, 1989.

[10] M. Burke. Liftings for Lebesgue measure. In *Set Theory of the reals*, volume 6 of *Israel Mathematical Conference Proceedings*, pages 119–150. 1993.

[11] P. Casevitz. *Dichotomies pour les espaces de suites reelles*. PhD thesis, Universite Paris 6, 1999.

[12] C.C. Chang and H.J. Keisler. *Model Theory*. North–Holland, 1973.

[13] J.P.R. Christensen. Some results with relation to the control measure problem. In R.M. Aron and S. Dineen, editors, *Vector space measures and applications II*, volume 645 of *Lecture Notes in Mathematics*, pages 27–34. Springer, 1978.

[14] W.W. Comfort and S. Negrepontis. *Theory of ultrafilters*. Springer, 1974.

[15] H.G. Dales and W.H. Woodin. *An Introduction to Independence for Analysts*, volume 115 of *London Mathematical Society Lecture Note Series*. Cambridge University Press, 1987.

[16] J. Diestel and J.J. Uhl, Jr. *Vector measures*. Amer. Math. Soc. Math. Surveys, Number 15, 1977.

[17] E.K. van Douwen. *Prime mappings, number of factors and binary operations*, volume 199 of *Dissertationes Mathematicae*. Warszawa, 1981.

[18] E.K. van Douwen. The automorphism group of $\mathcal{P}(\omega)/$ Fin need not be simple. *Topology and its Applications*, pages 97–104, 1990.

[19] E.K. van Douwen and J. van Mill. The homeomorphism extension theorem for $\beta\omega \setminus \omega$. In S. Andima et al., editors, *Papers on general topology and applications*, volume 704 of *Ann. New York Acad. Sci.*, pages 345–350. 1993.

[20] E.K. van Douwen, D. Monk, and M. Rubin. Some questions about Boolean algebras. *Algebra Universalis*, 11:220–243, 1980.

[21] A. Dow. Extending real-valued functions in $\beta\kappa$. *Fundamenta Mathematicae*, 152:21–41, 1997.

[22] A. Dow and I. Farah. Is $\mathcal{P}(\omega)$ a subalgebra? in preparation, 2001.

[23] A. Dow and K.P. Hart. ω^* has (almost) no continuous images. *Israel Journal of Mathematics*, 109:29–39, 1999.

[24] A. Dow and K.P. Hart. The measure algebra does not always embed. *Fundamenta Mathematicae*, 163:163–176, 2000.

[25] A. Dow, P. Simon, and J.E. Vaughan. Strong homology and the Proper Forcing Axiom. *Proceedings of the American Mathematical Society*, 106 (3):821–828, 1989.
[26] A. Ehrenfeucht, J. Kahn, R. Maddux, and J. Mycielski. On the dependence of functions on their variables. *Journal of Combinatorial Theory, ser. A*, 33:106–108, 1982.
[27] R. Engelking. *General Topology*. Heldermann, Berlin, 1989.
[28] P. Erdös. My Scottish Book "problems". In D. Mauldin, editor, *The Scottish Book*, pages 35–44. Birkhäuser, Boston, 1981.
[29] I. Farah. A coherent family of functions. *Proceedings of the American Mathematical Society*, 124:2845–2852, 1996.
[30] I. Farah. Embedding partially ordered sets into ω^ω. *Fundamenta Mathematicae*, 151:53–95, 1996.
[31] I Farah. OCA and towers in $\mathcal{P}(\mathbb{N})/$ Fin. *Commentationes Mathematicae Universitatis Carolinae*, 37:861–866, 1996.
[32] I. Farah. *Analytic ideals and their quotients*. PhD thesis, University of Toronto, 1997.
[33] I. Farah. Approximate homomorphisms. *Combinatorica*, 18:335–348, 1998.
[34] I. Farah. Cauchy nets and open colorings. *Publ. Inst. Math. (Beograd) (N.S.)*, 64(78):146–152, 1998. 50th anniversary of the Mathematical Institute, Serbian Academy of Sciences and Arts (Belgrade, 1996).
[35] I. Farah. Completely additive liftings. *The Bulletin of Symbolic Logic*, 4:37–54, 1998.
[36] I. Farah. Ideals induced by Tsirelson submeasures. *Fundamenta Mathematicae*, 159:243–258, 1999.
[37] I. Farah. Approximate homomorphisms II: Group homomorphisms. *Combinatorica*, 20:47–60, 2000.
[38] I. Farah. Liftings of homomorphisms between quotient structures and Ulam stability. In S. Buss, P. Hájek, and P. Pudlák, editors, *Logic Colloquium '98*, volume 13 of *Lecture notes in logic*, pages 173–196. A.K. Peters, 2000.
[39] I. Farah. Basis problem for turbulent actions I: Tsirelson submeasures. In *Proceedings of XI Latin American Symposium in Mathematical Logic, Merida, July 1998*, Annals of Pure and Applied Logic. to appear.
[40] I. Farah. Basis problem for turbulent actions II: c_0-equalities. *Proceedings of the London Mathematical Society*, 81, to appear.
[41] I. Farah. Dimension phenomena associated with $\beta\mathbb{N}$-spaces. *Topology and its Applications*, to appear.
[42] I. Farah. Functions essentially depending on at most one variable. *Journal of Combinatorial Theory, Ser. A*, to appear.
[43] Qi Feng. Homogeneity for open partitions of pairs of reals. *Transactions of the American Mathematical Society*, 339:659–684, 1993.
[44] G.L. Forti. Hyers–Ulam stability of functional equations in several variables. *Aequationes Math.*, 50:143–190, 1995.
[45] D.H. Fremlin. Measure algebras. In D. Monk and R. Bonnett, editors, *Handbook of Boolean algebras*, pages 877–980. North-Holland, Amsterdam, 1989.
[46] D.H. Fremlin. The partially ordered sets of measure theory and Tukey's ordering. *Note di Matematica*, Vol. XI:177–214, 1991.
[47] D.H. Fremlin. Notes on FARAH P99. preprint, University of Essex, June 1999.
[48] D.H. Fremlin. Embedding measure algebras in $\mathcal{P}(\mathbb{N})/\mathcal{Z}_0$. preprint, University of Essex, May 1999.
[49] D.H. Fremlin. Problems. University of Essex, version of April 21, 1994.
[50] F. Galvin. Letter of August 5, 1995.
[51] J. Hadamard. Sur les caracteres de convergence des series a termes positifs et sur les fonctions indefiniment croissantes. *Acta Mathematicae*, 18:319–336, 1884.
[52] K.P. Hart and J. van Mill. Open problems on $\beta\omega$. In J. van Mill and G.M. Reed, editors, *Open problems in topology*, pages 97–125. North-Holland, 1990.
[53] F. Hausdorff. Die Graduierung nach dem Endverlauf. *Abhandlungen der Königlich Sächsischen Gesellschaft der Wissenschaften; Mathematisch–Physiche Klasse*, 31:296–334, 1909.
[54] F. Hausdorff. Summen von $\aleph_1$ Mengen. *Fundamenta Mathematicae*, 26:241–255, 1936.
[55] G. Hjorth and A.S. Kechris. New dichotomies for Borel equivalence relations. *The Bulletin of Symbolic Logic*, 3:329–346, 1997.

[56] W. Hurewicz. Relativ Perfecte Teile von Punktmengen und Mengen (A). *Fundamenta Mathematicae*, 12:78–109, 1928.

[57] A. Ionescu Tulcea and C. Ionescu Tulcea. *Topics in the theory of lifting*. Ergebnisse der Mathematik und ihrer Grenzgebiete, Band 48. Springer-Verlag, New York, 1969.

[58] S.-A. Jalali-Naini. *The monotone subsets of Cantor space, filters and descriptive set theory*. PhD thesis, Oxford, 1976.

[59] B Jonsson and P. Olin. Almost direct products and saturation. *Compositio Mathematicae*, 20:125–132, 1968.

[60] W. Just. Nowhere dense P-subsets of ω^*. *Proceedings of the American Mathematical Society*, 106:1145–1146, 1989.

[61] W. Just. The space $(\omega^*)^{n+1}$ is not always a continuous image of $(\omega^*)^n$. *Fundamenta Mathematicae*, 132:59–72, 1989.

[62] W. Just. Repercussions on a problem of Erdös and Ulam about density ideals. *Canadian Journal of Mathematics*, 42:902–914, 1990.

[63] W. Just. A modification of Shelah's oracle chain condition with applications. *Transactions of the American Mathematical Society*, 329:325–341, 1992.

[64] W. Just. A weak version of AT from OCA. *Mathematical Science Research Institute Publications*, 26:281–291, 1992.

[65] W. Just and A. Krawczyk. On certain Boolean algebras $\mathcal{P}(\omega)/I$. *Transactions of the American Mathematical Society*, 285:411–429, 1984.

[66] N.J. Kalton. The Maharam problem. In G. Choquet et al., editors, *Séminaire Initiation à l' Analyse, 28e Année*, volume 18, pages 1–13. 1988/89.

[67] N.J. Kalton and J.W. Roberts. Uniformly exhaustive submeasures and nearly additive set functions. *Transactions of the American Mathematical Society*, 278:803–816, 1983.

[68] S. Kamo. Almost coinciding families and gaps in $\mathcal{P}(\omega)$. *J. Math. Soc. Japan*, 45:357–368, 1993.

[69] A. Kanamori. *The higher infinite*. Springer–Verlag, 1995.

[70] V. Kanovei and M. Reeken. New Radon–Nikodym ideals. *Mathematika*, to appear.

[71] V. Kanovei and M. Reeken. On Baire measurable homomorphisms of quotients of the additive group of the reals. *Mathematical Logic Quarterly*, to appear.

[72] V. Kanovei and M. Reeken. On Ulam's problem of stability of non-exact homomorphisms. *Proceedings of Moscow Steklov Mathematical Institute*, to appear.

[73] I. Kaplansky. Normed algebras. *Duke Math. Journal*, 16:399–418, 1949.

[74] A.S. Kechris. *Classical descriptive set theory*, volume 156 of *Graduate texts in mathematics*. Springer, 1995.

[75] A.S. Kechris. Rigidity properties of Borel ideals on the integers. In *8th Prague symposium on general topology and its relations to modern analysis and algebra, 1996*, volume 85 of *Topology and its applications*, pages 195–205, 1998.

[76] A.S. Kechris and A. Louveau. The structure of hypersmooth Borel equivalence relations. *Journal of the American Mathematical Society*, 10:215–242, 1997.

[77] A.S. Kechris, A. Louveau, and W.H. Woodin. The structure of σ-ideals of compact sets. *Transactions of the American Mathematical Society*, 301:263–288, 1987.

[78] H. J. Keisler. Logic with the quantifier "There exists uncountably many". *Annals of Mathematical Logic*, 1:1–93, 1970.

[79] S. Koppelberg. personal communication.

[80] K. Kunen. Some comments on box products. In A. Hajnal et al., editors, *Infinite and finite sets, Keszthely (Hungary), 1973*, volume 10 of *Coll. Math. Soc. János Bolyai*, pages 1011–1016. North-Holland, Amsterdam, 1975.

[81] K. Kunen. $\langle \kappa, \lambda^* \rangle$-gaps under MA. preprint, 1976.

[82] K. Kunen. *An Introduction to Independence Proofs*. North–Holland, 1980.

[83] K. Kunen. Where MA first fails. *The Journal of Symbolic Logic*, 53:429–433, 1988.

[84] C. Laflamme. Forcing with filters and complete combinatorics. *Annals of pure and applied logic*, 42:125–163, 1989.

[85] C. Laflamme. Combinatorial aspects of F_σ filters with an application to $\mathcal{N}$-sets. *Proceedings of the American Mathematical Society*, 125:3019–3025, 1997.

[86] C. Laflamme and J. Zhu. The Rudin–Blass ordering on ultrafilters. *Journal of Symbolic Logic*, 63:584–592, 1998.

[87] A. Louveau and B. Velickovic. A note on Borel equivalence relations. *Proceedings of the American Mathematical Society*, 120:255–259, 1994.

[88] A. Louveau and B. Velickovic. Analytic ideals and cofinal types. *Annals of Pure and Applied Logic*, 99:171–195, 1999.

[89] Н. Лузин. *О частях натурального ряда*. Изв. АН СССР, серия мат., 11, №5:714–722, 1947.

[90] S. Mardesic and A. Prasolov. Strong homology is not additive. *Transactions of the American Mathematical Society*, 307:725–744, 1988.

[91] D.A. Martin and R.M. Solovay. Internal Cohen extensions. *Annals of Mathematical Logic*, 2:143–178, 1970.

[92] A.R.D. Mathias. Solution of problems of Choquet and Puritz. In *Conference in Mathematical Logic–London '70*, volume 255 of *Springer Lecture Notes in Mathematics*, pages 204–210. 1972.

[93] A.R.D. Mathias. A remark on rare filters. In A. Hajnal et al., editors, *Infinite and finite sets, Vol. III*, volume 10 of *Coll. Math. Soc. János Bolyai*, pages 1095–1097. North Holland, 1975.

[94] K. Mazur. A modification of Louveau and Velickovic construction for F_σ-ideals. preprint, 199?

[95] K. Mazur. F_σ-ideals and $\omega_1 \omega_1^*$-gaps in the Boolean algebra $\mathcal{P}(\omega)/I$. *Fundamenta Mathematicae*, 138:103–111, 1991.

[96] K. Mazur. Towards the dichotomy for F_σ-ideals. preprint, 1996.

[97] J. van Mill. An introduction to $\beta\omega$. In K. Kunen and J. Vaughan, editors, *Handbook of Set-theoretic topology*, pages 503–560. North-Holland, 1984.

[98] J.D. Monk and R.M. Solovay. On the number of complete Boolean algebras. *Algebra Universalis*, 2:365–368, 1972.

[99] J.T. Moore. Linearly fibered Souslinean space under MA. *Topology Proceedings*, 24, Spring 1999.

[100] M.R. Oliver. The density ideal is equireducible with c_0. preprint, UCLA, 1997.

[101] I.I. Parovičenko. A universal bicompact of weight $\aleph$. *Soviet Mathematics Doklady*, 4:592–592, 1963.

[102] R.S. Pierce. Countable Boolean algebras. In J.D. Monk and R. Bonnett, editors, *Handbook of Boolean algebras*, volume 3, pages 775–886. North–Holland, Amsterdam, 1989.

[103] F. Rothberger. Sur la familles indenombrables de suites de nombres naturels et les problemes concernant la propriete C. *Proc. Cambridge Phil. Soc.*, 37:109–126, 1941.

[104] W. Rudin. Homogeneity problems in the theory of čech compactifications. *Duke Mathematics Journal*, 23:409–419, 1956.

[105] M. Scheepers. Gaps in $^\omega\omega$. In *Set theory of the reals*, volume 6 of *Israel Mathematical Conference Proceedings*, pages 439–561. 1993.

[106] S. Shelah. *Proper Forcing*. Lecture Notes in Mathematics 940. Springer, 1982.

[107] S. Shelah and J. Steprans. PFA implies all automorphisms are trivial. *Proceedings of the American Mathematical Society*, 104:1220–1225, 1988.

[108] S. Shelah and J. Steprans. Somewhere trivial automorphisms. *Journal of the London Mathematical Society*, 49:569–580, 1994.

[109] S. Shelah and J. Steprans. Martin's axiom is consistent with the existence of nowhere trivial automorphisms. preprint, Rutgers University, 2000.

[110] J.R. Shoenfield. The problem of predicativity. In Y. Bar-Hillel et al., editors, *Essays on the Foundations of Mathematics*, pages 132–142. The Magens Press, Jerusalem, 1961.

[111] P. Simon. Applications of independent linked families. In *Topology, theory and applications (Eger, 1983)*, pages 561–580. North-Holland, Amsterdam, 1985.

[112] S. Solecki. Analytic ideals. *The Bulletin of Symbolic Logic*, 2:339–348, 1996.

[113] S. Solecki. an untitled handwritten note. University of Indiana, Bloomington, 1997.

[114] S. Solecki. Analytic ideals and their applications. *Annals of Pure and Applied Logic*, 99:51–72, 1999.

[115] S. Solecki. Cofinal G_δ subsets of analytic ideals of compact sets. handwritten note, University of Indiana, Bloomington, 1999.

[116] S. Solecki and S. Todorcevic. Borel bases for analytic ideals. preprint, University of Toronto, 1998.

[117] R. Solovay. A model of set theory in which every set of reals is Lebesgue measurable. *Annals of Mathematics*, 92:1–56, 1970.
[118] R. Solovay. Discontinuous homomorphisms of Banach algebras. preprint, UC Berkeley, 1976.
[119] J. Steprans. The size of the autohomeomorphism group of the Čech-Stone compactification of the integers. preprint, York University, 2000.
[120] E. Szemeredi. On sets of integers containing no k elements in arithmetic progression. *Acta Arithmetica*, 27:199–245, 1975.
[121] M. Talagrand. Compacts de fonctions mesurables et filtres nonmesurables. *Studia Mathematica*, 67:13–43, 1980.
[122] M. Talagrand. A simple example of a pathological submeasure. *Mathematische Annalen*, 252:97–102, 1980.
[123] D. E. Talayco. Applications of Cohomology to Set Theory I: Hausdorff gaps. *Annals of Pure and Applied Logic*, 71:69–106, 1995.
[124] S. Todorcevic. Directed sets and cofinal types. *Transactions of the American Mathematical Society*, 290:711–723, 1985.
[125] S. Todorcevic. *Partition Problems in Topology*, volume 84 of *Contemporary mathematics*. American Mathematical Society, Providence, Rhode Island, 1989.
[126] S. Todorcevic. Analytic gaps. *Fundamenta Mathematicae*, 150:55–67, 1996.
[127] S. Todorcevic. Comparing the continuum with the first two uncountable cardinals. In M.L. Dalla Chiara et al., editors, *Proceedings of the 10th International Congress of Logic, Methodology and Philosophy of Science, Florence 1995*, pages 145–155. Kluwer Academic Publishers, 1997.
[128] S. Todorcevic. Definable ideals and gaps in their quotients. In C.A. DiPrisco et al., editors, *Set Theory: Techniques and Applications*, pages 213–226. Kluwer Academic Press, 1997.
[129] S. Todorcevic. The first derived limit and compactly F_σ-sets. *J. Math. Soc. Japan*, 50:831–836, 1998.
[130] S. Todorcevic. Gaps in analytic quotients. *Fundamenta Mathematicae*, 156:85–97, 1998.
[131] S. Todorcevic. Compact sets of Baire class-1 functions. *Journal of the American Mathematical Society*, 12:1179–1212, 1999.
[132] S. Todorcevic. Chain condition methods in topology. *Topology and its applications*, 101:45–82, 2000.
[133] S. Todorcevic. The orthogonal of every analytic P-ideal is countably generated. handwritten note, University of Toronto, July 1996.
[134] S. Todorcevic. A dichotomy for P-ideals of countable sets. *Fundamenta Mathematicae*, to appear.
[135] S. Todorcevic and I. Farah. *Some Applications of the Method of Forcing*. Mathematical Institute, Belgrade and Yenisei, Moscow, 1995.
[136] F. Topsøe. Some remarks concerning pathological submeasures. *Mathematica Scandinavica*, 38:159–166, 1976.
[137] B. Tsirelson. Not every Banach space contains an embedding of ℓ_p or c_0. *Functional Anal. Appl.*, 8:138–141, 1974. translated from the Russian.
[138] J.W. Tukey. *Convergence and uniformity in topology*. Princeton University Press, 1940.
[139] S.M. Ulam. *Problems in modern mathematics*. John Wiley & Sons, 1964.
[140] S.M. Ulam and D. Mauldin. Mathematical problems and games. *Advances in Applied Mathematics*, 8:281–344, 1987.
[141] B. Velickovic. Definable automorphisms of $\mathcal{P}(\omega)/\text{Fin}$. *Proceedings of the American Mathematical Society*, 96:130–135, 1986.
[142] B. Velickovic. OCA and automorphisms of $\mathcal{P}(\omega)/\text{Fin}$. *Topology and its Applications*, 49:1–12, 1992.
[143] P. Štěpánek and M. Rubin. Homogeneous Boolean algebras. In D. Monk and R. Bonnett, editors, *Handbook of Boolean algebras*, volume 2, pages 679–715. North–Holland, Amsterdam, 1989.
[144] W. Weiss. Partitioning topological spaces. In J. Nešetřil and V. Rödl, editors, *Mathematics of Ramsey Theory*, volume 5 of *Algorithms and Combinatorics*, pages 154–171. Springer–Verlag, Berlin, 1990.
[145] W.H. Woodin. Σ_1^2-absoluteness. Note of May 1985.
[146] W.H. Woodin. *The Axiom of Determinacy, forcing axioms and the nonstationary ideal*, volume 1 of *de Gruyter Series in Logic and Its Applications*. de Gruyter, 1999.

Index

additive lifting, 1
agree below k, 101
$\aleph_1$-dense, 88
$\aleph_1$-directed, 60
almost bijection, 123
almost disjoint, 7
almost disjoint sets, 73
almost equal, 7
almost includes, 7
almost lifting, 73, 132
almost refines, 135
amalgamation of homomorphisms, 12, 71
approximate homomorphism, 3
approximation
    δ-approximation for H, 21
    ε-approximation on $\mathcal{X}$, 93
asymptotic density zero, 9
asymptotically additive, 16
AT, 141
automorphism group, 78

Baire, see also Baire-measurable
Baire-measurable, 12
basic question, 1, 5, 74
$\beta\omega$-space, 120
Boolean algebra
    dual Hopfian, 82
    homogeneous, 83
    Hopfian, 82
    weakly homogeneous, 83
Borel cardinality, 12
bounding number, 61

ccc over Fin, 73
ccc over $\mathcal{J}$, 132
closed approximation to an ideal, 9
cofinal, 144
cofinality, 144
cofinally finer, 153
cofinally similar, 153
coherent family
    σ-directed, 66
    indexed by $\mathcal{A}$, 60
    indexed by $\mathbb{N}^\mathbb{N}$, 61
    modulo Fin, 85
    modulo $\mathcal{I}$, 66, 165
        a nontrivial one, 165
    of partial functions indexed by $\mathcal{A}$, 60
    of partitions indexed by $\mathbb{N}^\mathbb{N}$, 63
    trivial, 64
coloring, 55
    open, 55
comeager, 91
completely additive lifting, 1
conjecture
    Kechris–Mazur, 51, 53
    Todorcevic, 14
        version for arbitrary connecting maps, 114
        weaker form of, 51
    van Douwen's, 120, 123
countably saturated, 75
countably separated, 10
    families, 77
    pregap, 144

(d,l)-perfect, 125
d-bounded subset of $\alpha^{(d)}$, 132
d-compact, 117
δ-approximation for H, 21
dense ideal, 41
density ideal, 6
density ideal generated by a sequence of measures, 42
depends on at most one coordinate, 118
depends on at most one variable, 118
directed, 60
    κ-directed, 60
    σ-directed, 60
discrete sequence, 72
downwards closure, 91
dual filter, 16

elementary function, 123
    piecewise, 123
embeddable, 76
ε-approximate
    epimorphism, 52
    homomorphism, 21
ε-approximation on $\mathcal{X}$, 93
ε-tame, 93
eventual dominance, 61
everywhere nonmeager, 94
Extension Principle, 117

strong, 141
weak, 117

f-density zero, 8
family, *see also* coherent family
  neat, 99
  σ-trivial, 59, 62
  tree-like, 99
  trivial, 59, 62
finite-to-one reduction, 12
Fréchet ideal, 7
  generalized, 53
Fréchet property, 11, 83
Fubini product of two ideals, 8
function
  Baire-measurable, 12
  elementary, 123
  Erdös–Ulam function, 42
  incompatible, 59
  induced by h, 139
  monotonic, 20
  perfect, 124
  $\mathcal{R}$-measurable, 125
  σ-Borel, 101
  subadditive, 20

gap, *see also* nonlinear gap, 4, 76, 143
  analytic $\langle[\mathbb{R}]^{<\omega}, \mathcal{Q}\rangle$-gap, 156
  $\langle \mathfrak{b}, \mathfrak{b}\rangle$-gap, 158
  coherent, 165
  F_σ Hausdorff gap, 162
  Hausdorff, 145
    reflects, 148
  in $\mathcal{P}(\mathbb{N})/\mathcal{I}$, 144
  linear, 144, 154
  Luzin, 145
  of type $\langle \mathcal{P}, \mathcal{Q}\rangle$, 153
  $\langle \mathcal{P}, \mathcal{Q}\rangle$-gap, 153
  Rothberger, 145
gap spectrum, 144
  Hausdorff, 144
  linear, 144
groupwise density, 92

Hausdorff gaps in the quotient over Fin $\times \emptyset$
  reflect, 148
hereditary, 9, 86, 91
  closure, 91
homogeneous, 83, 122
  K_i-homogeneous, 55, 122
  weakly, 83
homomorphism
  ε-approximate, 21
  approximate, 3
  Baire, 5
  non-exact, 3
Hyers–Ulam stability, 25

$\mathcal{I}$-orthogonal, 144

$\mathcal{I}$-positive, 16
ideal
  analytic, 7
  analytic P-ideals, 5
  atomic, 36
  c_0-like, 51
  ccc over $\mathcal{J}$, 132
  ccc over Fin, 73
  countably generated, 74
  dense, 11, 41
  dense summable, 34
  density, 6
  density ideal, 42
  $\emptyset \times$ Fin, 8
  Erdös–Ulam, 8, 42
  $\mathcal{EU}_f$, *see also* ideal, Erdös–Ulam
  EU-ideals, 8
  F_σ P-ideal which is not summable, 33, *see also* [**36**]
  f-ideal, *see also* ideal, Erdös–Ulam
  Fin $\times$ Fin, 157
  Fin $\times \emptyset$, 8
  Fréchet, 7
  generalized Fréchet ideal, 53
  homogeneous, 53, 85
  $\mathcal{I}_{1/\sqrt{n}}$, 2
  $\mathcal{I}_{1/n}$, 1
  $\mathcal{I}_\alpha$, *see also* ideal, ordinal
  $\mathcal{I}_f$, *see also* ideal, summable
  isomorphic, 10
  $\mathcal{I} \oplus \mathcal{J}$, 8
  $\mathcal{I} \times \mathcal{J}$, 8
  Louveau–Velickovic, 31
  Mazur, 53, 154
  non-pathological analytic P-ideal, 25
  nontrivial, 10
  of d-bounded subsets of $\alpha^{(d)}$, 132
  of asymptotic density zero sets, 9
  of logarithmic density zero sets, 9
  ordinal ideal, 53
    topological ordinal ideal, 54
  P-ideal, 7
  pathological analytic P-ideal, 25
  summable, 8
  tall, 11
  Weiss, 54
  $\mathcal{Z}_\mu$, *see also* ideal, density
  $\mathcal{Z}_{\log}$, 9
  $\mathcal{Z}_0$, 9
incompatible, 59
indexed by $\mathcal{A}$, 62
induced by h, 139
induces, 125

K_i-homogeneous, 55, 122

lemma
  Prime Mapping Lemma, 120
  Todorcevic, 11

lifting, 1, 11
  additive, 1
  almost lifting, 73
  asymptotically additive, 16
  Baire-measurable, 5
  completely additive, 1
  of Φ on $\mathcal{J}$, 98
logarithmic density zero, 9, 42
lower semicontinuous submeasure, 7

meager, 11, 91

non-exact homomorphism, 3
non-pathological
  F_σ ideal, 53
  analytic P-ideal, 25
  submeasure, 21
nonlinear gap
  Hausdorff, 154
  Luzin, 145
  Rothberger, 154
nonmeager, 12
  everywhere, 94

OCA, 55, see also Open Coloring Axiom
OCA lifting theorem, 69
$\langle \omega, 1 \rangle$-limits, 144
open coloring, 55, 63
Open Coloring Axiom, 55
  the reflection component of, 56
  the evolution of, 56
open partition, 55
oracle-chain condition, 68
ordering
  Baire embeddability, 12
  Borel cardinality, 12
  dense, 34
  finite-to-one reduction, 6
  Rudin–Blass, 12
  Rudin–Keisler, 12
  Tukey, 153
ordinal
  additively indecomposable, see also indecomposable
  indecomposable, 53, 124
  multiplicatively indecomposable, 54
orthogonal, 10, 36, 144
  $\mathcal{I}$-orthogonal, 144
  of an ideal, 76
  submeasures, 42

P-ideal, 7
P-set, 77
  weak, 142
partial ordering
  directed, 60
  σ-directed, 60
partition
  isomorphic, 55

K_0^n, 100
L_0^n, 107
  open, 55
  Velickovic, 100
pathological, see also ideal
  submeasure, 21
perfect mapping, 124
piecewise elementary, 123
Polish space, 135
pregap
  countably separated, 144
  in $\mathcal{P}(\mathbb{N})/\mathcal{I}$, 144
  separated, 144
Principle of Open Coloring, see also Open Coloring Axiom, 63
Property of Baire, 12

question
  basic, 1
  Bell's, 141
  Dow's, 142
  Just–Krawczyk, 84

$\mathcal{R}$-measurable, 125
Radon–Nikodym property, 5, 15
  of $\emptyset \times$ Fin, 31
  of Fin, 19
  of Fin $\times \emptyset$, 19
  of density ideals, 42
  of Erdös–Ulam ideals, 31
  of generalized Fréchet ideals, 53
  of Louveau–Velickovic ideals, 31
  of Mazur ideals, 53
  of non-pathological F_σ-ideals, 53
  of non-pathological P-ideals, 31
  of summable ideals, 31
  of Weiss ideals, 54
rectangle, 62
reduced product, 157
reduction, see also ordering
refines, 135
restriction, 147

saturated quotient, 56
scale, 150
Schröder–Bernstein property, 41
sEP, see also strong Extension Principle, 141
separated, 76
separates, 144
sequential fan, 139
set
  $\aleph_1$-dense, 88
  almost disjoint, 73
  cofinal, 144
  comeager, 91
  d-compact, 117
  ε-tame, 93
  everywhere nonmeager, 94
  groupwise dense, 92

hereditary, 9, 86, 91
homogeneous, 122
$\mathcal{I}$-positive, 16
K_i-homogeneous, 55, 122
meager, 11, 91
nonmeager, 12
P-set, 77
    weak, 142
σ-K_1-homogeneous, 55
2-$\mathcal{I}$-small, 120
sides (of a gap), 153
σ-K_1-homogeneous, 55
σ-Borel function, 101
σ-directed, 60
    subset of $\mathcal{P}(\mathbb{N})$, 60
σ-trivial, 59
space
    $\beta\omega$-space, 120
    Parovičenko, 3
    Polish, 135
    Stone, 132
    Tsirelson, 34
stability program of Ulam, 3
stabilizer, 17
    (n, n')-ε-stabilizer, 95
    (n, n')-stabilizer, 17
stable, 3
    non-exact homomorphisms, 21
Stone space, 132
subadditive, 20
submeasure, 20
    absolutely continuous, 161
    Erdös–Ulam submeasure, 42
    Fubini, 53
    lower semicontinuous, 7
    non-pathological, 21
    norm of, 21
    orthogonal, 42
    pathological, 21
    supported by, 7
sum of two ideals, 8
summable ideal, 8
superperfect, *see also* tree

theorem
    Dow–Hart, 117
    Jalali–Naini, Talagrand, 92
    Jankov, von Neumann, 93
    Just, 76
    Just–Krawczyk, 115
    Kanovei–Reeken, 53
    Kechris, 47, 52
    Keisler's completeness theorem, 165
    lifting theorem for non-pathological analytic P-ideals, 31
    Mathias, 12
    Mazur, 9
    non-lifting theorem, 31
    OCA lifting theorem, 74
    OCA uniformization theorem, 64
    Olin, 14
    Parovičenko, 117
    Shelah, 68
    Shoenfield's absoluteness theorem, 57
    Σ_2-reflection for coherent families, 65
    Solecki, 9, 52
    Todorcevic, 10, 32, 76, 143, 145, 153
    Tukey, 153
    W. Rudin, 140
    Woodin, 58
Todorcevic separation property, 151
tree
    downwards closed, 102
    superperfect, 151
        $\mathcal{B}$-tree, 151
trivial
    autohomeomorphism of $(\omega^*)^{(d)}$, 123
    automorphism, 75
    coherent family, 64
    family, 62
    family of functions, 59
    isomorphism, 16
trivializes, 59, 62
    $\mathcal{F}$ on E_n, 64
TSP, *see also* Todorcevic separation property
Tukey map, 153
2-$\mathcal{I}$-small, 120
type
    of a linear gap, 144
    of a nonlinear gap, 153
    strong, 154

uncountable rectangle
    of compatible conditions, 88, 111
    of incompatible conditions, 88, 111
uniformization, 55, 67, 136

weak Extension Principle, 117
    for X, 139
wEP, 117
wEP(α, γ), 117
wEP(X), 140
working part, 88

zero density, 41

Index of special symbols

$\mathcal{I}_{1/n}$ 1
Φ_* 1
$\mathcal{I}_{1/\sqrt{n}}$ 2
Φ_h 5
$\mathcal{I}_f$ 6
$\emptyset$ 7
$\|A\|_\varphi$ 7
$\mathrm{Exh}(\varphi)$ 7
$\mathrm{Fin}(\varphi)$ 7
$A \perp B$ 7
$A \supseteq^* B$ 7
$A =^* B$ 7
$A \Delta B$ 7
$A \subseteq^\mathcal{I} B$ 7
$A =^\mathcal{I} B$ 7
A_m 8
$\mathcal{I} \oplus \mathcal{J}$ 8
$\mathcal{I} \times \mathcal{J}$ 8
$\mathcal{I} \upharpoonright A$ 8
$\forall^\infty$ 8
$\exists^\infty$ 8
$\mathrm{Fin} \times \emptyset$ 8
$\emptyset \times \mathrm{Fin}$ 8
$\mathbb{N}$ 8
$\mathcal{E}\mathcal{U}_f$ 8
φ_f 8
$\mathcal{Z}_0$ 9
$\mathcal{Z}_{\log}$ 9
$\mathcal{N}(\varphi)$ 10
$f''X$ 10
$\mathcal{I}^\perp$ 10
$\mathcal{P}(\mathbb{N})/\mathcal{I}$ 11
$[A]_\mathcal{I}$ 11
Φ_* 11
$\pi_\mathcal{I}, \pi_\mathcal{J}$ 11
$\mathcal{I} \leq_{\mathrm{RB}} \mathcal{J}$ 12
$\leq_f$ 12
$\mathcal{I} \leq_{\mathrm{RK}} \mathcal{J}$ 12
$\mathcal{I} \leq_{\mathrm{BE}} \mathcal{J}$ 12
$\mathcal{I} \leq^+_{\mathrm{BE}} \mathcal{J}$ 12
$\mathcal{I} \leq_{\mathrm{BC}} \mathcal{J}$ 12
$\Phi \oplus \Psi$ 12
$d_\varphi(A, B)$ 14
π 15
$\mathcal{I}^*$ 16
$\mathcal{I}^+$ 16

Ψ_H 16
Γ_f 19
$\mathrm{supp}(\varphi)$ 21
$\|\varphi\|$ 21
$\hat{\varphi}$ 21
$P(\varphi)$ 21
K_φ 21
μ_f 21
$s^\complement$ 22
$A[c, C]$ 35
$A[c, \cdot]$ 35
$A[\cdot, C]$ 35
$\mathcal{I}^\perp$ 36
a_k^f 38
$\mathcal{I}_{1/\sqrt{n}}$ 41
μ_f 42
φ_f 42
$\mathcal{E}\mathcal{U}_f$ 42
$\mathrm{at}^+(\varphi)$ 42
$\mathrm{at}^-(\varphi)$ 42
φ_μ 42
$\mathcal{Z}_\mu$ 42
F_μ 48
$G_{\mu\delta}$ 48
$\{0,1\}^{<\mathbb{N}}$ 50
$\mathcal{J}_{\mathrm{br}}$ 50
$[t]$ 50
$\mathcal{I}_\alpha$ 53
$P(\alpha)$ 54
$[X]^2$ 55
Π_1^2, Π_1 58
$f \leq g$ 61
$f \leq^* g$ 61
$\flat$ 61
$\Delta'(g_f, g_{f'})$ 62
$U \otimes V$ 62
$[A]^{<\mathbb{N}}$ 63
Σ_2 65
$\Phi \oplus \Psi$ 71
$\leq^+_{\mathrm{BE}}$ 72
$\mathcal{P}(\mathbb{N})/\mathcal{I} \hookrightarrow \mathcal{P}(\mathbb{N})/\mathcal{J}$ 75
$\mathcal{I} \leq_{\mathrm{EM}} \mathcal{J}$ 76
$\mathcal{I}^\perp$ 76
$\mathrm{Aut}(\mathcal{P}(\mathbb{N})/\mathcal{I})$ 78
S_∞ 79
G_f 79

INDEX OF SPECIAL SYMBOLS

H_f	79
Λ	80
$\mathcal{J}_{\text{cont}}$	85
h_A	85
$\mathcal{J}_1$	85
g_A	85
$L_0(D)$	86
Γ_f	87
$k(A)$	87
$\mathcal{P}$	87
$\mathcal{P}_{\omega_1}$	89
$\dot{\mathcal{X}}$	91
$[s;m]$	91
$\|A\|_\varphi$	92
$\text{Exh}(\varphi)$	92
$d_\varphi(A,B)$	92
$\mathcal{J}_{\text{cont}}$	99
$A \supseteq^* B$	99
$\{0,1\}^{<\mathbb{N}}$	99
$\mathcal{B}$	99
$\mathcal{A}$	100
$\mathcal{A}_0$	100
$K_0^{\bar{n}}, K_1^{\bar{n}}$	100
$\langle a,b \rangle \upharpoonright k = \langle a', b' \rangle \upharpoonright k$	101
$\mathcal{J}_\sigma^n$	103
$\mathcal{J}_{\text{cont}}^n$	103
C_1^a, C_0^a	106
$\mathcal{J}_1$	106
$\mathcal{J}_2$	106
$\mathcal{K}_a$	106
$x(b)$	108
$i(d,a)$	108
$k(d,a)$	108
T_a	108
$\mathcal{Y}_a$	108
$\mathcal{P}$	110
$\mathcal{P}_{\omega_1}$	112
$\mathbb{D}$	117
ω	117
$X^{(d)}$	117
$\mathcal{R}_\alpha^d$	117
$\mathcal{R}^d$	117
$X \twoheadrightarrow Y$	117
$(\mathcal{U}x)\tau(x)$	118
$\Delta A, [A]^2, \langle A \rangle^2$	121
$[A]^3$	122
$\{l,m,n\}_<$	122
G	131
$\text{cl}(\alpha)$	131
$\langle \alpha \rangle$	131
$\langle \alpha \rangle^{(d)}$	132
$\text{cl}(X)$	132
$\mathcal{K}(X)$	132
$\text{Stone}(\mathcal{B})$	132
π_i	134
$\mathbf{A} \prec \mathbf{B}$	135
$\mathbf{A} \prec^* \mathbf{B}$	135
$S(\omega)$	139
$\overline{A}$	140
$\mathcal{A} \perp_\mathcal{I} \mathcal{B}$	144
$\mathcal{A} \perp \mathcal{B}$	144
$\text{cf}(\mathcal{P})$	144
$A/\mathcal{I}$	144
$\subseteq^*$	144
$\mathfrak{b}$	145
$\mathcal{A} \upharpoonright C$	147
Γ_f	148
$\mathfrak{c}$	150
$\sqsubset$	151
$G(\mathcal{A},\mathcal{B})$	152
$\mathcal{P} \leq \mathcal{Q}$	153
$\mathcal{P}_1 \equiv \mathcal{P}_2$	153
$\mathcal{N}$	153
$\mathcal{A}/\mathcal{I}$	153
$\mathcal{B}^\perp/\mathcal{I}$	154
$[\mathbb{R}]^{<\omega}$	156
$\text{Exh}(\varphi)$	156
$\text{Fin} \times \text{Fin}$	157
$\prod_n \mathcal{P}_n / \text{Fin}$	157
$\prod_n \mathcal{P}_n$	159
$L^\omega(Q)$	165
$\text{add}(\mathcal{N})$	166

Editorial Information

To be published in the *Memoirs*, a paper must be correct, new, nontrivial, and significant. Further, it must be well written and of interest to a substantial number of mathematicians. Piecemeal results, such as an inconclusive step toward an unproved major theorem or a minor variation on a known result, are in general not acceptable for publication. Papers appearing in *Memoirs* are generally longer than those appearing in *Transactions*, which shares the same editorial committee.

As of July 31, 2000, the backlog for this journal was approximately 9 volumes. This estimate is the result of dividing the number of manuscripts for this journal in the Providence office that have not yet gone to the printer on the above date by the average number of monographs per volume over the previous twelve months, reduced by the number of volumes published in four months (the time necessary for preparing a volume for the printer). (There are 6 volumes per year, each containing at least 4 numbers.)

A Consent to Publish and Copyright Agreement is required before a paper will be published in the *Memoirs*. After a paper is accepted for publication, the Providence office will send a Consent to Publish and Copyright Agreement to all authors of the paper. By submitting a paper to the *Memoirs*, authors certify that the results have not been submitted to nor are they under consideration for publication by another journal, conference proceedings, or similar publication.

Information for Authors

Memoirs are printed from camera copy fully prepared by the author. This means that the finished book will look exactly like the copy submitted.

The paper must contain a *descriptive title* and an *abstract* that summarizes the article in language suitable for workers in the general field (algebra, analysis, etc.). The *descriptive title* should be short, but informative; useless or vague phrases such as "some remarks about" or "concerning" should be avoided. The *abstract* should be at least one complete sentence, and at most 300 words. Included with the footnotes to the paper should be the 2000 *Mathematics Subject Classification* representing the primary and secondary subjects of the article. The classifications are accessible from www.ams.org/msc/. The list of classifications is also available in print starting with the 1999 annual index of *Mathematical Reviews*. The Mathematics Subject Classification footnote may be followed by a list of *key words and phrases* describing the subject matter of the article and taken from it. Journal abbreviations used in bibliographies are listed in the latest *Mathematical Reviews* annual index. The series abbreviations are also accessible from www.ams.org/publications/. To help in preparing and verifying references, the AMS offers MR Lookup, a Reference Tool for Linking, at www.ams.org/mrlookup/. When the manuscript is submitted, authors should supply the editor with electronic addresses if available. These will be printed after the postal address at the end of the article.

Electronically prepared manuscripts. The AMS encourages electronically prepared manuscripts, with a strong preference for $\mathcal{A}_{\mathcal{M}}\mathcal{S}$-LaTeX. To this end, the Society has prepared $\mathcal{A}_{\mathcal{M}}\mathcal{S}$-LaTeX author packages for each AMS publication. Author packages include instructions for preparing electronic manuscripts, the *AMS Author Handbook*, samples, and a style file that generates the particular design specifications of that publication series. Though $\mathcal{A}_{\mathcal{M}}\mathcal{S}$-LaTeX is the highly preferred format of TeX, author packages are also available in $\mathcal{A}_{\mathcal{M}}\mathcal{S}$-TeX.

Authors may retrieve an author package from e-MATH starting from `www.ams.org/tex/` or via FTP to `ftp.ams.org` (login as `anonymous`, enter username as password, and type `cd pub/author-info`). The *AMS Author Handbook* and the *Instruction Manual* are available in PDF format following the author packages link from `www.ams.org/tex/`. The author package can be obtained free of charge by sending email to `pub@ams.org` (Internet) or from the Publication Division, American Mathematical Society, P.O. Box 6248, Providence, RI 02940-6248. When requesting an author package, please specify $\mathcal{AMS}$-LaTeX or $\mathcal{AMS}$-TeX, Macintosh or IBM (3.5) format, and the publication in which your paper will appear. Please be sure to include your complete mailing address.

Sending electronic files. After acceptance, the source file(s) should be sent to the Providence office (this includes any TeX source file, any graphics files, and the DVI or PostScript file).

Before sending the source file, be sure you have proofread your paper carefully. The files you send must be the EXACT files used to generate the proof copy that was accepted for publication. For all publications, authors are required to send a printed copy of their paper, which exactly matches the copy approved for publication, along with any graphics that will appear in the paper.

TeX files may be submitted by email, FTP, or on diskette. The DVI file(s) and PostScript files should be submitted only by FTP or on diskette unless they are encoded properly to submit through email. (DVI files are binary and PostScript files tend to be very large.)

Electronically prepared manuscripts can be sent via email to `pub-submit@ams.org` (Internet). The subject line of the message should include the publication code to identify it as a Memoir. TeX source files, DVI files, and PostScript files can be transferred over the Internet by FTP to the Internet node `e-math.ams.org` (130.44.1.100).

Electronic graphics. Comprehensive instructions on preparing graphics are available at `www.ams.org/jourhtml/graphics.html`. A few of the major requirements are given here.

Submit files for graphics as EPS (Encapsulated PostScript) files. This includes graphics originated via a graphics application as well as scanned photographs or other computer-generated images. If this is not possible, TIFF files are acceptable as long as they can be opened in Adobe Photoshop or Illustrator. No matter what method was used to produce the graphic, it is necessary to provide a paper copy to the AMS.

Authors using graphics packages for the creation of electronic art should also avoid the use of any lines thinner than 0.5 points in width. Many graphics packages allow the user to specify a "hairline" for a very thin line. Hairlines often look acceptable when proofed on a typical laser printer. However, when produced on a high-resolution laser imagesetter, hairlines become nearly invisible and will be lost entirely in the final printing process.

Screens should be set to values between 15% and 85%. Screens which fall outside of this range are too light or too dark to print correctly. Variations of screens within a graphic should be no less than 10%.

Inquiries. Any inquiries concerning a paper that has been accepted for publication should be sent directly to the Electronic Prepress Department, American Mathematical Society, P. O. Box 6248, Providence, RI 02940-6248.

Editors

This journal is designed particularly for long research papers (and groups of cognate papers) in pure and applied mathematics. Papers intended for publication in the *Memoirs* should be addressed to one of the following editors. In principle the Memoirs welcomes electronic submissions, and some of the editors, those whose names appear below with an asterisk (*), have indicated that they prefer them. However, editors reserve the right to request hard copies after papers have been submitted electronically. Authors are advised to make preliminary email inquiries to editors about whether they are likely to be able to handle submissions in a particular electronic form.

Algebra to CHARLES CURTIS, Department of Mathematics, University of Oregon, Eugene, OR 97403-1222 email: `cwc@darkwing.uoregon.edu`

Algebraic geometry and commutative algebra to LAWRENCE EIN, Department of Mathematics, University of Illinois, 851 S. Morgan (M/C 249), Chicago, IL 60607-7045; email: `ein@uic.edu`

Algebraic topology and cohomology of groups to STEWART PRIDDY, Department of Mathematics, Northwestern University, 2033 Sheridan Road, Evanston, IL 60208-2730; email: `priddy@math.nwu.edu`

Combinatorics and Lie theory to PHILIP J. HANLON, Department of Mathematics, University of Michigan, Ann Arbor, Michigan 48109-1003; email: `hanlon@math.lsa.umich.edu`

Complex analysis and complex geometry to DANIEL M. BURNS, Department of Mathematics, University of Michigan, Ann Arbor, MI 48109-1003; email: `dburns@math.lsa.umich.edu`

*__Differential geometry and global analysis__ to CHUU-LIAN TERNG, Department of Mathematics, Northeastern University, Huntington Avenue, Boston, MA 02115-5096; email: `terng@neu.edu`

*__Dynamical systems and ergodic theory__ to ROBERT F. WILLIAMS, Department of Mathematics, University of Texas, Austin, Texas 78712-1082; email: `bob@math.utexas.edu`

Geometric topology, knot theory, hyperbolic geometry, and general topoogy to JOHN LUECKE, Department of Mathematics, University of Texas, Austin, TX 78712-1082; email: `luecke@math.utexas.edu`

Harmonic analysis, representation theory, and Lie theory to ROBERT J. STANTON, Department of Mathematics, The Ohio State University, 231 West 18th Avenue, Columbus, OH 43210-1174; email: `stanton@math.ohio-state.edu`

*__Logic__ to THEODORE SLAMAN, Department of Mathematics, University of California, Berkeley, CA 94720-3840; email: `slaman@math.berkeley.edu`

Number theory to MICHAEL J. LARSEN, Department of Mathematics, Indiana University, Bloomington, IN 47405; email: `larsen@math.indiana.edu`

Operator algebras and functional analysis to BRUCE E. BLACKADAR, Department of Mathematics, University of Nevada, Reno, NV 89557; email: `bruceb@math.unr.edu`

*__Ordinary differential equations, partial differential equations, and applied mathematics__ to PETER W. BATES, Department of Mathematics, Brigham Young University, 292 TMCB, Provo, UT 84602-1001; email: `peter@math.byu.edu`

*__Partial differential equations and applied mathematics__ to BARBARA LEE KEYFITZ, Department of Mathematics, University of Houston, 4800 Calhoun Road, Houston, TX 77204-3476; email: `keyfitz@uh.edu`

*__Probability and statistics__ to KRZYSZTOF BURDZY, Department of Mathematics, University of Washington, Box 354350, Seattle, Washington 98195-4350; email: `burdzy@math.washington.edu`

*__Real and harmonic analysis and geometric partial differential equations__ to WILLIAM BECKNER, Department of Mathematics, University of Texas, Austin, TX 78712-1082; email: `beckner@math.utexas.edu`

All other communications to the editors should be addressed to the Managing Editor, WILLIAM BECKNER, Department of Mathematics, University of Texas, Austin, TX 78712-1082; email: `beckner@math.utexas.edu`.

Selected Titles in This Series

(Continued from the front of this publication)

672 **Yael Karshon,** Periodic Hamiltonian flows on four dimensional manifolds, 1999
671 **Andrzej Rosłanowski and Saharon Shelah,** Norms on possibilities I: Forcing with trees and creatures, 1999
670 **Steve Jackson,** A computation of δ^1_5, 1999
669 **Seán Keel and James McKernan,** Rational curves on quasi-projective surfaces, 1999
668 **E. N. Dancer and P. Poláčik,** Realization of vector fields and dynamics of spatially homogeneous parabolic equations, 1999
667 **Ethan Akin,** Simplicial dynamical systems, 1999
666 **Mark Hovey and Neil P. Strickland,** Morava K-theories and localisation, 1999
665 **George Lawrence Ashline,** The defect relation of meromorphic maps on parabolic manifolds, 1999
664 **Xia Chen,** Limit theorems for functionals of ergodic Markov chains with general state space, 1999
663 **Ola Bratteli and Palle E. T. Jorgensen,** Iterated function systems and permutation representation of the Cuntz algebra, 1999
662 **B. H. Bowditch,** Treelike structures arising from continua and convergence groups, 1999
661 **J. P. C. Greenlees,** Rational S^1-equivariant stable homotopy theory, 1999
660 **Dale E. Alspach,** Tensor products and independent sums of $\mathcal{L}_p$-spaces, $1 < p < \infty$, 1999
659 **R. D. Nussbaum and S. M. Verduyn Lunel,** Generalizations of the Perron-Frobenius theorem for nonlinear maps, 1999
658 **Hasna Riahi,** Study of the critical points at infinity arising from the failure of the Palais-Smale condition for n-body type problems, 1999
657 **Richard F. Bass and Krzysztof Burdzy,** Cutting Brownian paths, 1999
656 **W. G. Bade, H. G. Dales, and Z. A. Lykova,** Algebraic and strong splittings of extensions of Banach algebras, 1999
655 **Yuval Z. Flicker,** Matching of orbital integrals on $GL(4)$ and $GSp(2)$, 1999
654 **Wancheng Sheng and Tong Zhang,** The Riemann problem for the transportation equations in gas dynamics, 1999
653 **L. C. Evans and W. Gangbo,** Differential equations methods for the Monge-Kantorovich mass transfer problem, 1999
652 **Arne Meurman and Mirko Primc,** Annihilating fields of standard modules of $\mathfrak{sl}(2,\mathbb{C})^\sim$ and combinatorial identities, 1999
651 **Lindsay N. Childs, Cornelius Greither, David J. Moss, Jim Sauerberg, and Karl Zimmermann,** Hopf algebras, polynomial formal groups, and Raynaud orders, 1998
650 **Ian M. Musson and Michel Van den Bergh,** Invariants under Tori of rings of differential operators and related topics, 1998
649 **Bernd Stellmacher and Franz Georg Timmesfeld,** Rank 3 amalgams, 1998
648 **Raúl E. Curto and Lawrence A. Fialkow,** Flat extensions of positive moment matrices: Recursively generated relations, 1998
647 **Wenxian Shen and Yingfei Yi,** Almost automorphic and almost periodic dynamics in skew-product semiflows, 1998
646 **Russell Johnson and Mahesh Nerurkar,** Controllability, stabilization, and the regulator problem for random differential systems, 1998
645 **Peter W. Bates, Kening Lu, and Chongchun Zeng,** Existence and persistence of invariant manifolds for semiflows in Banach space, 1998

For a complete list of titles in this series, visit the
AMS Bookstore at **www.ams.org/bookstore/**.